Thixoforming

Edited by
Gerhard Hirt and Reiner Kopp

Further Reading

Herlach, D. M. (ed.)

Phase Transformations in Multicomponent Melts

2009
ISBN: 978-3-527-31994-7

Riedel, R., Chen, I-W. (eds.)

Ceramics Science and Technology

Volume 1: Structures

2008
ISBN: 978-3-527-31155-2

Breme, J., Kirkpatrick, C. J., Thull, R. (eds.)

Metallic Biomaterial Interfaces

2008
ISBN: 978-3-527-31860-5

de With, G.

Structure, Deformation, and Integrity of Materials

Volume I: Fundamentals and Elasticity / Volume II: Plasticity, Visco-elasticity, and Fracture

2006
ISBN: 978-3-527-31426-3

Kainer, K. U. (ed.)

Metal Matrix Composites

Custom-made Materials for Automotive and Aerospace Engineering

2006
ISBN: 978-3-527-31360-0

Schumann, H., Oettel, H.

Metallografie
14th Edition

2004
ISBN: 978-3-527-30679-4

Thixoforming

Semi-solid Metal Processing

Edited by
Gerhard Hirt and Reiner Kopp

WILEY-VCH

WILEY-VCH Verlag GmbH & Co. KGaA

The Editors

Prof. Dr. G. Hirt
Institute of Metal Forming
RWTH Aachen
Intzestrasse 10
52072 Aachen
Germany

Prof. Dr. Reiner Kopp
Institute of Metal Forming
RWTH Aachen
Intzestrasse 10
52072 Aachen
Germany

Library of Congress Card No.: applied for

British Library Cataloguing-in-Publication Data
A catalogue record for this book is available from the British Library.

Bibliographic information published by the Deutsche Nationalbibliothek
The Deutsche Nationalbibliothek lists this publication in the Deutsche Nationalbibliografie; detailed bibliographic data are available on the Internet at http://dnb.d-nb.de.

© 2009 WILEY-VCH Verlag GmbH & Co. KGaA, Weinheim

Typesetting   Thomson Digital, Noida, India
Printing   Strauss GmbH, Mörlenbach
Binding   Litges & Dopf GmbH, Heppenheim

Printed in the Federal Republic of Germany
Printed on acid-free paper

ISBN: 978-3-527-32204-6

Contents

Thixoforming: Semi-solid Metal Processing. Edited by G. Hirt and R. Kopp
Copyright © 2009 WILEY-VCH Verlag GmbH & Co. KGaA, Weinheim
ISBN: 978-3-527-32204-6

Preface

Semi-solid forming of metals is a fascinating technology offering the opportunity to manufacture net-shaped metal components of complex geometry in a single forming operation. At the same time, high mechanical properties can be achieved due to the unique microstructure and flow behaviour. Successful semi-solid forming processes require narrow tolerances in all process steps, including feedstock generation, reheating and the forming process. It is this strong and highly nonlinear interrelation between the parameters of each process step on the one hand and the material microstructure and flow behaviour on the other which still causes a challenge for scientific understanding and economic mass production.

This book first gives a substantial general overview of worldwide achievements of semi-solid metal (SSM) technology to date. The main part then presents latest research results concerning the material fundamentals and process technology and material and process modelling. In addition to contributions from internationally recognized scientists elsewhere, most of the results presented were obtained within the activities of the collaborative research centre SFB289, 'Forming of metals in the semi-solid state and their properties', at RWTH Aachen University. This research centre was funded at RWTH from 1996 to 2007 by the Deutsche Forschungsgemeinschaft (DFG) involving nine institutes and with a total budget of about €19.2 million. The centre has contributed more than 250 publications in various international journals and conferences. Even though this book briefly covers the whole period, it is especially intended to make the results of the most recent activities, dating from 2004 to 2007, available as a whole to the international community.

The book covers semi-solid forming of aluminium alloys and steels from feedstock generation to part properties and also process control and die technologies. It is intended for engineers and scientists in industry and academia who want to achieve a general overview of the technology involved and a deeper understanding of the fundamental basics of this innovative technology.

Aachen, November 2008

Gerhard Hirt
Reiner Kopp

Thixoforming: Semi-solid Metal Processing. Edited by G. Hirt and R. Kopp
Copyright © 2009 WILEY-VCH Verlag GmbH & Co. KGaA, Weinheim
ISBN: 978-3-527-32204-6

List of Contributors

Dirk Abel
RWTH Aachen University
Institute of Automatic Control
Steinbachstrasse 54
52056 Aachen
Germany

Carsten Afrath
RWTH Aachen University
Foundry Institute
Intzeststrasse 5
52056 Aachen
Germany

Mahmoud Ahmadein
RWTH Aachen University
Foundry Institute
Intzestrasse 5
52056 Aachen
Germany

Alexander Arnold
RWTH Aachen University
IME – Process Metallurgy
and Metal Recycling
Intzestrasse 3
52072 Aachen
Germany

René Baadjou
RWTH Aachen University
Institute of Metal Forming
Intzestrasse 10
52072 Aachen
Germany

Régis Bigot
Arts et Metiers Paristech
ENSAM
Cedex 3
Laboratoire de Génie Industriel et
Production Mécanique
57078 Metz
France

Wolfgang Bleck
RWTH Aachen University
Institute of Ferrous Metallurgy
Intzestrasse 1
52072 Aachen
Germany

Kirsten Bobzin
RWTH Aachen University
Institut für Oberflächentechnik
Templergraben 55
52056 Aachen
Germany

Thixoforming: Semi-solid Metal Processing. Edited by G. Hirt and R. Kopp
Copyright © 2009 WILEY-VCH Verlag GmbH & Co. KGaA, Weinheim
ISBN: 978-3-527-32204-6

Andreas Bührig-Polaczek
RWTH Aachen University
Foundry Institute
Intzestrasse 5
52072 Aachen
Germany

Matthias Bünck
RWTH Aachen University
Foundry Institute
Intzestrasse 5
52072 Aachen
Germany

Pierre Cézard
ASCOMETAL-CREAS
BP 70045
57301 Hagondange
France

Sebastian Dziallach
RWTH Aachen University
Institute of Ferrous Metallurgy
Intzestrasse 1
52072 Aachen
Germany

Véronique Favier
Arts et Metiers Paristech
UMR CNRS 8006
Laboratoire d'Ingénierie
des Matériaux ENSAM
151 Boulevard de l'Hôpital
75013 Paris
France

Bernd Friedrich
RWTH Aachen University
IME – Process Metallurgy and Metal
Recycling
Intzestrasse 3
52072 Aachen
Germany

Rainer Gasper
RWTH Aachen University
Institute of Automatic Control
Steinbachstraße 54
52056 Aachen
Germany

David Hajas
RWTH Aachen University
Materials Chemistry
Kopernikusstraße 16
52056 Aachen
Germany

Bengt Hallstedt
RWTH Aachen University
Materials Chemistry
Kopernikusstrasse 16
52074 Aachen
Germany

Gerhard Hirt
RWTH Aachen University
Institute of Metal Forming
Intzestrasse 10
52072 Aachen
Germany

Markus Hufschmidt
RWTH Aachen University
Aachener Verfahrenstechnik
Turmstrasse 46
52056 Aachen
Germany

Philipp Immich
RWTH Aachen University
Institut für Oberflächentechnik
Templergraben 55
52056 Aachen
Germany

Liudmila Khizhnyakova
RWTH Aachen University
Institute of Metal Forming
Intzestrasse 10
52072 Aachen
Germany

Frederik Knauf
RWTH Aachen University
Institute of Metal Forming
Intzestrasse 10
52072 Aachen
Germany

Reiner Kopp
RWTH Aachen University
Institute of Metal Forming
52072 Aachen
Germany

Fabian Küthe
RWTH Aachen University
Foundry Institute
Intzestrasse 5
52072 Aachen
Germany

Erich Lugscheider
RWTH Aachen University
Institut für Oberflächentechnik
Templergraben 55
52056 Aachen
Germany

Heike Meuser
V & M Deutschland GmbH
Schützenstrasse 124
45476 Mülheim a. d. Ruhr
Germany

Michael Modigell
RWTH Aachen University
Aachener Verfahrenstechnik
Turmstrasse 46
52056 Aachen
Germany

Simon Münstermann
RWTH Aachen University
Institut für Gesteinshüttenkunde
Mauerstraße 5
52056 Aachen
Germany

Tony Noll
RWTH Aachen University
IME – Process Metallurgy
and Metal Recycling
Intzestrasse 3
52072 Aachen
Germany

Lars Pape
RWTH Aachen University
Aachener Verfahrenstechnik
Turmstrasse 46
52056 Aachen
Germany

Wolfgang Püttgen
McKinsey & Company
Königsallee 60c
40026 Düsseldorf
Germany

Roger Sauermann
RWTH Aachen University
Process Metallurgy and Metal
Recycling
Intzestrasse 3
52072 Aachen
Germany

Jochen M. Schneider
RWTH Aachen University
Materials Chemistry
Kopernikusstrasse 16
52074 Aachen
Germany

Alexander Schönbohm
RWTH Aachen University
Institute of Automatic Control
Steinbachstraße 54
52056 Aachen
Germany

Ingold Seidl
RWTH Aachen University
Institute of Metal Forming
Intzestrasse 10
52072 Aachen
Germany

Hideki Shimahara
RWTH Aachen University
Institute of Metal Forming
Intzestrasse 10
52072 Aachen
Germany

Rainer Telle
RWTH Aachen University
Mauerstraße 5
Institut für Gesteinshüttenkunde
52056 Aachen
Germany

Peter J. Uggowitzer
ETH Zurich
Department of Materials
Laboratory of Metal Physics and
Technology
8093 Zurich
Switzerland

Dirk I. Uhlenhaut
ETH Zurich
Department of Materials
Laboratory of Metal Physics and
Technology
8093 Zurich
Switzerland

Ksenija Vasilić
RWTH Aachen University
Aachener Verfahrenstechnik
Turmstrasse 46
52056 Aachen
Germany

1

Semi-solid Forming of Aluminium and Steel – Introduction and Overview*

Gerhard Hirt, Liudmila Khizhnyakova, René Baadjou, Frederik Knauf, and Reiner Kopp

1.1
Introduction

The origins of semi-solid metal forming date back to the early 1970s, when Flemings and co-workers studied the flow behaviour of metals in a semi-solid state [1]. Soon the first industrialization by Alumax and ITT-TEVES was achieved for automotive applications such as chassis components, brake cylinders, rims and so on. Many patents, particularly concerning the production of the required specific primary material, hindered wider development during that time. At the end of the 1980s, an intensive development period started also in Europe. With the alternative electromagnetic stirring methods implemented by Pechiney (France), Ormet (USA) and SAG (Austria), primary material in variable dimensions and quality became available. New heating technologies and online-controlled pressure casting machines constituted the basis for appropriate production equipment. This led to various impressive mass production applications in the field of chassis (e.g. Porsche, DaimlerChrysler, Alfa Romeo) and car body components (e.g. Audi, Fiat, DaimlerChrysler, in addition to lower volume production for other fields. However, due to the increasing contest with the rapidly improving quality of highly cost-effective fully liquid casting processes, some of these activities have been terminated in the meantime. One reason, namely the cost disadvantage caused by the use of special primary material, could be reduced by the introduction of new rheocasting processes, which helped to keep several applications of forming in the semi-solid state. For example, STAMPAL (Italy) changed the production of some components from 'classical thixocasting' to 'new rheocasting'. Other challenges are the narrow process windows for billet production, reheating and forming, which have to be maintained to achieve highly repeatable production. This requires a fundamental and detailed understanding of the physical basics of each process step, taking into account details of the material behaviour and microstructure development.

---

* A List of Abbreviations can be found at the
end of this chapter.

Thixoforming: Semi-solid Metal Processing. Edited by G. Hirt and R. Kopp
Copyright © 2009 WILEY-VCH Verlag GmbH & Co. KGaA, Weinheim
ISBN: 978-3-527-32204-6

The worldwide intensive efforts in industry and science to develop semi-solid metal forming are motivated by the significant technological and scientific potential which these innovative technologies can provide:

- *Compared with conventional casting*, the high viscosity of semi-solid metal allows macroscopic turbulence during die filling to be avoided and subsequently reduces part defects that could arise from air entrapment. A second advantage is that due to high solid fraction of about 40% during die filling the loss of volume during complete solidification is reduced, leading to correspondingly reduced shrinkage porosity or allowing higher cross-sectional changes than are possible in conventional castings. Furthermore, the low gas content leads to microstructures which are suitable for welding and heat treatment even in very filigree components, which among others is an essential argument for the existing thixocasting serial production of aluminium alloys for the automotive industry. In addition, the lowered process temperature of the semi-solid metal can lead to a significant increase in tool life compared with conventional die casting.

- *Compared with conventional forging*, thixoforming offers significantly reduced forming loads and the opportunity to produce complex geometry components, which could not be produced by forging. In addition the near net shape capabilities of thixoforming reduce machining to a minimum. However, the usually wrought high-strength aluminium alloys, which are typically used in forging, are not well suited for semi-solid forming, especially because they have a tendency for hot cracking during solidification. Therefore, the superior mechanical properties of forged components cannot completely be achieved by thixoforming. Also, the production cycles in forging are much shorter than in thixoforming, which requires time for solidification. Accordingly, a substitution of forging by semi-solid forming will only offer economic benefits if parts with significant added value can be produced. This could be achieved by increased geometric complexity, by weight saving when substituting steel by aluminium or by production of composite components. Additional benefits may arise if some final machining or joining operations can be avoided.

Industrial thixocasting of aluminium alloys is based on the benefits in comparison with casting processes, which have led to various serial production methods in spite of the additional costs for primary material and investment costs for heating equipment. However, there is strong competition with highly economical and improved casting processes such as vacuum casting, squeeze casting, fully automated die casting and optimized casting aluminium alloys such as AlMg5SiMn. This strictly limits the range of economically feasible semi-solid forming applications to such components, which fully exploit the technical advantages listed above. In addition, the narrow process window and the complexity of the process chain require significant experience in semi-solid metal series production, which is today available for example at SAG (Austria), Stampal (Italy), AFT (USA), and Pechiney (France).

Thixoforming of higher melting iron-based alloys has also been investigated in early work at MIT [2, 3], Alumax [4] and Sheffield University [5]. However, these activities, which showed impressive success, nevertheless faced significant technological challenges and they were not in the centre of general interest, which at that time was much more focused on aluminium alloys. Not before the middle of the 1990s did semi-solid forming of steel become the focus of various research activities worldwide [6–14]. In these projects, the general feasibility of semi-solid forming of higher melting point alloys was demonstrated. The major challenges, however, which still need to be overcome, are mainly related to the high temperature range, which causes high thermal loading of tools and dies and difficulties in achieving a homogeneous temperature distribution in the billet. Also, the tendency for oxidation and scale formation and in some alloys the complex microstructure evolution during reheating cause difficult problems to solve.

Concerning the competition between semi-solid forming of steels and conventional casting and forging technologies, the situation differs significantly from that discussed above for aluminium. The reason is that cost-efficient permanent mould casting processes for steel do not exist. This is currently restricted by the high thermal loading that the dies and injection systems would have to face. Thixoforming, however, is performed at lower temperatures and, even more important, the internal energy of semi-solid metals is significantly reduced compared with the fully liquid state. This means that using the high fluidity of semi-solid steels, it might for the first time become possible to manufacture near net shape steel components with complicated geometry in a single-step permanent mould process.

Conventionally forged steel components may also be geometrically complicated, but the relation between flow length and wall thickness is much smaller than in semi-solid forming. Also, complex geometries require multi-stage forging processes and significantly more final machining. This indicates that the application potential for semi-solid forming of steels could be significantly higher than for aluminium alloys. However, there are still challenging technical and scientific problems to solve.

The goals of this book are as follows:

- to summarize fundamental knowledge and technological applications for semi-solid forming;
- to contribute to a better understanding of the governing relations between the process parameters and the achieved part quality and to present selected technological solutions to overcome existing process challenges.

Since several books and longer reviews have already been written focusing on semi-solid forming of aluminium alloys and with respect to the market potential described above, most of this book is dedicated to semi-solid forming of steels. However, the methods applied and most of the results achieved are also valid for other alloy systems.

The rest of this introductory chapter will give a brief overview of the development and state of the art of the technology of semi-solid metal forming. Part One covers the material science aspects such as microstructure evolution and characterization in the semi-solid condition. Part Two is dedicated to numerical modelling of the constitutive material behaviour and its application in process modelling. This involves both two-phase modelling and micro–macro coupling, since both must be taken into account for an adequate representation of the material behaviour. Part Three is related to tool and die technologies, which are one of the major challenges for semi-solid forming of high-temperature alloys such as steel. Part Four presents technological applications and process alternatives of semi-solid forming.

1.2
Early Work on Flow Behaviour and Technology Development

1.2.1
Basic Findings Concerning the Rheology of Metals in the Semi-solid Condition

The development of semi-solid forming of metals dates back to early work of Flemings in the 1970s [15]. He and his co-workers studied the behaviour of solidifying metallic melts under conditions in which they form a suspension of globular primary solid particles in a liquid metallic melt (Figure 1.1b). At that time, this was achieved by stirring the slurry while cooling it to the desired temperature. The initial findings clearly showed that under such conditions the viscosity of metals in the semi-solid state depends on the solid fraction, shear rate and time history. Even though this behaviour is to be expected for two-phase emulsions, it was a fairly new discovery for metallic systems, which under normal conditions would have shown a dendritic microstructure (Figure 1.1a).

The observations concerning the flow behaviour, which were achieved in Couette-type rheometers, may be summarized as follows:

- *Influence of solid fraction*
  The viscosity strongly depends on the volume fraction of solid particles, typically showing a steep increase when the solid fraction becomes higher than 35–50%.

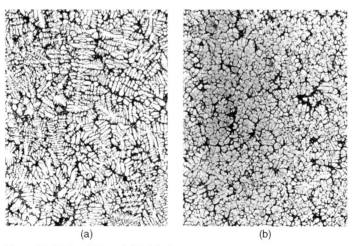

(a)   (b)

Figure 1.1 (a) Dendritic and (b) globular microstructure.

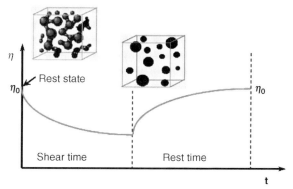

Figure 1.2 Time-dependent thixotropic behaviour.

- *Influence of shear rate*
  Varying the shear rate in the rheometer revealed that the apparent viscosity also depends strongly on the actual shear rate. For example, in an Sn–15Pb alloy at a solid fraction of 40%, the viscosity varies from about 10^6 Pa s at a shear rate $0.001\,s^{-1}$, which is typical for glass working, to 10^{-2} Pa s at $200\,s^{-1}$, which is similar to bicycle oil [15].

- *Influence of time history*
  If semi-solid slurries are allowed to stand, the globular particles tend to agglomerate and the viscosity increases with time. If the material is sheared, the agglomerates are broken up and the viscosity falls. This time-dependant thixotropic behaviour is shown schematically in Figure 1.2.

Many studies have been performed since then to investigate this behaviour in more detail for different alloys, to explain the governing mechanisms and to derive models to describe the material response. One mechanism, which distinguishes semi-solid metal slurries from other suspensions, is that the particle shape and size vary irreversibly with time, as shown in Figure 1.3.

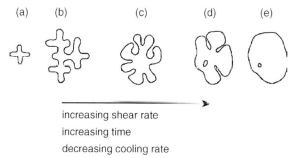

increasing shear rate
increasing time
decreasing cooling rate

Figure 1.3 Schematic illustration of evolution of structure during solidification with vigorous agitation: (a) initial dendritic fragment; (b) dendritic growth; (c) rosette; (d) ripened rosette; (e) spheroid [15].

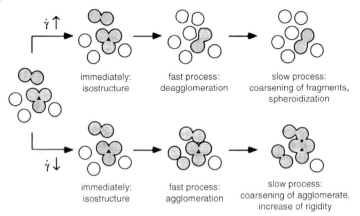

Figure 1.4 Schematic model describing the fast and slow processes in a semi-solid material's structure after up and down jumps of shear rate [16].

This mechanism is combined with agglomeration of particles and breakdown of agglomerates with increasing and decreasing shear rate (Figure 1.4). Thus the transient material behaviour is governed by a combination of fast, reversible and slow, irreversible mechanisms.

This combination and the tendency that under the influence of shear and pressure gradients a separation of liquid phase and solid particles may occur cause difficulties for appropriate material modelling. One-phase models, which assume a shear rate and time-dependent viscosity, can be sufficient to describe the die filling in semi-solid casting, especially in cases where metal flow is mainly governed by the imposed boundary conditions due to the geometry of the part. However, the part quality is also determined by the distribution of the chemical composition, which is dominated by the occurrence of liquid–solid separation. This can only be modelled using two-phase models, which solve the governing flow equations for the liquid and solid phase separately while imposing appropriate coupling terms. A comprehensive review concerning the rheology of semi-solid metal slurries and their material modelling has recently been prepared by Atkinson [16]. Here, Chapter 3 will be dedicated to this topic, describing experimental procedures and modelling approaches on both the micro- and macro-scales.

1.2.2
First Steps Towards a Semi-solid Metal-forming Technology

Already at a very early stage, the general idea arose that this specific flow behaviour could be beneficial for forming processes. Most of the potential advantages, which have been presented in Section 1.1, were already seen from the very beginning [15]. With respect to market potential and temperature range, it was straightforward that the technological exploitation concentrated on aluminium alloys, being of interest for many structural applications while offering a moderate temperature range below 600 °C for semi-solid processing. However, the idea of producing the slurry by cooling from liquid turned out not to be a practical solution at that time and the break-through was not achieved until an alternative route was found. This new route was based on the idea that a suitable suspension of globular solid particles in a liquid melt could also be achieved by reheating solidified billets to the desired liquid fraction, if these billets were prepared to have a fine-grained globular microstructure. Accordingly, the emerging technology, which was used to demonstrate the feasibility to manufacture complex parts, consisted of three major steps [18]:

- *Billet production:* Proprietary technologies were developed to cast billets with a fine-grained globular microstructure. These technologies included direct chill casting with active (mechanical or electromagnetic) or passive stirring and thermomecha-nical processing of solid feedstock (see Sections 1.3.1.1 and 1.3.1.2).

- *Reheating to the semi-solid condition:* The necessity to heat billets accurately and homogeneously to the semi-solid condition required specific heating strategies and the development of adequate equipment and control systems. Inductive heating was preferably used, but also conventional radiation/convection-type furnaces were applied in series production (see Sections 1.3.2.1 and 1.3.2.2).

- *Forming operations and applications:* After reheating, the semi-solid billet is usually formed into the final shape in just one forming operation. Two basic forming principles are generally applicable:

  – In thixocasting the billet is injected into a closed die, as in high-pressure die casting.

  – In thixoforging, the billet is formed by using an upper and a lower die which move against each other. Using these principles, relatively complex and chal-lenging demonstrator parts were already manufactured in the 1980s, such as automotive brake master cylinders, automobile wheels, electrical connectors, valve bodies and plumbing fittings [18].

1.3
Today's Technologies of Semi-solid Metal Forming

Over the years, a family of process routes and process alternatives for semi-solid forming have evolved from this early work. Reviews concerning these technologies and their industrial application have been published [18–21]. Some of them have succeeded in being applied in industrial mass production; others are still limited to the laboratory scale. Generally, these process routes can be structured according to Figure 1.5.

Following the early idea of separating the slurry preparation from the forming operation, various methods to provide solid billets with a fine-grained globular microstructure have been developed. These will be briefly discussed in general in Section 1.3.1, whereas specific details are given later in Chapter 4. Before forming,

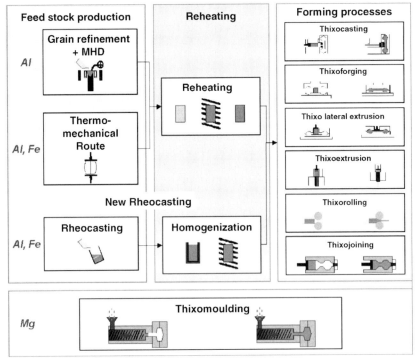

Figure 1.5 SMM process routes.

the billet has to be brought to a homogeneous semi-solid condition. Depending on the process route, this can be a reheating stage (see Section 1.3.2) or a homogenization and/or consolidation stage in the more recently developed routes which create the slurry directly from liquid stage (see the overview in Section 1.3.3 and discussion of specific aspects in Chapters 4 and 9). The further processing of the billet or slurry is then performed using modifications of conventional processes such as high-pressure die casting, forging, lateral extrusion, bar extrusion and rolling. Of these, only semi-solid casting and to a lesser extent modifications of semi-solid forging have so far been applied to industrial production (see Sections 1.3.4 and 1.3.5 for an overview and Part Four for specific details).

A completely different approach is the thixomoulding of magnesium alloys. This process uses magnesium chips and a modified injection moulding machine to produce magnesium components in a process similar to polymer injection moulding. The solid chips are fed through the screw system of the machine where they are heated to the semi-solid condition. The slurry is accumulated in the casting chamber and injected into the die. The process is commercially applied mainly for electronic products such as laptop housings and so on where magnesium substitutes polymers. The advantage lies in the fact that the problems which are related to the handling of liquid magnesium, as in hot-chamber high-pressure die casting, can be avoided.

1.3.1
Preparation of Billets for Semi-solid Forming

1.3.1.1 Direct Chill Casting

Direct chill casting is an established process for the production of aluminium billets for forging, extrusion and rolling. Accordingly, it was an obvious idea to use this process to produce billets with a fine-grained globular microstructure that were required for the thixoforming route. Transferring the observations from the early rheometer tests initially resulted in laboratory concepts using active mechanical stirring to produce a slurry before feeding it into the direct chill (DC) casting mould (Figure 1.6a). This, however, was soon replaced by DC casting with magnetohydro-dynamic (MHD) stirring in the mould, thus directly influencing the solidification by vigorous agitation to avoid the formation of dendritic structures. Patents [22–24] introduced variations of this idea, such as circumferential or vertical stirring (Figure 1.6 c, d) and a combination of both stirring types.

Alternatives to electromagnetic or active mechanical stirring are the so-called passive stirring and chemical grain refinement. In passive stirring (Figure 1.6b), the liquid metal is forced to flow through a system of obstacles (i.e. ceramic balls) while being cooled into the semi-solid range. The shear generated by this forced flow prohibits the formation of large dendrites and creates 'nuclei' of crystals by dendrite fragments. This process was developed by Moschini [25] and has been used to supply feedstock for the mass production of pressure-tight fuel rails by semi-solid casting.

Also in conventional DC casting of forging or extrusion feedstock, the goal is to produce a fine-grained and homogeneous microstructure. For this purpose, various methods of chemical grain refinement have been developed, providing a large number of 'nuclei' by addition of Al–5Ti–B particles. It has been shown that these procedures in some cases can be adopted to produce feedstock which exhibits the typical microstructure and flow behaviour for semi-solid forming [26, 27].

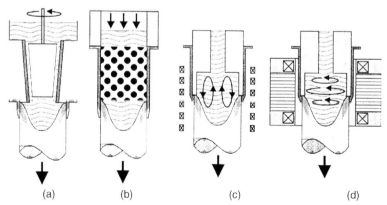

(a)   (b)   (c)   (d)

Figure 1.6 Stirring modes: (a) mechanical stirring; (b) passive stirring; (c) electromagnetic 'vertical' stirring; (d) electromagnetic 'horizontal' stirring.

1.3.1.2 Thermomechanical Treatment

In addition to casting, there also exist solid working routes to produce feedstock for semi-solid forming. These routes typically consist of plastic deformation followed by recrystallization during the reheating stage. If reheating is then performed into the semi-solid range, the fine grains start to melt at their boundaries, resulting in globular particles surrounded by liquid. Depending on whether the plastic deformation has been performed as hot working or cold working, the process is referred to as the strain induced melt activated (SIMA) route [22, 28].

Economically, these thermomechanical routes cannot compete with MHD casting of the typically applied aluminium alloys since they involve an additional extrusion step which is usually limited to smaller diameters. However, some high-performance alloys and composites are generally delivered in extruded condition (i.e. after spray forming) and can be used in this condition for semi-solid forming.

However, regarding thixoforming of steels, the thermomechanical route has so far been the most suitable method to provide adequate feedstock from various steel grades. With respect to large-scale production units for steel, there is not much chance of economically casting relatively small amounts of small-diameter feedstock by special DC casting processes.

1.3.2
Reheating of Billets and Alternative Slurry Production

1.3.2.1 Heating Furnaces and Strategies

In the classical thixoforming route, the billets are reheated to the semi-solid condition prior to forming. This reheating step is critical, because it defines the microstructure and flow behaviour of the billet and, depending on the alloy, the reheating may also directly influence the mechanical properties of the final part. For example, when reheating the typically used AlSi7Mg alloy (A357), it is not only important that the final temperature of 580 °C is reached, but also that the holding time allows one to bring the coarsened silicon particles completely into solution, which would otherwise reduce the elongation to fracture since they form relatively large brittle inclusions [29]. On the other hand, undesired grain growth will occur during slow reheating and the mechanical stability of the billet decreases the longer it is maintained in the semi-solid condition.

With respect to these effects reheating of the billets must be:

- quick to avoid grain growth and for economic reasons;
- precise to achieve the desired liquid fraction very reproducibly;
- homogeneous throughout the billet volume to avoid property gradients.

In an industrial environment, mostly inductive heating or radiation/convection heating are used. Although both processes have been applied for industrial heating of billets to semi-solid forming, the use of a classical radiation/convection furnace has only been reported by Moschini [25]. However, this installation was one of the first to perform successfully the mass production of pressure-tight automotive components. The advantage of this concept is the relatively cheap furnace and a robust process control structure. The main disadvantage is the long heating time, which is a consequence of the necessity to transfer the heat through the billet surface.

Typically, inductive heating is used due to the advantages given by shorter heating time and the possibility of flexible process control. Induction allows faster heating, because the heat is generated by eddy currents inside the billet volume even though the so-called skin effect leads to a higher power density close to the billet surface. The skin effect can be characterized by the so-called penetration depth, which is the distance from the billet surface within which approximately 86% of the total power is induced into the billet. For a given billet and induction coil geometry, this penetration depth decreases with increasing frequency of the oscillator. This suggests using low-frequency heating if only the homogeneity of the temperature distribution would have to be considered. On the other hand, low frequency means high electromagnetic forces, which are in contradiction to heating a soft semi-solid billet. Accordingly experience has shown that for typical billet diameters from 76 to 150 mm, induction furnaces in the medium frequency range from 250 to 1000 Hz are to be preferred.

Due to the skin effect, the requirements of homogeneous and fast reheating to the desired semi-solid state are contradictory. Fast reheating would require a high power input, which leads to significant radial temperature gradients. Accordingly, typical heating cycles as shown in Figure 1.7 consist of several stages starting with high power for rapid heating until the surface reaches the target temperature. Consequently, the power is reduced in order to compensate for heat losses due to convection and radiation and in addition allows temperature homogenization by heat conduction inside the billet. In addition to the radial gradients due to the skin effect there are also corner effects at the billet's ends. These can be reduced by special coil design or other standard measures to guide the electromagnetic field [13, 30].

To realize these heating cycles and to reach the cycle time of typical forming operations such as high-pressure die casting, two types of machines have been developed: carousel-type machines and individual billet heating systems. Carousel-type machines consist of a certain number of induction coils (12–16) which are placed around the circumference of a billet carousel (Figure 1.8). During heating, the billets are placed on ceramic pedestals and are indexed from coil to coil by rotating the carousel. Thus a large number of billets are heated at the same time. Typically, several

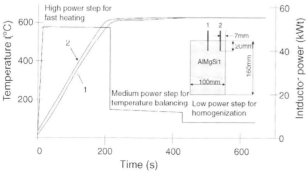

Figure 1.7 Typical heating cycle for AlMgSi1.

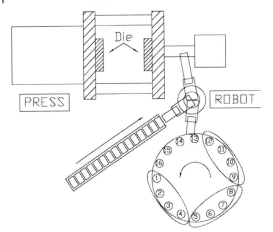

Figure 1.8 Carousel-type machine billet heating system.

groups of coils are created and each group is connected to one frequency generator supplying the group of coils with a certain power input. The heating cycle is realized by passing the billets from group to group. The main advantage of this system is a relatively cheap power supply, which typically consists of three frequency generators each of which is designed for either high, medium or low power input. Also, the mechanical design is relatively simple, requiring a drive for the turntable and a simple pick and place device for billet handling.

The main disadvantage is a lack of flexibility and control, and especially that there is no possibility of controlling the heating cycle of each billet individually. This causes problems at least if there is an unexpected interruption of the forming machine, since it is not possible to put such a system into a 'hold' condition.

This disadvantage led to the development of the individual coil concept, which also heats several billets at the same time. However, in this case each billet is placed in its individual coil, each of which is connected to a separate frequency generator (Figure 1.9). This requires that the number of generators is as large as the number of coils (i.e. 12) and that each generator is designed for the maximum power input required during the initial fast heating period. The advantage is that the heating cycle of each billet can now be controlled individually. The disadvantage is high cost, because in such a case a large number of high power generators are required. Also, the mechanical handling of the billets is more expensive because each coil requires its individual loading device and additionally a robot or other more complex feeding system is required to handle the billets in the different coil locations.

There are two types of individual billet heating set-ups, one where the billet stands in an upright position on the pedestal and the other in a lying position in a transport shell. The first concept has the advantage of easier handling because of the unhindered reachability with regard to automation, but is limited to relatively short billets and lower liquid fraction due to the decrease in the billet's stability. The second concept offers higher flexibility regarding the liquid fraction and the billet's size but has the disadvantage that a container is required for handling.

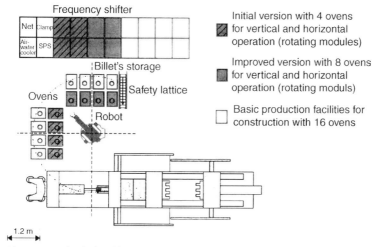

Figure 1.9 Individual coil heating system [31, 32].

In the case of heating steel, protection against the oxidation of the surface by flushing with a protective gas such as Ar or N_2 is required, since otherwise inclusions in finished parts might occur during the subsequent forming operation. This also reduces radiation heat loss, since an oxidized dark steel surface causes increased radiation loss compared with a blank surface.

1.3.2.2 Control of the Billet Condition During and After Reheating

A general problem is how to identify whether the heating process has reached the desired condition. A very popular way is to cut the billets with a knife by hand, which gives a lot of information about the general condition and the local 'softness' distribution if it is performed by an experienced person. However, this test does not give quantitative numbers, it destroys the billet and, because it is performed after the process, it cannot be used to control the billet condition during the heating process. The measurement of temperature by thermocouples can be used as an aid in process setup, but it is not precise enough to adjust the liquid fraction accurately because in the semi-solid interval very small changes in temperature can lead to significant changes in liquid fraction (Figure 1.10).

Accordingly, also in the individual billet heating the heating process in many cases is controlled by giving a predetermined power–time curve without control of the billet condition. This results in limited reproducibility if there are some other disturbing factors such as variations in billet composition, billet volume or ambient temperature. An improvement that has been introduced into industrial production plants is to measure the energy that has been induced into the billet. This can basically be achieved by measuring electric parameters which are available in the system [34]. However, even in this case some assumptions concerning efficiency and losses have to be made, which are not completely independent of the mentioned variations of the other process parameters.

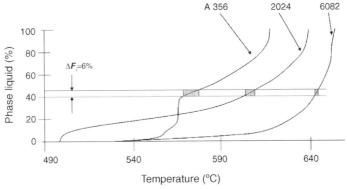

Figure 1.10 Liquid fraction as a function of temperature for various aluminium alloys [33].

In order to set the desired target liquid fraction reproducibly despite the given process fluctuations, some different control concepts have been tested successfully, for instance applying an electromagnetic sensor [33]. This sensor is introduced in the pedestal and detects the change in electric conductivity in aluminium billets resulting from melting. It could be demonstrated that feedback control based on this sensor could manage holding periods within the heating cycle. However, also for this sensor the output values depend on the chemical composition of the billet, which may vary slightly from batch to batch, so that initial adjustments are always necessary. In Part Four, a more detailed discussion concerning the heating strategy for steel will be given and the use of a hybrid strategy will be presented, where a pyrometer is used to detect the onset of melting and from then on a defined power–time curve is used.

1.3.3
Alternative Routes for Slurry Preparation

A major factor contributing to the higher costs of the thixoforming process is the high premium associated with the costs of producing the primary billets with globular microstructure by MHD casting, billet cutting and reheating them to the semi-solid condition. In trying to avoid these costs and also associated metal losses, cheaper and shorter routes to achieve the slurry would be very attractive. In recent years, a family of processes have been developed, which create the slurry during cooling from the melt: cooling slope, low superheat casting, single slug production method, continuous rheoconversion process, SEED process, a method of billet production directly using liquid electrolysed aluminium. Most of these processes make use of the high nucleation rate associated with low-temperature casting and chill cooling, but chemical reactions and other principles are also used.

1.3.3.1 The UBE New Rheocasting Process
The new UBE rheocasting process is based on the principle of feedstock production by manipulating the solidification conditions [35, 36]. The molten metal at near-liquidus temperature is poured into a tiled crucible and grain nucleation occurs on

the side of the crucible. The grain size is fine because the temperature is near liquidus. There is no need for specially treated thixoformable feedstock and scrap can be readily recycled within the plant. This new rheocasting route has a lower unit cost than thixoforming, due to the lower starting material cost [35].

1.3.3.2 The Cooling Slope Method

The slurries in the cooling slope method are made by the simple process of pouring the slightly superheated melt down a cooling slope and subsequent solidification in a die. Granular crystals nucleate and grow on the slope wall and are washed away from the wall by fluid motion. The melt, containing a large number of these nuclei crystals, solidifies in the die, resulting in a fine globular microstructure. The size of the ingot is determined by the weight of the molten metal and the die diameter [37]. The ingots can be used directly for rheo- or thixoprocessing after the appropriate reheating. The running costs of the process are significantly reduced in comparison with DC casting with MHD or mechanical stirring, and no special equipment is needed.

1.3.3.3 Low Superheat Casting

The low superheat casting route is shorter than the cooling slope route, using only a die. A slightly superheated melt is poured directly into the die and solidified. For aluminium alloy A356, low superheat casting into a cooper die with melt superheat of more than $10\,^{\circ}C$ results in a dendritic microstructure after reheating the billet to a semi-solid state ($580\,^{\circ}C$). Two conditions are necessary when casting ingots appropriate for thixoforming: one is that the superheat is lower than $10\,^{\circ}C$, and the other is that the molten metal is rapidly solidified because the cooling rate of the material in the die affects the shape of the primary crystals [37].

1.3.3.4 Single Slug Production Method (SSP Method)

The globulitic structure is generated in the same way as in the MHD-method. A magnetic field causes stirring of the molten mass. The growing dendrite structure is destroyed by shear forces generated by the flow. In comparison with MHD, the microstructure of the single slug production (SSP)-processed billet has a fine globular α-phase [38]. Using the SSP process, billets can be produced in near net shape quality directly in the heating device. This method provides the flexibility to change alloys rapidly and the capability to process alloys that are usually difficult to cast. For more details, see [38].

1.3.3.5 The Continuous Rheoconversion Process (CRP)

The continuous rheoconversion process (CRP) is based on a passive liquid mixing technique in which the nucleation and growth of the primary phase are controlled using a specially designed 'reactor'. The reactor provides heat extraction, copious nucleation and forced convection during the initial stage of solidification, thus leading to the formation of globular microstructure. The CPR reactor is mounted above the shot sleeve of a die casting machine. During each run, a dosing furnace is used to pump melt from the holding furnace to the inlet of the reactor. The melt flows through the reactor into the shot sleeve. The slurry fraction solid is adjusted by changing the temperature and flow rate of the cooling water.

Recently, the CRP has been scaled up for industrial applications. Experimental results with various commercial aluminium alloys indicated that the CRP is effective for manufacture high-quality feedstock. Process advantages include the simplicity of the process, a wide process window and the feasibility of recycling scrap metal within the process flowstream [39].

1.3.3.6 The SEED Process

A liquid-based slurry-making process known as the SEED technology for semi-solid forming is currently entering the industrial stages. The SEED process helps to overcome problems with high costs of feedstock. A large range of foundry and wrought alloy compositions (e.g. A206, A319, A356/357, AA6061 and AA6082) can be processed by this technology. In addition, the process can produce different slug dimensions and weights up to approximately 18 kg.

The technology involves two main steps: (1) heat extraction to achieve the desired liquid–solid mixture and (2) drainage of excess liquid to produce a self-supporting semi-solid slug that is formed under pressure. The principle is based on achieving rapid thermal equilibrium between the metallic container and the bulk of the metal by proper process parameter selection such as pouring temperature, eccentric mechanical stirring and drainage of a portion of eutectic liquid. For more details, see [40].

1.3.3.7 Low Superheat Pouring with a Shear Field (LSPSF)

The process named low superheat pouring with a shear field (LSPSF) uses solidification conditions to control nucleation, nuclei survival and grain growth by means of low superheat pouring, vigorous mixing and rapid cooling during the initial stage of solidification, combined with a much slower cooling thereafter. So far, the investigations on A356, A201 and A380 alloys have revealed no eutectic entrapped within the primary phase. The advantages of the LSPSF include process simplicity with high efficiency, easy incorporation into existing metal forming installations and a wide process window for pouring temperature [41].

1.3.3.8 The Gas Bubbles Technique

Flowing of gas bubbles in the melt is another effective means to provide agitation during the initial stages of solidification. The experiments with aluminium alloy A357 showed the possibility of creating a globular semi-solid metal microstructure. With a gas diffuser placed in the bottom of the melt, fine argon gas bubbles are produced. The process is carried out until the predetermined target temperature or solid fraction in the melt is achieved. Then, bubbling is stopped and the melt is allowed to cool, thus leading to a fine globular microstructure [42].

1.3.3.9 Method of Billet Production Directly Using Liquid Electrolysed Aluminium

Another possibility for producing billets with a globular microstructure on an industrial scale is a production line of semi-solid aluminium alloy billets directly using liquid electrolysed aluminium. The line can produce various aluminium alloy billets for the SSP industry (e.g. for an existing production line in China: withdrawal rate 1000 mm min^{-1} or productivity 25 tons per day, diameter range 60–90 mm and set length 4 m [43]).

After the electrolysed aluminium refinement, the liquid alloy is continuously poured into the mould system at a temperature 0–10 °C above the liquidus. Through the electromagnetic stirring, mechanical vibration and cooling, the liquid alloy is formed into a semi-solid slurry. More detailed information is given in [43].

1.3.4
Process Alternatives for Semi-solid Forming

Of the forming process alternatives listed at the beginning of Section 1.3, only the production of individual components has achieved industrial application, semi-solid rolling and extrusion being limited to laboratory trials. For semi-solid forming of shaped parts, the process variants shown in Figure 1.11 utilize the specific flow and forming properties of semi-solid materials with specific modifications in terms of process design and the installed equipment.

Thixocasting (a) describes a process where an ingot billet with liquid fraction in the range 40–60% is squeezed into a closed die by a shot piston, comparable to high-pressure die casting. The process is mainly performed on conventional real-time controlled die casting machines where the shot chamber system is adapted to the semi-solid billet insert. In thixocasting, the die filling velocity is significantly faster than in thixoforging but lower than in conventional high-pressure die casting. The flow properties of the semi-solid material lead to a laminar die filling. By the closed flow front, air enclosed in the die cavity is evenly conducted through venting channels and the division plane. The use of a faceplate within the gating system reduces the entry and inclusion of damaging surface oxides from the reheated billet [14]. Since the highly advanced die technology from high-pressure die casting can be used in this

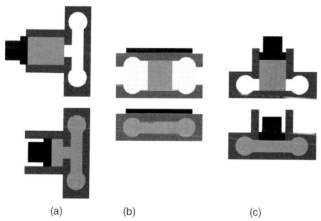

(a) (b) (c)

Figure 1.11 Schematic view of different thixoforming processes: (a) thixocasting; (b) thixoforging; (c) thixo lateral extrusion.

process, it is possible to manufacture very complex components. A potential disadvantage is that feeding of shrinkage porosity in some cases might be limited because it mainly has to be performed through the gating system [17].

In thixoforging (b), the semi-solid billet of significantly lower liquid fraction (30–40%) is inserted straight in the lower half of the horizontally sectioned tool analogous to the conventional drop forging process. The forming operation is performed by closing either one or both of the die halves. In contrast to thixocasting, the force transmission for the forming and densification step is applied over the whole tool surface so that hydrostatic pressure is affected evenly during solidification. However, the geometric complexity is limited to geometries which allow forging without flash and oxides on the billet surface must be reduced to a minimum, because there is the risk of including surface during forming.

The process alternative of thixo lateral extrusion (c) is characterized by squeezing the semi-solid material into an already closed die. Using servo-hydraulic presses, the forming velocity is of the same order of magnitude as in thixoforging. However, in terms of the diversity of parts, it allows an increased degree of geometric freedom, for example undercut sections in tooling [17].

1.3.5
Industrial Application Potential

This chapter is intended to give a brief overview of industry-oriented applications of semi-solid metal forming. While industrial mass production is basically limited to semi-solid casting of AlSi7Mg, there have been numerous industrially driven prototype developments with other alloy systems related to automotive and other applications. Most of these applications have been reported in the biennial series of the International Conference on Semi-Solid Processing of Alloys and Composites and there is a review dedicated to light metal alloys [20]. Therefore, this section will only discuss some basic aspects of semi-solid casting of light metals and will give an overview concerning demonstration projects on semi-solid casting and forming of iron-based alloys.

1.3.5.1 AlSi7Mg (A356, A357)
The volume fraction of eutectic in these alloys is about 45%. This and the distribution of liquid fraction as a function of temperature (Figure 1.10) reduce the difficulties to reheating these alloys to a defined liquid fraction just above the eutectic temperature. They also show a good casting behaviour in terms of flow length, joining of flow fronts and little tendency to form hot cracks during solidification. These advantages significantly widen the process window during reheating and forming, allowing defect-free components to be produced. In addition, these alloys show good corrosion behaviour and they offer a wide range of mechanical properties, including conditions of high strength or high elongation to fracture, which can be adjusted by heat treatment procedures and by the choice of the Mg content (0.25–0.45% in A356 and 0.45 to 0.60% in A357). Accordingly, these alloys are dominant in industrial production by semi-solid casting. Since industrial applications are mostly driven by a

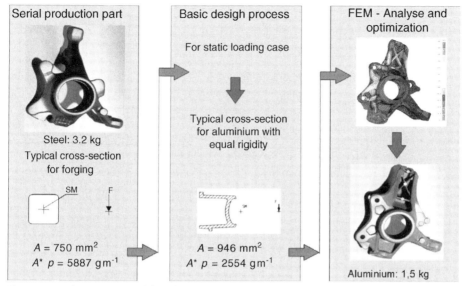

Serial production part	Basic desigh process	FEM - Analyse and optimization
Steel: 3.2 kg	For static loading case	

Figure 1.12 Substituting original forged steel steering knuckle (weight 3.2 kg) by A357 semi-solid casting (weight 1.5 kg) [44].

cost–performance relation, successful serial applications make use of the specific superior properties of semi-solid formed components to compensate for the additional costs involved. Examples of such applications include pressure-tight components such as fuel rails (Magneti Marelli), automotive structural chassis and suspension components such as steering knuckles and connecting rods (Stampal), taking advantage of the high mechanical properties, or front door pillars (SAG), requiring close tolerances and high weldability.

One difficulty when introducing the semi-solid metal (SSM) process is that usually it is not sufficient to substitute only the process for a given part geometry. In most cases a process-specific redesign will be required to achieve substantial benefits. This especially holds if the goal is to achieve significant weight savings, that is, by substituting steel forging by semi-solid forming of aluminium. As an example, Figure 1.12 shows the redesign of a forged steering knuckle [44]. The new design uses the process capabilities of semi-solid casting and reduces weight by creating 'hollow' cross-sections with high stiffness, which is not possible in forging. The component, which has been semi-solid cast in A357 alloy, fulfilled all dynamic testing procedures that were required for this component while reducing the part weight to 50% of the original steel solution.

1.3.5.2 Other Aluminium Alloys

Even though the broad range of mechanical properties achievable by the AlSi7Mg alloys covers many technical fields, some applications require specific properties which could be better achieved by other alloy systems. This has led to various alloy developments and demonstration tests to exploit the potential of semi-solid forming also for such cases. These developments show that in most considered aluminium alloys it is in principle possible to create the fine-grained globular microstructure required for semi-solid forming [45, 46]. However, during heating and forming the difficulties in achieving sound parts vary significantly depending on, for example, reheating behaviour, flow behaviour and hot crack sensitivity. Witulski *et al.* have given a list of criteria which can be used to judge the suitability of a specific alloy for semi-solid forming [46]. Selected examples and results are given below:

Applications Requiring High Wear Resistance   High wear resistance can be achieved by semi-solid forming of hypereutectic silicon alloys, such as AlSi17Cu4Mg (A390), or by using metal matrix composites (MMCs). Both materials are difficult to cast by conventional processes and may also cause high tool wear in machining. In semi-solid forming, A390 has proved to allow good die filling while achieving a micro-structure in which the primary silicon particles are evenly distributed (Figure 1.13).

Feedstock billets from SiC particle-reinforced metal matrix composites such as Duralcan [47] can be produced by MHD casting if particle sedimentation is avoided in

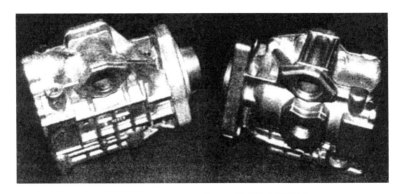

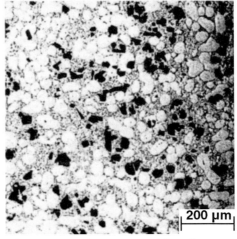

200 µm

Figure 1.13 Diesel injection pump housing semi-solid cast from A390 and microstructure with even distribution of primary silicon.

AlMg5Si2MnCr	very good casting behaviour
AlMg5Si2	same, but larger variation
AlSi7Mg0,3	tendency towards sticking
AlMg3,5Si1,4Mn	tendency towards sticking
AlMg2Si0,8	severe sticking / welding in the die

Figure 1.14 Space frame node demonstrator and alloy casting performance [49].

the furnace by appropriate stirring. Then the reheating and semi-solid casting behaviour of these billets is relatively easy due to the stabilizing effect of the non-melting particles. A significant advantage of semi-solid forming instead of casting these materials is that the particles do not agglomerate or separate during the solidification of the parts.

Usually Wrought Alloys (*i.e.* 2*xxx*, 5*xxx*, 7*xxx* Series)   These alloys, which are usually forged, are well known for their superior mechanical properties at room and elevated temperature. There have been various attempts to use these alloys also in semi-solid casting [45, 46], but they suffer from a very narrow temperature window, limited 'flow length' and a tendency to form hot cracks. Some success has been achieved using modified alloy compositions of the type AlMg5Si2Mn [48, 49], for example the space frame node for an Audi car structure, which requires a yield strength between 120 and 150 MPa, an ultimate strength above 180 MPa and elongation above 15% (Figure 1.14). This thin-walled structure must achieve these properties without heat treatment to avoid distortion, so that AlSi7Mg was not a possible solution. It has been semi-solid cast successfully from MHD cast billets of AlMg5Si2Mn [49].

1.3.5.3   High Melting Point Alloys (*i.e.* Steels)
In contrast to aluminium, SSM forming of higher melting alloys such as steel has not yet reached the state of industrial applicability due to technological problems, mostly due to the higher process temperature. The investigation of the semi-solid processing of steel started in the 1970s at MIT [2, 3] and was followed by Alumax [4] and the University of Sheffield [5]. These investigations showed by means of successfully produced parts that it is possible to apply SSM to the production of steel components. In the late 1990s, the semi-solid forming of steel again became the focus of great interest because of the expected market potential which was already described in Chapter 1.1. Since then, several Japanese and European projects have been carried out [6–14]. The target of these projects was the development of the necessary technology for semi-solid production of steel components and their common focus was the investigation of suitable steel grades and tool materials. To give an overview of the past research work on semi-solid forming of steels, representative parts of the projects are shown in Table 1.1. It can be seen that part weights from less then 200 g up to more than 3 kg have been produced using carbon steels, tool steels and cast iron.

Table 1.1 Overview of semi-solid processed steel part examples.

Year	1992	1992	1996	1997	2003
Process	Thixoforging	Thixoforging	Thixoforging	Thixocasting	Thixo lateral extrusion
Weight	<200 g	~1.3 kg	137 g	1010 g (2 kg*) 379 g (870 g*)	~250 g
Alloy	X105CrMo17, X5CrNi18-10	HS6-5-2, CoCr28MoNi4	FC-10/20/30, FCD-45 (cast iron)	C70S6, 100Cr6, HS6-5-2	100Cr65, HS6-5-3
Tool material	—	Graphite	—	X38CrMoV5-3, TZM	NiCu20TiAl9
No. of part	>1000	~30	—	~250	<30
Reference	[4]	[5]	[6]	[7, 8]	[7, 8]

Year	2003	2003	2004	2004	2004
Process	Thixoforging	Thixocasting	Thixoforging	Thixocasting	Thixo-, rheoforging
Weight	~630 g	—	3.5 kg	385 g (2985 g) 66 g (3 p. in a shot)	360 g
Alloy	C38, C60, C80, HS6-5-3	FCD450-10 (cast iron)	49MnVS3, 70MnVS5	X210CrW12	X210CrW12, HS6-5-2, 100Cr6
Tool material	Si3N4 ceramics, X38CrMoV5-1, X38CrMoV5-3	Co-base alloy (NGK Insulators, MC-9)	Si3N4 ceramics, X45MoCrV5-3-1, Ni-base alloys	X38CrMoV5-1, Laser sintered direct steel powder	TZM, X38CrMoV5-1, X45MoCrV5-3-1
No. of part	>100	>1200	>100	~30	>100
Reference	[10, 11]	[12]		[17]	[13, 14]

Billet weight.

The temperature window for these alloys is well above 1250 °C and may reach up to 1500 °C depending on the material chosen (see Chapter 3). This means that the heating procedures must take into account severe heat loss due to radiation and also surface degradation caused by scale formation. Accordingly, specific technologies have been developed to achieve acceptable precise, homogeneous and reproducible billet heating results (see Chapter 9). However, as a result of the high temperature differences between the billet and the surroundings, it would require extensive efforts to bring temperature gradients within the billet down to the same homogeneity as in the processing of aluminium alloys.

The excessive cyclic thermal loading of tools and dies causes another challenge for the process. Even though the energy content of semi-solid steel is far below that of fully liquid steel, it causes high temperature gradients and thermal stresses within the tools and dies. So far no efficient tool material that can survive several thousand semi-solid forming operations has been developed. Therefore, the selection of tool materials has to be made by considering given load profiles during the forming operation depending on the forming concept and material. Characteristic properties of several tool materials, which have already been applied in different studies (see Table 1.1), and their suitability regarding semi-solid processing are summarized in Table 1.2. Specific heat capacity, thermal conductivity and density (not included in Table 1.2) are decisive for the thermal balance between workpiece and tool material. On the one hand, high values of these quantities lead to accelerated extraction of heat from the workpiece, which causes a premature solidification, thus resulting in a poor flow length. On the other hand, low values of these quantities lead to an elevated tool temperature on the contact surface, which can cause severe material damage depending on its temperature resistance. The mechanical damage on the tool surface occurs if the local forming pressure exceeds the strength of the tool material at a given local temperature (hot strength). Damage can also occur as a form of corrosion, abrasion or adhesion due to the contact with workpiece material, especially with the steel melt when it comes to semi-solid state. Another typical type of tool damage is propagation of a crack. For a metallic tool material, hot cracking occurs after a large number of thermal shock cycles as a consequence of thermal fatigue. For ceramic tool materials, cracks occur instantaneously when a certain magnitude of thermal shock is exceeded. Even though silicon nitride ceramics possess high thermal shock resistance compared with other ceramics, crack propagation can only be inhibited by maintaining sufficient compressive stress, for example via reinforcement, since it possesses very poor tensile strength, contrary to its high compressive strength. Generally, the combination of high heat capacity, high heat conductivity and low thermal expansion leads to a good thermoshock resistance, for example for Mo-base alloys. It is possible to combine some of these tool materials within a set of forming dies in order to adjust the tool to the local load profile. In this case, the different expansion coefficients of the tool materials have to be carefully considered for tool construction. Furthermore, an improvement of the applicability of metallic tool materials may be realized by surface coating technologies. For this purpose, different coating concepts are currently being investigated. As tool life is one of the decisive criteria for the economic performance of semi-solid forming of steels, this topic is dealt with in detail in Chapter 8.

Table 1.2 Rating matrix of different tool materials for semi-solid forming of steel.

Material properties and criteria	Hot-work tool steels [51, 52]	Ni-base alloys [51, 52]	Mo-base alloys [53]	Cu-base alloys [51]	Silicon nitride ceramics [50]	Graphite [54, 55]
Specific heat capacity ($J\,kg^{-1}\,K^{-1}$)	480–710	440–550	~310	400–430	700–1200	700–2200
Heat conductivity ($W\,m^{-1}\,K^{-1}$)	25–40	12–55	90–135	250–350	10–20	75–80
Strength ($N\,mm^{-2}$) at 20 °C	~1500	~1200	~800	440–500	~1000	50–100
at 600 °C	~1000	~1200	~500	—	~900	65–130
at 1000 °C	—	~200	~450	—	~600	75–150
Expansion coefficient ($10^{-6}\,K^{-1}$)	11–13	~13	~5.3	~19	2.7–3	3.9–5.3
Temperature resistance	−	0	++	−	++	++
Thermoshock resistance	0	0	+	0	−	+
Adhesion resistance	−	−	+	+	++	−
Flow length	+	+	0	−	++	+

List of Abbreviations

DC	direct chill
MHD	magnetohydrodynamic
MMC	metal matrix composite
SIMA	strain induced melt activated
SSM	semi-solid metal
SSP	single slug production
CRP	continuous rheoconversion process
LSPSF	low superheat pouring with a shear field

References

1 Spencer, B., Mehrabian, R. and Flemings, M.C. (1972) *Metallurgical Transactions*, **3**, 1925.

2 Flemings, M.C., Riek, R.G. and Young, K.P. (1976) *Materials Science and Engineering*, **25**, 103–117.

3 Flemings, M.C. and Young, K.P. (1978) *Yearbook of Science and Technology*, McGraw-Hill, New York, pp. 49–58.

4 Midson, S.P., Nicholas, N.H., Nichting, R.A. and Young, K.P. (1992) Proceedings of the 2nd S2P Conference, Cambridge, USA, pp. 140.

5 Kapranos, P., Kirkwood, D.H. and Sellars, C.M. (1993) *Journal of Engineering Manufacture, B1a*, **207** (4), 1.

6 Kiuchi, M., Sugiyama, S. and Arai, M. (1996) *Journal of the Japan Society for Technology of Plasticity*, **37** (430), 1219–1224.

7 Püttgen, W., Bleck, W., Seidl, I., Kopp, R. and Bertrand, C. (2005) *Advanced Engineering Materials*, **7** (8), 726.

8 Puttgen, W., Pant, M., Bleck, W., Seidl, I., Rabitsch, R. and Testani, C. (2006) *Steel Research International*, **77** (5), 342–348.

9 Omar, M.Z., Palmiere, E.J., Howe, A.A., Atkinson, H.V. and Kapranos, P. (2005) *Materials Science and Engineering A*, **395**, 53–61.

10 Behrens, B.A., Haller, B. and Fischer, D. (2004) *Steel Research International*, **75** (8/9), 561.

11 Rassili, A., Robelet, M. and Fischer, D. (2006) Proceedings of the 9th ESAFORM Conference, Glasgow, pp. 819–822.

12 Tsuchiya, M., Ueno, H. and Takagi, I. (2003) *JSAE Review*, **24**, 205–214.

13 Küthe, F., Schönbohm, A., Abel, D. and Kopp, R. (2004) *Steel Research International*, **75** (8/9), 601–606.

14 Hirt, G., Shimahara, H., Seidl, I., Küthe, F., Abel, D. and Schönbohm, A. (2005) *CIRP*, **54/1**, 257–260.

15 Flemings, M.C. (1991) *Metallurgical Transactions A*, **22** (5), 957.

16 Atkinson, H.V. (2005) Modelling the semisolid processing of metallic alloys. *Progress in Materials Science*, **50**, 341–412.

17 Hirt, G., Bleck, W., Bührig-Polaczek, A., Shimahara, H., Püttgen, W. and Afrath, C. (2006) Proceedings of the 9th S2P Conference, Korea.

18 Kenney, M.P., Courtois, J.A., Evans, R.D., Farrior, G.M., Kyonka, C.P., Koch, A.A. and Young, K.P. (1988) *Metals Handbook,* Vol. 15, 9th edn, ASM International, Metals Park, OH, pp. 327–338.

19 Fan, Z. (2002) *International Materials Reviews*, **47**, 49–85.

20 de Figueredo, A. (ed.) (2001) *Science and Technology of Semi-solid Metal Processing*, North American Die Casting Association, Rosemont, IL.

21 Midson, S.P. (1996) Proceedings of the 4th S2P Conference, England.

22 Young, K.P., Kyonka, C.P. and Courtois, J.A. (1982) Fine grained metal composition, US Patent 4,415,374, 30 March.

23 Young, K.P. (1987) US Patent 4,687,042.

24 Kenney, M.P., Young, K.P. and Koch, A.A. (1984) US Patent 4,473,107.

25 Moschini, R. (1992) Manufacture of Automotive Components by Semi-Liquid Forming Process. Proceedings of the 2nd S2P Conference, USA.

26 Gabatuler, J.-P. (1992) Proceedings of the 2nd S2P Conference, USA.

27 Wan, G., Witulski, T. and Hirt, G. (1994) Thixoforming of aluminium alloys using modified chemical grain refinement for billet production. *La Metallurgia Italiana*, **86** (1), 29–36.

28 Kirkwood, D.H., Sellars, C.M. and Elias Boyed, L.G. (1992) Thixotropic materials. European Patent 0305375, 28 October.

29 Young, K.P. and Fitze, R. (1994) Proceedings of the 3rd S2P Conference, Japan.

30 Behrens, B.-A., Fischer, D., Haller, B., Rassili, A., Klemm, H., Flüss, A., Walkin, B., Karlsson, M., Robelet, M. and Cucatto, A. (2004) Proceedings of the 8th S2P Conference, Cyprus, Session 1.

31 Hirt, G., Cremer, R. and Sommer, K. (1997) Sensor controlled induction heating for semi solid forming of aluminium alloys. Proceedings of Congrès International de l'Induction dans les Procédés Industriels, Paris, 26–29 May, Vol. 1, pp. 489ff.

32 Hirt, G., Cremer, R., Sommer, K. and Witulski, T. (1997) Advances in thixoforming – plant technology, component manufacturing and simulation. Proceedings of Nadca's 19th International Die Casting Congress and Exposition, Minneapolis, MN, 3–6 November, pp. 377–382.

33 Cremer, R., Winkelmann, A. and Hirt, G. (1996) Sensor controlled induction heating of aluminium alloys for semi solid forming, Proceedings of the 4th S2P Conference, England.

34 Gräf, T., Jürgens, R. and Gies, J. (2000) Controlled inductive heating for thixotropic materials into the semi-solid state. Proceedings of the 6th S2P Conference, Italy.

35 Hall, K., Kaufmann, H. and Mundl, A. (2000) Proceedings of the 6th S2P Conference, Italy (eds G.L. Chiarmetta and M. Rosso).

36 UBE Industries (1996) Method and apparatus of shaping semisolid metals. European Patent 0745694.

37 Haga, T., Kapranos, P., Kirkwood, D.H. and Atkinson, H.V. (2002) Proceedings of the 7th S2P Conference, Japan.

38 Müller-Späth, H., Achten, M. and Sahm, P.R. (1996) Proceedings of the 4th S2P Conference, England.

39 Pan, Q., Wiesner, S. and Apelian, D. (2006) Proceedings of the 9th S2P Conference, Korea.

40 Langlias, J. and Lemieux, A. (2006) Proceedings of the 9th S2P Conference, Korea.

41 Guo, H. and Yang, X. (2006) Proceedings of the 9th S2P Conference, Korea.

42 Wannaish, J., Martinez, R.A. and Flemings, M.C. (2006) Proceedings of the 9th S2P Conference, Korea.

43 Xing, S., Zhang, L., Zhang, P., Du, Y., Yao, J., Wu, C., Wang, J., Zeng, D. and Li, W. (2002) Proceedings of the 7th S2P Conference, Japan.

44 Hirt, G., Cremer, R., Tinius, H.-C. and Witulski, T. (1998) Lightweight near net shape components produced by thixoforming, *Materials and Design*, **18** (4/6), 315–321.

45 Kapranos, P. and Atkinson, H.V. (2002) Proceedings of the 7th S2P Conference, Japan.

46 Witulski, T., Morjan, U., Niedick, I. and Hirt, G. (1998) Proceedings of the 5th S2P Conference, USA.

47 Dualcan Composites – Mechanical and Physical Property Data, Duralcan Bulletin, February (1990).

48 Garat, M., Maenner, L. and Sztur, C. (2000) Proceedings of the 6th S2P Conference, Italy.

49 Hirt, G. (2000) Proceedings of the 6th S2P Conference, Italy.

50 Kopp, R., Shimahara, H., Schneider, J.M., Kurapov, D., Telle, R., Münstermann, S., Lugscheider, E., Bobzin, K. and Maes, M. (2004) *Steel Research International*, **75** (8/9), 570.

51 Hirt, G. and Nohn, B. (1999) Proceedings of the DVM-Congress, Berlin.

52 Wegst, W. (1995) *Key to Steel*, 17th edn. Stahlschlüssel Wegst.

53 Plansee-Group, Products Information Brochure, Material Properties and Applications of Molybdenum, http://www.plansee.com/hlw/im_materials_ENG_HTML.htm.

54 SGL Carbon Group, Products Information of Specialty Graphite Grades for Industrial Applications, http://www.sglcarbon.com/gs/grades/industrial.html.

55 Carbon-Industrie-Produkte GmbH, Products Information Brochure of Graphite for Die and Mould Making http://www.carbon-graphit.de/files/pdf/Funkenerosion.pdf.

Part One
Material Fundamentals of Metal Forming in the Semi-solid State

Thixoforming: Semi-solid Metal Processing. Edited by G. Hirt and R. Kopp
Copyright © 2009 WILEY-VCH Verlag GmbH & Co. KGaA, Weinheim
ISBN: 978-3-527-32204-6

2
Metallurgical Aspects of SSM Processing*

Peter J. Uggowitzer and Dirk I. Uhlenhaut

2.1
Introduction

With the discovery of shear thinning and the thixotropic behaviour of partially solidified alloys under vigorous agitation, a new era in forming technology was started, namely semi-solid metal (SSM) processing. The new technology promises several important advantages: improved die filling, less air entrapment and less oxide inclusions due to the higher viscosity compared with fully liquid melts, longer die life, shorter solidification time, reduced cycle time and therefore higher productivity due to lower heat content and lower process temperature and reduced shrinkage, and thus near net shape or even net shape production due to the partially solidified slurry.

Two basic routes of SSM processing – termed 'rheocasting' and 'thixocasting' – proved their feasibility in industrial trials. The rheo-route involves the preparation of an SSM slurry from liquid alloys and transfer of the prepared slurry directly to a die or mould for component shaping. The thixo-route is basically a two-step process, involving the preparation of a feedstock material with thixotropic characteristics, reheating the solid feedstock material to semi-solid temperature and shaping the semi-solid slurry into components. Both routes aim at the formation of an 'ideal' slurry that exhibits an accurately specified volume fraction of fine and spherical solid particles uniformly distributed in the liquid matrix [1]. Such microstructural tuning requires specific properties of the alloys used, namely sufficient width of the freezing range and adequate temperature sensitivity S^* [2]. For all possible alloying systems, the liquid and solid phases in the freezing range will differ in their chemical composition due to near-equilibrium element partitioning. During complete solidification in the die, however, for specific alloying systems, that is, single-phase systems, this element partitioning might not be outweighed by diffusion processes and can cause severe deterioration of properties. Such systems are less capable at SSM processing. A further aspect concerning the selection of appropriate alloying

*A List of Abbreviations and Symbols can be found at the end of this chapter.

Thixoforming: Semi-solid Metal Processing. Edited by G. Hirt and R. Kopp
Copyright © 2009 WILEY-VCH Verlag GmbH & Co. KGaA, Weinheim
ISBN: 978-3-527-32204-6

systems is the formation of intermetallic phases (IMPs) in the residual liquid, and the width of the terminal freezing range (TFR), which is the non-equilibrium partial solidification range near termination of solidification that indicates the alloy's proneness to hot tearing.

In the following sections, the above-mentioned criteria of temperature sensitivity, proneness to segregation and hot tearing, and IMP-formation are discussed for selected alloying systems. Both light metals (aluminium and magnesium) and iron-based alloys are considered. In addition, some metallurgical aspects are discussed with regard to the characteristics of slurry formation in the rheo- and thixo-routes, and the impact of variations in the alloy compositions on the structure and properties of SSM components.

2.2
Temperature Sensitivity S^* and Solid–Liquid Fraction

For the thixo-formability of alloys, the correct adjustment of the solid–liquid fraction is of crucial importance. In general, systems with a wide freezing rage are beneficial. In a real thixo-process, the temperature will always be subject to some error. The impact of temperature variations on the present initial solid fraction, f_S^*, can be expressed by the (negative) slope of the equilibrium solid fraction curve:

$$S^* = -\frac{df_S^*}{dT} \tag{2.1}$$

S^* depends on the alloy's composition and also on the initial amount of solid or liquid. It is advisable to consider the average slope in the partial freezing range $f_S^* = 0.4$–0.6. The average sensitivity is inversely proportional to this partial freezing range. Therefore, as for selection of appropriate alloys, the temperature sensitivity S^* should be small.

Table 2.1 indicates that for the light metal cast alloys A356 and AZ91 a temperature error of ± 5 K results in a variation of solid fraction of only about ± 0.04, but for the wrought alloy AA6082 the variation is about ± 0.135 or roughly 25% of the initial fraction. As for the Fe-based alloys, the tool steel X210CrW12 proves to be more suitable for SSM processing than the bearing steel 100Cr6.

Table 2.1 Relevant temperatures and temperature sensitivities of various alloys [2, 3].

Alloy	T_S (°C)	T_L (°C)	T_{50} (°C)	ΔT^{40-60} (°C)	S^* (K^{-1})
AlSi7Mg (A356)	557	614	574	17	0.0083
AlSi1Mg (AA6082)	557	647	637	7	0.027
AlSi9Cu3 (A380)	548	603	566	10	0.039
AZ91	470	600	570	22	0.0087
100Cr6	1348	1461	1427	19	0.010
HS6-5-2	1175	1432	1360	35	0.0055
X210CrW12	1221	1366	1291	50	0.0045

It is important to note that only the interglobular liquid phase contributes to shear thinning and therefore thixotropic behaviour. Intraglobular liquid phase that may form during reheating of the thixo-feedstock does not contribute to the viscosity of the slurry and therefore the numbers in Table 2.1 are applicable only for 'ideal' slurries.

While the usability of the criteria 'solid fraction' and 'temperature sensitivity' has been confirmed for Al and Mg alloys, the situation with Fe-based alloys is more complex. Here, due to technological difficulties (tooling problems and oxidation), the absolute temperature level must also be taken into account. High-alloyed steels with a complex melting behaviour mostly exhibit a wide solidification interval at relatively low temperatures and are therefore less sensitive to temperature changes and easier to thixo-form. Figure 2.1 illustrates the situation for different steel grades by means of solidification curves (DTA measurements) [3]. It is worth noting that the complex steels (X220CrVMo134, X210CrW12, HS6-5-3) also exhibit eutectic solidification. The benefit of such phase transformation is addressed in Section 2.4.

As for the solid fraction adjustment, it is recommended also to take the absolute temperature at $f_S = 50\%$ into account, which is $T_{50} \approx 850$ K for AlSi7Mg, $T_{50} \approx 910$ K for AlSi1Mg, $T_{50} \approx 1565$ K for X210CrW12 and $T_{50} \approx 1700$ K for 100Cr6. As a rough appraisal for the evaluation of the thixo-ability of an alloy, we propose a simple multiplication of S^* and T_{50}. For AlSi7Mg $S^* T_{50}$ is 7.5, for X210CrW12 it is 7.0, for 100Cr6 it is 17.0 and for AlSi1Mg it is 24.6. According to this simple rule, AlSi7Mg and X210CrW12 are of comparable capability, whereas the adjustment with 100Cr6 and AlSi1Mg is much more complicated.

As mentioned above, from the temperature sensitivity a wide freezing rage can be considered to be beneficial. A wide freezing rage is achieved in systems with a low element partitioning ratio, k (element content in solid phase/element content on

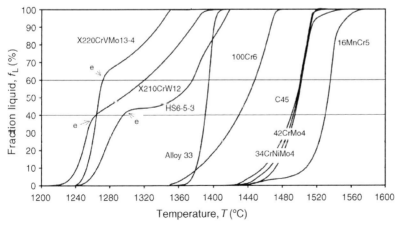

Figure 2.1 Solidification curves for various steel grades; the complex steels exhibit eutectic solidification at relatively low temperatures [4].

liquid phase) and a high slope of the liquidus line, m_l. According to Easton and StJohn [4], a high product $m_l(k-1)$ reflects on the one hand a high grain growth restriction, which can be considered to be beneficial, but on the other a high constitutional undercooling ΔT_C. A high ΔT_C, however, results in increased proneness to dendritic solidification, which is a disadvantage for SSM processing.

2.3
Slurry Formation in the Rheo- and Thixo-routes

As already mentioned, two major categories of SSM slurry-making processes exist. The first group comprises all processes in which slurries are produced directly from the liquid phase. These routes are termed the rheo-route or rheocasting, or 'slurry-on-demand' (SoD) technology. The second category of SSM processing – the thixo-route – deals with solid feedstock that was agitated during solidification and exhibits a rosette-like structure. It is then reheated into the freezing range in order to form a non-dendritic, globular structure of the primary phase.

From the microstructural point of view, there is a significant difference between the thixo- and rheo-materials. Spheroidization of the thixo-material occurs during reheating and final holding at the casting temperature. During this process, liquid pockets are entrapped in the primary globules and cannot contribute as a 'lubricant' to the final forming step. In SoD processes, this entrapment never occurs during solidification of the primary aluminium phase, and therefore the full amount of liquid phase can contribute to the forming operation (Figure 2.2).

When talking about the globular 'primary phase', it implies that also a 'secondary phase' exists. For most light metal casting alloys this is in fact the case; the secondary phase is eutectic, as for A356, A319 and AZ91. But complex steel grades also exhibit pronounced eutectic solidification; see Figure 2.1, label 'e'. During reheating, the eutectic transforms to liquid and no anomalous remelting occurs. In contrast to this

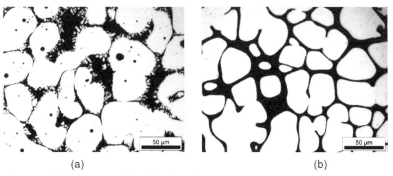

(a) (b)

Figure 2.2 Microstructure of the alloy A356; thixo-route (a) and rheo-route (b). Entrapped liquid pockets are visible in the thixo-sample.

desired behaviour, an extruded or rolled single-phase feedstock material features completely different remelting characteristics. Due to dynamic recrystallization during hot deformation, an extruded or rolled feedstock material does not exhibit microsegregation (coring), that is, there is no gradient in the alloying content within a single grain. When during reheating the temperature reaches the solidus temperature, remelting starts at grain boundaries. As the liquid dissolves more alloying elements than the solid, the adjacent solid becomes depleted in alloying elements. The solidus temperature within this depleted seam is increased, that is, further remelting into the solid grain can only occur when the temperature is raised. This phenomenon is called self-blocking remelting [5]. Within the grains, where the original alloying content has been preserved, the solidus temperature stays unchanged and remelting may start simultaneously everywhere inside the grains, that is, many tiny liquid droplets develop. Once again, the vicinal area of every small liquid inclusion becomes depleted in solute content, the local solidus temperature is raised and remelting stops. This is the reason for the formation of many small liquid droplets within each grain in the early stage of reheating into the semi-solid state. These liquid droplets become fewer and larger by growth and coalescence with increasing holding time in the semi-solid state seeking to reduce the interfacial energy. Such an entrapped liquid is shown in Figure 2.3 for single-phase AZ80 and 100Cr6 feedstock material.

Self-blocking remelting is an exclusive feature of hot deformed or homogenized feedstock material and does not occur in microsegregated feedstock materials that are produced by casting techniques.

Extruded or rolled feedstock material may not be ideal in respect of entrapped liquid, but it has its advantage in superior sphericity of the solid particles compared with most other feedstock materials. As it does not contain any dendrites or rosette-like structures, little time is needed to spheroidize the grains in the semi-solid state. The excellent sphericity of the solid particles is helpful for good fluidity and may counterbalance the detrimental effect of the entrapped liquid.

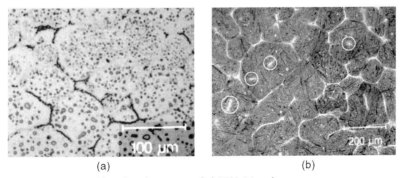

(a)   (b)

Figure 2.3 Entrapped liquid pockets in extruded AZ80 (a) and rolled 100Cr6 (b) (circles) feedstock material, reheated to the semi-solid state.

2.4
Proneness to Segregation and Hot Tearing

A significant difference in microstructural development of partly eutectic alloys on the one hand and single-phase alloys on the other is observed not only for slurry formation but also for the solidification after the forming process. Figure 2.4 illustrates the freezing behaviour for various alloying systems, calculated for non-equilibrium solidification using the Scheil model. Commercial forming processes do not allow equilibrium solidification, which is why a eutectic is also often found in 'single-phase' alloys (see Figure 2.4: AA6082, AZ91 and 100Cr6).

The presence and morphology of the eutectic phases significantly influence the properties of the components. A comparison of the Al alloy A356 with the Mg alloy AZ91 illustrates this peculiarity. A356 is SSM processed just above the eutectic temperature of 575 °C where it features an adequate liquid fraction f_L of 51%. During cooling, this liquid solidifies as a eutectic and the final microstructure consist of 49% α-Al and 51% eutectic. AZ91 has to be processed at a temperature of about 570 °C ($f_L \approx 50$%) and non-equilibrium solidification ends with alloying element enrichment in the remaining

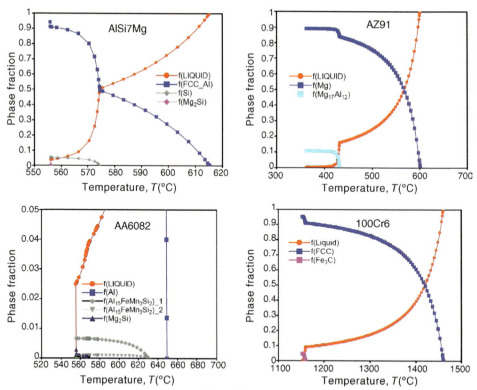

Figure 2.4 Non-equilibrium (Scheil) solidification behaviour of various alloying systems; 'partly eutectic' alloy A356 and 'single-phase alloys' AA6082, AZ91 and 100Cr6.

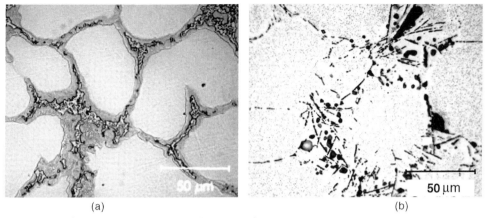

(a)   (b)

Figure 2.5 IMP formation along the grain boundaries. $Mg_{17}Al_{12}$ in AZ91 (a), Si and AlFeMnSi in AA6082 (b).

liquid and thus the formation of about 16% eutectic. However, it is not the total amount of eutectic that most controls the properties but the amount of the brittle phase, Si in A356 and $Mg_{17}Al_{12}$ in AZ91, and also its distribution and morphology. The Al–Si eutectic contains 11.3% Si and the eutectic in AZ alloys contains as much as 66% of the brittle IMP $Mg_{17}Al_{12}$. The globular shape of the primary Mg phase facilitates the formation of a contiguous network of eutectic. As a result, the microstructure is characterized by about 10% of coarse and very brittle intermetallic $Mg_{17}Al_{12}$ concentrated along the grain boundaries. Non-equilibrium segregation into the liquid phase occurs not only for the main alloying elements but also for trace elements such as Fe. In AA6082, for example, Fe enrichment in the liquid leads to the formation of eutectic Si and brittle Fe-rich IMPs along the grain boundaries. Figure 2.5 illustrates the IMP morphology in SSM-processed AZ91 and AA6082.

Such IMP formation due to non-equilibrium segregation significantly reduces the properties of the SSM-processed alloys. A356 with ~50% eutectic exhibits an elonga-tion to fracture of typical $12 \pm 5\%$ in the T6 condition, whereas it is only about $7 \pm 3\%$ for AZ91 (F and T4; ~15% eutectic) and $7 \pm 3\%$ for AA6082 (T6; ~2.5% eutectic) [6].

Element segregation is also an important issue in SSM processing of 'single-phase' steels such as 100Cr6. As already indicated in Figure 2.4 (Note: the appearance of a eutectic phase should be treated with caution since the calculation was performed without considering C to be a fast-diffusing element [7]), and clearly disclosed by micrographs, X-ray maps and diffusion simulations in Figures 2.6–2.8, pronounced segregation of the alloying elements (Cr, Mn and Si) and slight C enrichment towards the intergranular regions occur during cooling from the semi-solid range. In these regions, the Cr content reaches a value well above 3% and therefore the martensite start temperature decreases to below room temperature [8, 9] . This generates the formation of a unique microstructure: martensite with an intergranular network of retained austenite (Figure 2.3b), accompanied by the sporadic formation of carbides at the grain boundaries.

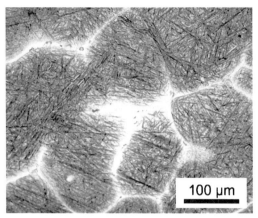

Figure 2.6 Optical image of the 100Cr6 microstructure after quenching from 1425 °C ($f_L = 0.57$) [8].

The presence of austenite would normally be expected to improve the toughness properties. However, it was shown that the impact energy of such a condition is low and fracture occurs in an intergranular manner. Intergranular fracture of carbon steels is commonly observed after overheating during austenitizing and is called 'burning' when the temperature exceeds the solidus temperature and local melting occurs at the austenite grain boundaries [10]. The intergranular fracture is then attributed to a segregation of sulfur into the liquid phase and subsequent sulfide formation during solidification. Since SSM can be viewed as an extreme form of burning, it seems reasonable to assume that the intergranular fracture is caused by segregation of sulfur. As shown in Figure 2.9, the formation of MnS along the grain boundaries can in fact be observed. Such thin, chain-like MnS precipitates are known to cause intergranular 'burning' failure [10]. The sporadic presence of carbides in the austenite grain boundaries might additionally contribute to the intercrystalline character of the fracture.

Another very important criterion concerning alloy selection and alloy modification in SSM technology is the terminal freezing range (TFR), that is, the extent of non-

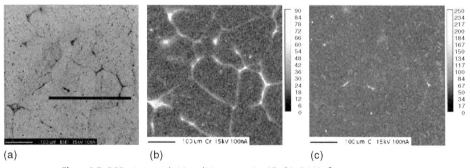

(a)   (b)   (c)

Figure 2.7 BSE micrograph (a) and X-ray mapping [Cr (b); C (c)] of a 100Cr6 sample quenched from 1400 °C [9].

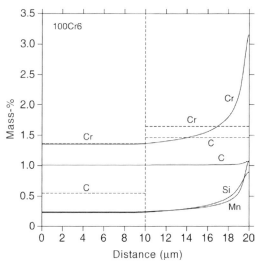

Figure 2.8 DICTRA-calculated composition profiles after quenching with $50 \, K \, s^{-1}$ from $1418 \, °C$ ($f_L = 0.57$). The dashed vertical line shows the original solid–liquid interface before quenching. The dashed horizontal lines show the C and Cr contents before quenching [8].

equilibrium freezing range near termination of solidification, $\Delta T^{88–98}$, proposed by Djurdjevic and Schmid-Fetzer [11]. A pronounced TFR may cause hot tearing and should therefore be avoided. Since only 'single-phase' alloys may develop a distinct TFR, the applicability of SSM should be assessed also from this point of view. For Mg–Al alloys, for example, it was shown that an increase in Al and Ca content might

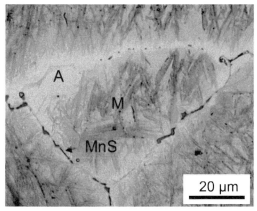

Figure 2.9 Optical microscopy image of the intergranular region of 100Cr6 (quenched from $1425 \, °C$; $f_L = 0.57$); martensite (M), retained austenite (A), and manganese sulfides (MnS) along the grain boundaries [8].

be helpful in reducing TFR. At first sight, it is the increased amount of eutectic that helps in reducing the alloy's susceptibility to hot tearing.

In summary, from a simplified metallurgical view, choosing SSM material with a significant amount of eutectic can reduce the proneness to segregation and hot tearing. For this reason, Al and Mg wrought alloys are considered to be rather unfavourable. Among the Fe-base alloys, use of the steel grades X210CrW12 and HS6-5-2 can be recommended [12].

As for segregation, only the microscopic aspect was considered in this section. Macroscopic segregation due to long-range decomposition of the liquid and solid phases during the SSM forming process is mostly caused by inappropriate technological conditions and is due to only to a minor extent solely to metallurgical circumstances. In this respect, one important aspect is discussed in the next section.

2.5
Impact of Variations in Alloy Composition

Fluctuations in alloy composition affect all casting processes, from gravity casting to high-pressure die casting. While this is a problem in fully liquid casting processes, it is even more so in semi-solid casting processes. With process stability in mind, the solidification range and in particular the ΔT^{40-60} temperature interval and the sensitivity to temperature fluctuations S^* in this temperature range have already been described as important factors influencing the choice of alloys for SSM production. For the alloys AlSi7Mg and AZ91 it was shown that ΔT^{40-60} is 17 and 22 °C, respectively, and that the sensitivity to temperature change, S^*, at $f_L = 0.5$ is 0.83 and 0.87% K^{-1}. International alloy standards tolerate significant fluctuations in the content of major and minor alloying elements. For the alloy A356, the tolerance field for Si is 6.5–7.5% and for Mg 0.25–0.45%. In the magnesium alloy AZ91, the aluminium and zinc contents can vary between 8.5 to 9.5% and 0.45 to 0.9%, respectively. With the upper and lower limits of these alloy compositions, the solidification curves shift for by degrees (Figure 2.10).

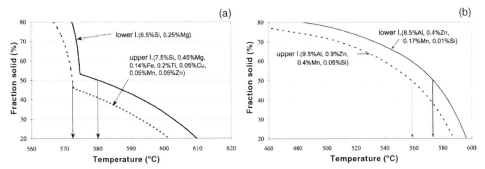

Figure 2.10 Calculated cooling curves for AlSi7Mg (a) and AZ91 (b) at the upper and lower limits of tolerated alloy composition, indicating the shift of temperature at $f_L = 0.5$ [13].

For A356, there is a temperature shift of ~7.5 °C, and for AZ91 the shift is ~12 °C. For the alloy AlSi7Mg, it must also be noted that a 50% solid fraction at the upper compositional limit cannot be reached without any eutectic solidification in the slug. If eutectic is present during the holding period of the slug, a very coarse eutectic phase will be formed. This reduces the mechanical properties, especially elongation values. However, in general, it can be stated that both alloys are well suited for semi-solids. Let us assume that SSM forming takes place at 570 °C, aiming for a 50% solid fraction in the alloy AZ91. If the alloy is at the upper limit of the tolerance field, the actual solid fraction would only be around 44%, whereas if the alloy is at the lower limit f_S would be about 55%, and only a little influence on product quality can be assumed.

For the steel X210CrW12, fluctuations of 2.0–2.3% C, 11–13% Cr and 0.6–0.8% W are tolerated, resulting in a temperature shift of about 35 °C. This corresponds to a solid fraction variation from about 42% to 57%, which is also considered to be not very harmful.

The situation is more complicated for the Al wrought alloy AA6082, with tolerated fluctuations of 0.7–1.3% Si, 0.6–1.2% Mg and 0.1–0.45% Mn. Here the temperature shift is approximately 11 °C, but with the high S^* value of 0.027 K^{-1} this corresponds to an intolerable solid fraction variation of about 30%. Similar problems are expected for the steel grade 100Cr6. The accepted chemical variations are 0.97–1.1% C, 1.3–1.6% Cr, 0.15–0.3% Si and 0.25–0.45% Mn. This implies a possible temperature shift of about 18 °C or a variation in solid fraction of almost 20%.

2.6
Conclusion

The impressive bundle of advantages mentioned in the Introduction has been the driving force for continued SSM research worldwide. In contrast to what was expected several years ago, the acceptance of SSM technology in industry is still at a relatively low level. The essential demand is to produce components at a constant high-quality level. From a metallurgical point of view, only a few alloy systems fulfil the requirements for high-quality SSM production. However, also in case of 'optimal' alloy selection, unintentional fluctuations in solid fraction due to fluctuations in alloy composition may cause quality problems. For a given set of forming parameters, an alloy with a composition at the lower or upper end of its standard may generate severe defects such as macro-segregation. This does not mean that SSM processing at high or low solid fractions is generally bad, or that it cannot yield high-quality components. If variations in solid fraction occur, the forming parameters have to be adjusted properly. On the other hand, if the process parameters are kept constant, only a limitation in compositional fluctuation can help to reduce scrap production. In order to stabilize the SSM process, the use of temperature control in combination with chemical monitoring is highly recommended.

List of Abbreviations and Symbols

SoD	slurry-on-demand
SSM	semi-solid metal
TFR	terminal freezing range
f_S^*	initial solid fraction
f_L	liquid fraction
f_S	solid fraction
k	element content in solid phase/element content on liquid phase
m_l	slope of the liquidus line
S^*	temperature sensitivity
T_{50}	temperature at which there is 50% liquid
ΔT_C	constitutional undercooling

References

1 Fan, Z. (2002) *International Materials Reviews*, **47** (2), 49–84.

2 Liu, D., Atkinson, H.V. and Jones, H. (2005) *Acta Materialia*, **53**, 3807–3819.

3 Püttgen, W. and Bleck, W. (2004) *Steel Research International*, **75**, 531–536.

4 Easton, M.A. and StJohn, D.H. (2001) *Acta Materialia*, **49**, 1867–1878.

5 Kleiner, S., Beffort, O. and Uggowitzer, P.J. (2004) *Scripta Materialia*, **51**, 405–410.

6 Kaufmann, H. and Uggowitzer, P.J. (2007) *Metallurgy and Processing of High-integrity Light Metal Pressure Castings*, Schiele & Schön, Berlin.

7 Hallstedt, B. and Schneider, J.M. (2009) In *Thixoforming – Semi-solid Materials Processing* (ed. G. Hirt), Wiley-VCH Verlag GmbH, Weinheim, Chapter 4.

8 Püttgen, W., Hallstedt, B., Bleck, W., Löffler, J.F. and Uggowitzer, P.J. (2007) *Acta Materialia*, **55**, 6553–6560.

9 Püttgen, W., Hallstedt, B., Bleck, W. and Uggowitzer, P.J. (2007) *Acta Materialia*, **55**, 1033–1042.

10 Hale, G.E. and Nutting, J. (1984) *International Materials Reviews*, **29**, 273–298.

11 Djurdjevic, M.B. and Schmid-Fetzer, R. (2006) *Materials Science and Engineering A*, **41**, 24–33.

12 Uhlenhaut, D.I., Kradolfer, J., Püttgen, W., Löffler, J.F. and Uggowitzer, P.J. (2006) *Acta Materialia*, **54**, 2727–2734.

13 Kaufmann, H., Fragner, W. and Uggowitzer, P.J. (2005) *International Journal of Cast Metals Research*, **18** (5), 279–285.

3
Material Aspects of Steel Thixoforming

Wolfgang Bleck, Sebastian Dziallach, Heike Meuser, Wolfgang Püttgen, and Peter. J. Uggowitzer

3.1
Introduction

For thixoforming, the alloy has to fulfil two material-specific requirements:

- First, distinct two- or multiphase areas are essential for a potential treatment of metals in the partial liquid state, so that a metallic suspension of solid-phase particles and the liquid phase can be adjusted within a wide temperature interval. For process control, low temperature sensitivity during the adjustment of the designated solid-phase fraction is of particular importance. Metals exhibiting a high temperature sensitivity cannot, or only with great effort, be heated homogeneously and can accordingly be processed poorly.

- Second, globular solid particles should be embedded within a liquid matrix. By this, the separation of solid and liquid elements can be reduced or even completely suppressed. The deformation of a suspension of slab-shaped, columnar or dendritic solid-phase particles within a liquid medium inevitably leads to intense decomposition effects of the solid and liquid phases, resulting in pronounced macro-segregation and impaired mechanical properties.

For the determination of the temperature intervals, various and easily applicable standard methods exist. In contrasts, the analysis of the structural formations of the partial liquid state is far more complex. Since no direct procedures for online measurement exist, the characterization of the microstructure of the partial liquid state is in general carried out metallographically on quenched, partial liquid samples. Generally, it is assumed that the microstructure does not change significantly during quenching and the quenched microstructure matches the microstructure shortly before quenching in the partial liquid state. New results indicate that nevertheless microstructural changes in the material are caused by the quenching procedure.

Thixoforming: Semi-solid Metal Processing. Edited by G. Hirt and R. Kopp
Copyright © 2009 WILEY-VCH Verlag GmbH & Co. KGaA, Weinheim
ISBN: 978-3-527-32204-6

In Section 3.2, the requirements for successful thixoforming are summarized from the materials and processing points of view. In Section 3.3, the alloying systems which have been investigated in detail are introduced.

In Section 3.4, the phenomena during quenching from the partial liquid state and their influence on the liquid state concentration and the structural morphology are examined more closely. Furthermore, the analysis procedures for the appraisal of the microstructural parameters and the material choice are evaluated. An estimation and evaluation concerning the accuracy of the determination of the structural parameters have to be conducted.

In Section 3.5, investigations of the regulation of the casting behaviour are described. In addition to the structural parameters, the adjustment of a fluid metallic suspension is also of importance for the process optimization of the thixoforming of steel.

In Section 3.6, the potential of adjusted heat-treatment strategies based on the partial liquid state are examined, because completely new material properties can be expected due to the multiphase structure and resulting different element-content distributions in the solid and liquid phases. The liquid phase is highly enriched with alloying elements, in which the considerable carbon enrichment plays a specific role during the subsequent transformation behaviour. Therefore, the emphasis is on the structure examination and the development of adjusted heat-treatment strategies based on the partial liquid state of the steel grades X210CrW12 and 100Cr6.

The research presented in this chapter aims at a reassessment of the potential and the suitability of the different steel grades for thixoforming.

3.2
Background

3.2.1
Material Prerequisites for the Thixoforming of Steel

The thixotropic behaviour of partial liquid metallic suspensions was analysed for the first time in the mid-1970s and developed further during the ensuing years [1–3]. In the examination of tin–lead alloys, which were sheared during solidification within their solidification-interval, remarkably low viscosities were detected despite their high solid fraction. Based on these observations, different semi-solid metal (SSM) technologies for aluminium, magnesium, copper and steel were developed during the following decades. A precondition for SSM technologies is a sufficiently large solidification interval, in which a globular grain structure in the partial liquid state can develop. It was proved to be advantageous if the primary material already exhibited a globular grain structure. Therefore, various primary material creation routes for the different materials were developed.

Whereas thixoforming of low-melting metals such as aluminium and magnesium is already applied industrially, thixoforming of steels is still being developed. Especially the high process temperatures of the steels pose challenges concerning suitable tools, the heating technology and the process conduct [4]. Due to the decreasing process temperatures with increasing carbon concentration, high-carbon steels are advantageous pertaining to the tool strain. In addition to the location of the solidification interval, the size or rather the development of the solids fraction with temperature is of importance. To allow easy adjustment of the liquid fraction, the increase of the cooling curve should be as low as possible. A prerequisite for a successful semi-solid forming is a particular microstructural formation in the semi-liquid state, so that the material is available as a solid–liquid suspension with a low viscosity during the forming.

3.2.1.1 Rheological Behaviour of Thixotropic Metal Suspensions

In contrast to Newtonian fluids, for instance water, which shows an increase in viscosity with increase in shear stress, metal alloys in the temperature interval between the solidus and the liquidus temperature exhibit thixotropic flow behaviour, when solid particles exist in a globular form within a liquid matrix and are able to move freely against each other(Figure 3.1). Partial liquid metals in the unstressed state can be treated similarly to solids, because of the developing coalescence of the solids particles.

In the primary examination concerning the behaviour of metallic suspensions, an increase in shear stress and a decrease in cooling rate during the process lead to denser and rounder primary particles. The mechanisms that occur in globulitic metal suspensions are not entirely resolved, but according to the theory developed from experiments agglomerations of particles leading to a higher viscosity accumulate in the quiescent state due to interfacial tensions [2, 3]. The shape of the solid globules is a crucial parameter for the description of the microstructure in the partial liquid state. The morphology of the solid particles alters, for example, during stirring with a change in the stir level (Figure 3.2). Whereas the material in the unstressed state is characterized by cohesive grain clusters, the particles developed with increasing shearing rate, increasing shearing time and decreasing cooling rate are no longer dendritic but rosette-shaped and ultimately globular. Partial liquid bodies can be handled like solids due to the developed solid network, provided that the shearing forces are too small to

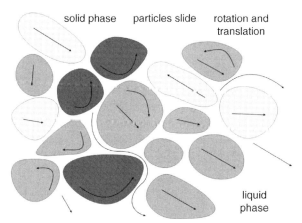

Figure 3.1 Depiction of the different mechanisms to explain the flowability of partial liquid metallic suspensions [5].

Increase in shear rate and intensity of turbulence

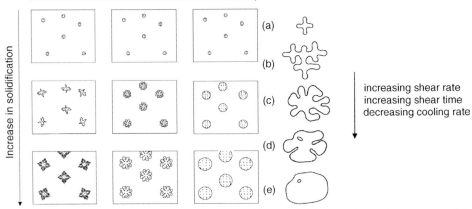

Increase in solidification

increasing shear rate
increasing shear time
decreasing cooling rate

Figure 3.2 Structural changes during the solidification and shearing of metallic suspensions [14].

destroy them. The development of the network permits a relatively easy insertion of partial liquid billet into the casting or forging tools. On applying a sufficient shear stress, the solid framework breaks open and the previously formed agglomerations can be dissolved. A fluid suspension of solid particles within a liquid matrix-phase develops, resulting in a reduction in viscosity. The structure formation of partial liquid casts under convection, or generally under external strain, is therefore the result of the interplay of agglomeration and destruction of agglomerates through induced shear forces. This interplay can be examined with different methods, where, depending on the particular method, shearing rates of up to $1\,s^{-1}$ with shear rate leaps of up to $1500\,s^{-2}$ can be realized [6]. With low shearing rates, large agglomerates are generated, which disintegrate with higher shearing rates and exist as isolated particles in the liquid phase. The proportion of the intra-globular, immobile liquid phase rises with increasing agglomerate size, so that in this case higher viscosities are measured due to the lower liquid-phase fraction available for sliding.

As a result of the multitude of different rheological phenomena, it is usually not reasonable to describe a material using only one measurement. Therefore, most materials are characterized using viscosity curves, that is, viscosities subject to the shearing incline, or using flow curves or the shearing incline subject to shear stress. For metallic alloys, these curves have to be determined for each temperature in the semi-solid state. Therefore, three approaches are in particular used for the analysis of the rheological properties of partially solidified metal alloys [7]:

1. isothermal and stationary viscosity measurements for the identification of the flow curve;
2. experimental study of the time-dependent behaviour under isothermal conditions (ramp experiments to investigate hysteresis loops, leap experiments and measurements of different initial viscosities); and
3. cooling or heating experiments to measure the viscosity under a constant cooling or heating rate and shear rate.

Due to the multitude of different experimental equipment and rheometer designs for the determination of the rheological characteristics of partial liquid metallic suspensions, these are not discussed here in detail and readers are rather referred to the literature [4, 8–12]. A model with which the structural changes in the semi-liquid state can be simulated is presented in [13].

3.2.1.2 The Sponge Effect

As already mentioned, in addition to the phase fraction, the morphology of the phases is of crucial importance for the flow behaviour of partial liquid suspensions. In this way, the particles are able to slide past each other or to move in addition to that by means of translation and rotation movements. The microstructural development during the reheating and the holding in the semi-liquid state are furthermore characterized by an initial spheroidization and a subsequent diffusion controlled microstructural coarsening. This can be described according to the Ostwald ripening mechanism and the coalescence process [15]. Figure 3.3a shows an increase in the carcass degree. The different material currents and diffusion processes occurring during the coalition of two equally sized particles are illustrated in Figure 3.3b.

Parallel to the spheroidization and the growth of the solid phase, the connectivity increases, that is, the strength of the steric carcass increases, and this bestows solid material character and a very high viscosity on the material. If this cannot be broken up with a high enough shearing stress, no flowable suspension develops. A structural viscous, thixotropic flow can only be observed if the operating shearing strengths

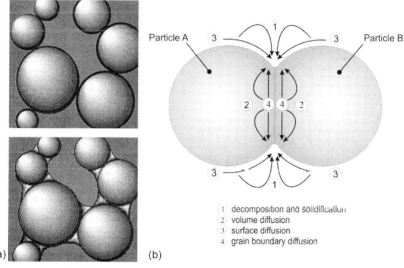

Particle A   Particle B

1 decomposition and solidification
2 volume diffusion
3 surface diffusion
4 grain boundary diffusion

(a)   (b)

Figure 3.3 Growth of the carcass volume and the viscosity by coagulation of the globulites during the isothermal halt or temperature decreasing in the two-phase area (a) and diffusion activity during coagulation (b) [16].

exceed the cohesiveness and stiffness of the solid-phase carcass. If the externally applied forces are too low, the body is only deformed elasto-plastically and no thixotropy occurs. The 'frame firmness' of the solid phase can be described by the degree of carcass creation. If the shearing forces operating during the transformation of a partial liquid material are not sufficient to destroy the steric cohesive carcass, a mere plastic deformation and a compression of the solid phase carcass result. The reheated billet will be plastically compressed and no thixotropy occurs, so that the liquid phase, which is embedded in the spaces between the solid particles, is extruded from the compacted solid phase. Because this is similar to the wringing out of a water-filled sponge, this effect is also referred to as the 'sponge effect'. It can, for instance, be quantified by quantitative measurements of the chemical composition within the unit [15]. In re-extrusion experiments with the alloy AIMgSi1, a homogeneous deformation without noteworthy phase separation occurred only with a liquid-phase fraction between 40 and 50%. The segregation due to the sponge effect should generally be avoided because of the resulting poor component-properties. In particular cases it can be positively used as an alternative option for the production of gradient components.

3.2.1.3 Temperature Interval Size and Temperature Sensitivity

For thixoformability of a metal, the possibility of the defined adjustment of the solid–liquid fraction in the solidification-interval is a fundamental criterion. This is independent of whether the partial liquid state of an alloy is reached by heating a solid body as in conventional thixoforming or by directed cooling of a melt as in the rheo-routes. Because pure metals and alloys of eutectic composition do not have a solidification interval but rather a specific transformation temperature, these are only restrictedly or not at all suitable for thixoforming. Due to the different requirements of the thixo-procedures on the primary material, no explicit threshold for the solid- or liquid-phase portion is given in the literature. Recent studies show that for thixoforging solid-phase fractions of 60–80% and for thixocasting 30–50% are used [4]. The determination of the optimal concentrations is subject of current research. It cannot be finally indicated independent of the structure of the primary metal and the resulting microstructural formation.

Apart from the plant-specific calibration of an exact and homogeneous temperature distribution within the to-be-moulded billet, further factors such as the temperature sensitivity of the solid-phase fraction S^* (Equation 3.1) and the temperature sensitivity of the enthalpy of the solid–liquid system L^* (Equation 3.2) are relevant for the adjustments of the designated phase fractions.

$$S^* = \mathrm{d}f_s/\mathrm{d}T \tag{3.1}$$

$$L^* = \mathrm{d}H/\mathrm{d}T \tag{3.2}$$

where f_s = solid-phase fraction, H = enthalpy and T = temperature.

Table 3.1 shows some values for technically interesting aluminium and magnesium alloys and steels [17, 18]. It has to be considered here that the L^* value is strongly dependent on the observed temperature interval. It is, for instance, 58.8 for the alloy

Table 3.1 Summary of relevant temperatures and characteristics for aluminium alloys and steels [18]. The values for the temperature sensitivity of the enthalpy L^* are averaged values for the interval $0.4 < f_s < 0.6$ [17].

Alloy	T_s (°C)	T_L (°C)	T_{50} (°C)	df_L^{50}/dT	ΔT^{40-60} (°C)	T_{30} (°C)	df_L^{30}/dT (°C)	ΔT^{20-40} (°C)	L^* (J g^{-1} K^{-1})
A356	557	614	574	0.0083	17	572	0.092	3	11.5
AA6082	557	647	637	0.027	7	627	0.015	14	13.0
A319	500	603	562	0.039	10	554	0.015	14	10.4
100Cr6	1348	1461	1427	0.011	19	1405	0.007	27	3.2
HS6-5-2	1175	1432	1360	0.006	35	1308	0.002	59	2.0
X210CrW12	1221	1366	1291	0.004	50	1254	0.028	13	1.6

A356 with a 40% liquid phase and 4.6 for a 60% liquid phase. Special attention has to be paid to the particular observed interval.

Especially steels with a complex casting behaviour and a resulting distinctive solidification interval are suitable for thixoforming, because of their multiphase microstructure and their advantageous liquid-phase interval size. The curve progression of the solid-phase fraction f_s as a function of the temperature $[f_s = f(T)]$ is of technical relevance for thixoforming processes, because the temperature sensitivity of the phase fractions increases strongly with an increasingly steep curve progression and accurate adjustments become more difficult [17]. Figure 3.4 shows the development of the liquid-phase concentrations of different steels [19]. It is advantageous for the interval-size if the liquid-phase development overlaps with the disaggregation of carbides.

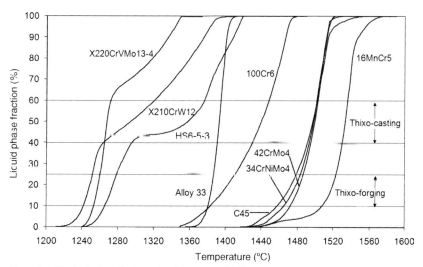

Figure 3.4 Particular liquid phase developments of different steels at a heating rate of 10 K min^{-1} determined using DTA [19].

The examined steels solidify in one-phase (100Cr6) or multi-phase form (X210CrW12). For the steel 100Cr6, the results of thermodynamic calculations under equilibrium terms (Thermo-Calc) and with diffusion (DICTRA) show slight differences, so that during solidification a low segregation aptness exists for this steel [18]. Due to its relatively large semi-solid interval, the steel 100Cr6 should be in general suitable for thixoforging and thixocasting. For the steel X210CrW12, a comparison of equilibrium calculations and calculations with diffusion shows that a tendency to segregate exists until the completion of solidification, so that the M_7C_3 carbide is formed, as indicated in Figure 3.5a. The steep incline of the liquid phase concentration up to 40% suggests that this steel could be rather more suitable for thixocasting than for thixoforging, due to its lower temperature sensitivity with higher liquid-phase fraction [18].

A further crucial factor for the position and size of the semi-solid interval of different aluminium alloys and steels is the variation of the chemical composition within the specified boundary concentrations of the alloy standards [18, 20]. In steels, the elements carbon and chromium in particular have a crucial influence on the liquid-phase progression (Figure 3.5b). For the steel X210CrW12, the interval-size for thixocasting thus decreases in the extreme case from 50 to 5 K, so that a batch with disadvantageous composition is no longer processable in the partial liquid state [18].

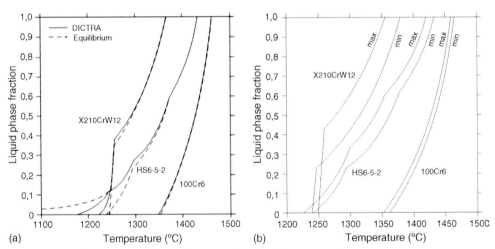

(a)   (b)

Figure 3.5 Liquid-phase fracture as a function of temperature for 100Cr6, HS6-5-3 and X210CrW12 steels. The solid lines show the diffusion evaluation by DICTRA and the dashed lines show the equilibrium evaluation by Thermo-Calc (a). Equilibrium evaluation of the determined liquid phase fraction as a function of temperature with maximum and minimum composition of alloying elements corresponds to the valid standards (b) [18].

3.2.1.4 **Microstructure in the Partial Liquid State**

Influences on Microstructural Development During Solidification   The compositions
of the liquid and solid phases for different temperatures are given in equilibrium
by the course of the solidus and liquidus lines of an alloying system (Figure 3.6).
For eutectic alloys with chemical compositions between C_{max} and C_E, the liquid
phase is constituted by eutectic transformations. For single-phase alloys with chemical
compositions below C_{max}, no eutectic is formed under equilibrium conditions. The
aluminium alloy AlSi7Mg0.3 (A356), used predominantly for thixoforming, belongs to the eutectic type and exhibits at 580 °C a favourable liquid-phase fraction
of about 45%, convenient for thixotropic forming with a eutectic temperature of
577 °C [17].

In technical casting processes, the equilibrium condition according to the phase
diagram is not achieved. The equilibrium condition is only reached if, during solidification, enough time is available, so that by means of diffusion in the solid or liquid
phase concentration differences of the alloying elements can be compensated. The
dashed line in Figure 3.6b shows that solidification in non-equilibrium conditions
causes not only a broadening of the solidification interval but also the occurrence of
an 'unexpected eutectic'. In this case, melts and crystals appear with compositions
which do not comply with those in the finite state diagram. The composition at the

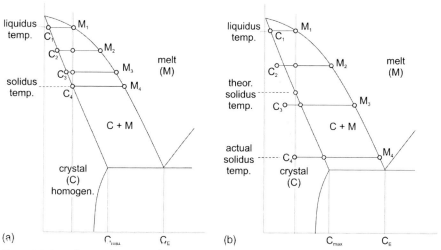

(a)

(b)

Figure 3.6 Crystallization by means of a schematic phase diagram
of a binary alloy or rather a quasi-binary cut of a multi-component
system with eutectic and a high solubility in the solid state:
solidification of a homogeneous alloy in equilibrium (a) and
solidification of a heterogeneous alloy in non-equilibrium (b).
With more rapid cooling and/or low diffusion potential the solidus
line is shifted to lower temperatures.

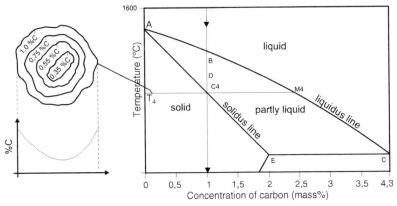

Figure 3.7 Simplified schematic solidification or concentration development of steel with 1% carbon [21].

centre of the primary crystals changes only slightly during solidification, so that the average concentration is below equilibrium. For aluminium alloys with an element concentration below the maximum solubility, the formation of brittle secondary phases even has to be expected. This is the case, for example, for the forming of the magnesium alloys AZ91 or AM60 by means of thixo- or pressure die casting. Despite a maximum solubility of 12.6% Al, these alloys show a significant fraction of brittle, eutectic β-phase ($Mg_{17}Al_{12}$) [17].

For the development of the conventional thixoforming process of steel, especially the carbon concentration and the local carbon distribution exert a crucial influence on the fusion behaviour during reheating. In Figure 3.7, micro-segregations for a steel with a carbon concentration of 1% are depicted schematically under the assumption that the carbon does not diffuse in the solid phase. Hence micro-segregations also results in a shift of the solidus line to lower temperatures and a broadening of the solidification interval. The chemical homogeneity adjusted during the primary material production therefore has a considerable influence on the fusion behaviour and the suitability of a material for thixoforming.

For the development of rheo-processes, accurate knowledge of the nucleation and nucleus growth process is necessary. Typically, a dendritic structure of the solid phase is formed during the solidification process of metals due to the constitutional undercooling of the liquid phase. In most cases, the temperature gradient in the liquid state is not sufficient or the solidification rate is not high enough to avoid dendritic solidification. To obtain a globular instead of a dendritic microstructure, the solidifying alloy has to be specially treated. For all primary material routes, local concentration differences have to be avoided because they can lead to undercooling and superheating in different areas of the liquid, leading to an inhomogeneous distribution of the solid and liquid phases.

In order to understand the nucleation and crystal growth processes, numerical modelling techniques are applied [22]. Here local undercooling is seen to be responsible for the formation of a dendritic structure. The growth rate of a dendrite barb of an undercooled liquid can be calculated with the KTG (Kurz–Giovanola–Trivedi) model [23]. The complete undercooling of the solid–liquid boundary surface ΔT is given by the sum of the following contributions:

$$\Delta T = \Delta T_c + \Delta T_t + \Delta T_r + \Delta T_k \tag{3.3}$$

where ΔT_c, ΔT_t, ΔT_r and ΔT_k are the undercooling contributions concerning the concentration, the local temperature, the curvature of the dendrite barb and the attachment kinetics, respectively. The kinetic undercooling term, ΔT_k, is only relevant to very high cooling rates and can be neglected under the conditions of rheo- and thixoforming [24]. In general, lower undercooling is advantageous for the formation of globular structures, so that the Gibbs–Thomson fraction, ΔT_r, gains importance. It incorporates the decrease in the liquidus temperature due to the bent boundary surface and counteracts the contributions resulting from the temperature gradient and the difference in the chemical composition. For small globulite radii, this leads to the dissolution of solid-phase protrusions which prevent the formation of a dendrite and benefit further globular growth. This self-healing effect only occurs if the Gibbs–Thomson fraction is high in comparison with the other two contributions. With increasing growth of the globulites, protrusions with increased radii can also be formed. These protrusions do not degenerate and can lead to the formation of dendrites. With increasing solidification, the contributions $|T_c + T_t|$, therefore, become increasingly important, so that these have to be reduced by appropriate means to allow further globular solidification. This can be achieved by forced convection or by lower cooling rates. In conclusion, three conditions for the development of a small-grained, globular microstructure during the cooling to the partial liquid state result from these deliberations [25]:

1. existence of a sufficiently high number of nuclei;
2. boundaries around the nucleus intervals;
3. slow cooling of the cast.

A general overview of the numerical simulation of solidification processes concerning the prediction of the microstructure can be found in [26–28].

Structural Development During the Heating Process Particularly, apart from the liquid phase content, the grain size and the grain shape, also the steric arrangements of the solid and liquid phases are of critical importance for a treatment in the partial liquid state. During heating, these parameters are decided or rather adjustable through the homogeneity or inhomogeneity of the element distribution. In steels, the fusion behaviour and the distribution of the solid and liquid phase can be controlled through the carbide distribution. The steric arrangement of the solid-phase particles can be described by the degree of carcass development, the contiguity and the contiguity volume.

No universally valid boundary value for the grain size is currently given in the literature because such a threshold can be dependent on the wall thickness of the to-be-filled component segment. For conventional aluminium alloys, it could already be shown that the grain coarsening mechanisms that occur during semi-solid reheating can be strongly reduced through specific alloy modification (addition of barium to the conventional aluminium alloy AA6082). This can be ascribed to the decreasing effect of barium on the contiguity, because barium raises the contact penetration of the cast due to its influence on surface energy and thereby confines the cohesion mechanism of the solid phase [15].

For aluminium alloy A356, analyses concerning the influence of primary material during conventional thixoforming show that dendritic grains with grain radii of more than 800 μm during isothermal holding in the partial liquid state exhibit an irregular, round grain form, whereas dendritic primary material with grain radii from 200 to 600 μm with a relatively long holding time show a homogeneous, coarse-grained globular structure. Primary material with grain sizes of less than 200 μm exhibits a globular structure already after short holding times. Cast material with grain sizes of less than 110 μm requires no isothermal holding time to reach a globular structure for semi-solid forming. Based on these results, Equation 3.4 concerning the connection of the grain size in the primary material and the size of the globulites after remelting was derived [29]:

$$\text{Size of globulites after remelting}\,(\mu m) = 1 + 0.09 \times \frac{\text{particle size}_{\text{as cast}}\,(\mu m) - 110}{\text{time}\,(min) + 0.5} \tag{3.4}$$

3.2.2
Structural Examination

3.2.2.1 Structural Parameters
All structural components can be characterized by their crystal structure and their chemical composition and also by their geometry. While the structure of the structural components is generally relatively easy to specify, the description of its geometry and its proximity relationships (except for simple quantification) is considerably more complicated and is, therefore, usually carried out qualitatively with the aid of simplifying parameters. For the characterization of the microstructure of partial liquid samples, these are usually quenched and interpreted metallographically. The production and analysis of level sections of the to-be-examined material provide only indirect information about the steric structural parameters. The stereology provides the necessary procedures to reconstruct the structure of the three-dimensional space based on two-dimensional sections or projections on to a plain. Procedures of geometric probability computation and differential geometry can be utilized. By means of two-dimensional grindings, thixoforming process-relevant parameters can be detected or rather estimated. These are described in detail in the following sections.

Volume Fraction in the Solid and Liquid Phases   In partially liquid metals, the most important parameters are the volume fractions of the solid and liquid phases, because they crucially influence the viscosity and therefore the flow and form-filling behaviour of the material. The microstructure development in the partial liquid state is also caused by the phase fractions, so that these parameters are of special importance for the basic analysis and also for process control. Depending on the process examined, the volume fraction of the solid phase should amount to between 30 and 80 for thixoforming. On the one hand, it should not be too small to avoid turbulent form filling so that a sufficiently high viscosity is achieved. On the other, it should not be too high to ensure complete form filling. The phase fractions are given for a certain temperature between the solidus and liquidus temperatures under equilibrium conditions and can be read off from the appropriate phase diagram. In all other cases, they are dependent on the homogeneity of the material and the previous thermal history. If equilibrium conditions are assumed, the solid-phase fraction can be derived thermodynamically from the lever rule. For non-equilibrium conditions, the calculation is possible with various methods. A well-known method is the Scheil–Gulliver model, for which it is assumed that no diffusion in the solid phase occurs. This assumption is only valid, however, for short holding times or high cooling rates. For the determination of the solid-phase fraction, various processes are used which either define the solid-phase fraction directly or by means of its effect on special physical properties [30]:

- thermodynamic data (equilibrium phase diagrams);
- thermal processes [differential thermal analysis (DTA), differential scanning calorimetry (DSC)];
- quantitative metallography by means of quenched samples from the semi-solid interval;
- measurement of the propagation rate of ultrasound within the partial liquid medium;
- measurement of the electrical resistance or the magnetic permeability;
- measurement of mechanical properties (e.g. penetration tests, flow experiments).

In practice, the first three methods are mostly used, as the others not only need extensive calibration but also do not exhibit a definite relationship between the measured characteristic parameters (ultrasound speed rate, electrical resistance, deformation resistance) and the phase fraction. Furthermore, these methods are dependent on the microstructure, so that especially the distribution of the solid-phase particles and their connectivity exert an essential influence on the result and lead to an inaccurate measurement.

The experimental determination of the volume fractions is mostly done with DTA or by means of quenching experiments from the partial liquid state, where for aluminium alloys measurements with micro-tomography are also possible [31]. An extensive discussion of possible error sources in solid-phase determination with DTA, quenching experiments and thermodynamic calculations can be found in [30]. In summary, it can be said that all three methods allow for a rough prediction of the phase fraction. The use of thermodynamic data provides information on the maximum width of the semi-solid interval. However, consideration of the prior thermal history and of the microstructure is currently not yet possible. The development of adequate software is the object of on-going research studies and, in concert with constantly rising computation capacity, this might become possible in the future.

The advantage of the thermal procedures is the easy and cost-efficient specimen preparation, but a systematic error has to be considered because of the use of peak reference surface integration. In comparison with industrial application, the disadvantages of thermal procedures concerning the direct transferability of the results for semi-solid technologies (considerably quicker inductive heating) is the lower feasible heating rate of DTA and DSC devices.

For the metallographic determination of the average phase contents by means of quenching experiments, the random and uniform distribution of the liquid and solid phases in the sample volume has to be ensured. Furthermore, the phase concentrations should not change significantly during quenching. This is generally the case if the liquid phase transforms in such a way that it can be easily distinguished from the already present solid-phase components. If the alloys are quenched from temperatures slightly above the eutectic temperature, the contour of the round globulites and the two-phase eutectic can easily be distinguished. The characterization of the microstructure is more difficult in cases in which the quenching temperature is much higher than the eutectic temperature. In that case, secondary, solid particles are formed, which can hardly be differentiated from the existing globulites, while the remaining cast solidifies eutectically. Remarkably, research effort is focused on this issue. In special cases, solid particles present in the partial liquid state can be differentiated from secondary phase fractions by means of special etching techniques due to their chemical composition. However, the boundary of the different solid phases is often very blurred so that no precise characterization is feasible.

Grain Size and Grain Size Distribution in the Solid Phase   The influence of the grain size on the rheological properties of the partial liquid material is the object of current research [4]. It should, on the one hand, be large enough to build a dimensionally rigid solid-phase carcass. On the other, it should be small enough that also solid material can flow into thin component areas. It is usually assumed that the smallest to-be-filled component widths without causing noteworthy separations of solid- and liquid-phase components should not be less than 20–30 times the grain radius [31]. For a laminar mould filling without pore or entrapment formation, the globulites should preferably be round and separated from each other, so that the formation of segregations due to separations of the solid and liquid phases can be reduced to a minimum.

The grain size distribution by means of 2D analysis is difficult to define because of the random view of the grain sections. To take this influence into consideration, a Saltykov correction can be conducted. Furthermore, by means of extensive serial sections, the three-dimensional solid-phase carcass can be reconstructed. An overview of the hitherto conducted research with different materials can be found in [32]. A largely error-free analysis of the grain size distribution can, however, only be determined by means of 3D examination methods. X-ray micro-tomography permits a destruction-free determination of the structural parameters with dissolution of 2 µm within 15 min [31–34].

Form Factor of the Solid Phase   The globulite form factor, *F*, is an important parameter for thixoforming because it strongly influences the flowability and the viscosity of the material. In general, longer holding times, higher temperatures and higher shear rates in the partial liquid state result in rounder grains. With equal holding times and temperatures, electromagnetically stirred material exhibits rounder particles than conventionally cast material [29, 33]. The form factor is generally defined by the equation

$$F = \frac{4\pi A}{U^2} \qquad (3.5)$$

where A is the grain section and U is the grain circumference of the globulites. For ideal round globulites, the form factor takes the value 1. A low value for this parameter indicates dilation or a complex (e.g. dendritic) grain structure. Occasionally the form factor is also defined as $1/F$, so that in this case large values (>1) yield a complex grain structure. In the analyses conducted, the form factor according to Equation 3.5 is exclusively used and this form factor is also denoted the object-specific form factor. With two-dimensional sections of complex grains, especially of dendrites, the form factor is strongly dependent on the particular magnification utilized so that a simple, object-specific appraisal by means of image analysis is subject to a large error. The complexity of the solid phase actually present is not reflected in this case. Therefore, a dimensionless 'grain-specific' form factor, F_G, according to Equation 3.6 was introduced in the literature [33]:

$$F_G = \frac{1}{6\pi f_s} \cdot \frac{S_v^2}{N_A} \qquad (3.6)$$

where f_s is the volume fraction of the solid phase, S_v is the solid–liquid boundary surface per volume unit and N_A is the amount of globulites per volume unit of the sample. This form factor takes the value 1 for perfect, round globulites and values >1 for complex structures.

Volume Fraction of the Intra-Globular Liquid Phase In contrast to the inter-globular liquid phase, which can contribute to the sliding of the solid-phase particles during forming, the intra-globular liquid phase is unwanted. Three phenomena can essentially be held responsible for the development of the intra-globular liquid phase:

- With the existence of a very homogeneous primary structure, an obstruction of the solid–liquid phase boundary surface can be observed due to the self-blocking-remelting effect. In this process, the material melts locally at the grain boundaries due to the marginally enriched alloying elements there and the associated lower solidus temperature. If now further alloying elements (e.g. carbon) diffuse over short distances into the already liquid zone, this leads in the depleted areas to higher solidus temperatures and with constant temperature to an obstruction of the phase boundaries. In the grain interior, however, the nominal composition remains, so that areas within the solid grains melt and form intra-globular liquid phases with an increase in temperature [35].

- During remelting of highly segregated, banded structured, hot-rolled material, coarse, clouded melting behaviour is exhibited, as the carbon-rich carbide bands possess lower solidus temperatures [36].

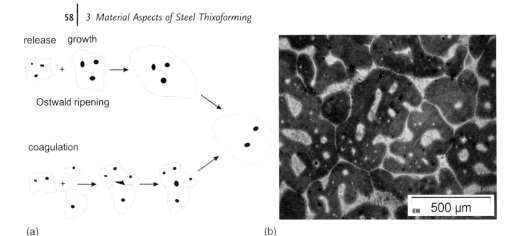

(a)   (b)

Figure 3.8 Schematic influence of Ostwald ripening and coalescence on the fraction of intra-globular liquid phase (a) [37]. Micrograph with intra- and inter-globular liquid phase (b) [36].

- With the existence of dendritic structures, enclosed liquid phase areas are formed during heating in areas with locally higher alloying element contents.

For thixoforming, the problem with the intra-globular liquid phase is that it cannot contribute to the sliding of the globulites during forming and therefore the liquid-phase volume fraction is lower than the theoretical value. From this follows on the one hand an undesirably raised viscosity and on the other the sponge effect is promoted. The solid-phase particles adhere more to each other and higher forces are necessary to obtain a flowable suspension. The development of the enclosed liquid-phase fractions during partial melting can be described by the two coarsening phenomena of the globulites, Ostwald ripening and coalescence (Figure 3.8). They have a reverse influence on the intra-globular liquid-phase fraction. Ostwald ripening leads to the aggregation of enclosed liquid-phase areas due to the dissolution of smaller particles and the growth of larger particles. In contrast, the aggregation and coalescence of different particles lead to an increase in the enclosed liquid phase [37].

During reheating of dendritic structures, intra-globular liquid phases are often formed in discrete globulites which do not have a connection to the liquid phase network due to the coalescence of the secondary dendrite arms. As a result of the tendency to save interfacial energy, the intra-globular liquid-phase fractions are nearly globular. An unambiguous determination of its fraction is only possible by means of 3D analysis, because with a 2D polished section areas can appear to be isolated even when they have a connection to the liquid-phase network (inter-globular or inter-dendritic liquid phase) in deeper levels [38].

Contiguity, Matrix Character and Contiguity Volume   In addition to the liquid-phase content, the grain size and the grain shape, the steric arrangement of the solid phase is of crucial importance for processing in the partial liquid state. To warrant a dimensionally stable handling of the partial liquid billet, the solid-phase particles have to form a network to absorb the forces. The average contact quantity to other particles per particle is referred to as convectivity, whereas the contiguity (Cf_s) characterizes the steric arrangement or rather the cohesion of solid-phase particles and can be described by the average proportion of solid–solid and solid–liquid boundary surfaces. The contiguity has a major influence on the dimensional stability of a partial liquid billet and, therefore, also on the viscosity of the material. Simplified, contiguity can be understood as the degree of carcass formation of neighbouring particles, where three different types of particle contacts are distinguished: not wetted grain boundaries, partially wetted grain boundaries and completely wetted grain boundaries [39–42]. For the characterization of neighbouring relationships of partial liquid suspensions, the network character of a material is defined by means of the degree of carcass formation or rather the contiguity of the solid phase or the matrix character (M_{fl}) of the liquid phase. The contiguity is characterized by the proportion of common boundary surfaces within one component compared with all boundary surfaces. The contiguity results for partial liquid metals according to Equation 3.7 from the proportion of the grain boundary surface in the solid phase to its complete surface [39]:

$$Cf_s = \frac{2 S_v^{ss}}{2 S_v^{ss} + S_v^{sl}} \tag{3.7}$$

where Cf_s is the contiguity value of the solid phase, S_v^{ss} is the boundary surface of the solid-phase particles, S_v^{sl} is the boundary surface of the solid (s = solid) and the liquid phase (l = liquid). The determination of the contiguity takes place on metallographic images by measuring the solid–solid and solid–liquid boundary surfaces. It can take values between 0 and 1. If Cf_s takes the value zero, the particles of the solid phase are completely surrounded by melt and are isolated from the liquid matrix.

As the carcass strength of the solid phases depends on both the contiguity and the volume fraction of the solid phase, the product $C_s f_s$ might be used instead of the contiguity. This factor is referred to as the contiguity volume and describes the volume of connected phase areas [41]:

$$C_s f_s = V_s Cf_s \tag{3.8}$$

where $C_s f_s$ is the contiguity volume, V_s is the volume fraction of the solid phase and Cf_s is the contiguity value of the solid phase. Examinations of the aluminium alloy AA6082 showed that the contiguity volume is an important factor for thixoforming [43].

The matrix character can be established according to the following equations [44]:

$$M_{fs} = \frac{L_{f_s}}{\sum L_i}; \quad M_{fl} = \frac{L_{fl}}{\sum L_i}; \quad \sum M_i = 1 \tag{3.9}$$

where M_{fs} is the matrix character of the solid phase, M_{fl} is the matrix character of the liquid phase, L_{fs} is the carcass line length of the solid phase, L_{fl} is the carcass line length of the liquid phase, $\sum L_i$ is the sum of the carcass line lengths and $\sum M_i$ is the sum of the matrix characters of the two phases. For the matrix character, by definition,

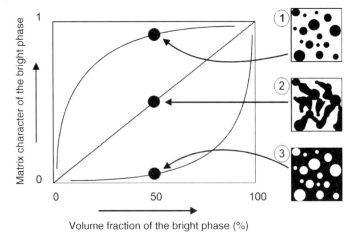

Figure 3.9 Schematic explanation of the matrix character depending on the volume fraction [44].

values between 0 and 1 are possible. It is only dependent on the position of the phase boundaries, not on the grain boundaries. The matrix character is a measure of the degree of connectivity of the investigated structural constituent. By means of this, it is possible to evaluate the position, specifically the transition between 'entrapment character' and 'matrix character' of a phase. If the matrix character is depicted depending on the volume fraction of the observed phase, the structural arrangements shown schematically in Figure 3.9 can be explained as follows. In image 3, the matrix character is observed in the light phase. Despite a volume fraction of 50% of light phase, the matrix character is low, as the globulites are located isolated in the dark phase. In image 2, the matrix character of the dark phase in about 50% because this phase pervades the structural arrangement like at network. The matrix character in image 1 amounts to nearly 90%, as the matrix of the structural arrangement is formed by the light phase.

When metals recrystallize, a large amount of high-angle grain boundaries are formed, which melt preferably during heating. The higher the deformation degree and, resulting from that, the smaller the grain size, the smaller the grain size of the globulites becomes. To intensify this effect, elements can be alloyed which decrease the solid–liquid interfacial energy (γ_{sl}). The melting of a grain boundary takes place as soon as the condition $2\gamma_{sl} < \gamma_{gb}$ (where γ_{gb} is the interfacial energy of the grain boundary) is fulfilled. Therefore, the reduction of the solid–liquid interfacial energy leads to increased melting of the grain boundary, which leads to increased isolation of the particles in the liquid matrix. For example, the connectivity of the solid-phase particles and thereby also the contiguity volume in aluminium alloys can be influenced by adding the alloying element barium, because this element has a strong influence on the solid–liquid interfacial energy. Adding barium leads to the formation of a liquid-phase film for grains with a disorientation of about 4°, whereas only grain boundaries with a disorientation of more than 10° are melted in a batch without barium [15].

3.3
Alloying Systems

Mainly highly alloyed materials are suitable for thixoforming, because of their broad melting interval and their reduced solidus temperatures. Therefore, model alloys were chosen, which belong to the group of tool steels. According to the temperature of their main application field, these are divided into cold, warm and high-speed working steels (EN ISO 4957) [45]. The division of cold working steels into three groups takes place according to their carbon content and the resulting microstructure: hypoeutectoidic, hypereutectoidic and hypoeutectoidic–ledeburitic (Figure 3.10). In the diagram, it can be seen that for a chromium content of about 13 mass% a ledeburitic eutectic can be already formed with a carbon content of 0.8 mass%. The probably most essential property of cold working steels is hardness, which is dependent on the alloying elements and the microstructural components. The hardness is influenced by the matrix, which is characterized by the amounts of martensite and retained austenite and also by the included carbides, for which the constitution and the arrangements are relevant. The attainable hardness as a function of the tempering temperature serves essentially as characteristic attribute [46].

In hypoeutectoidic steels (0.4–0.7%), the carbon is nearly completely dissolved in the matrix during holding at hardening temperature and the accomplishable hardness increases with rising C content. However, for steels in this group, it remains below 62 HRC, whereas the retained austenite content is low. In hypereutectoidic steels (0.8–1.5%), the carbon content is chosen to be so high that on the one hand the full martensite hardness of at least 64 HRC is reached, but on the other still up to about 5 vol.% unsolved carbides remain in the matrix. In hypoeutectic or ledeburitic steels, the C content is over 1.5%, so that towards the end of the solidification a carbide eutectic is formed, which will nearly completely be spit out during the usually

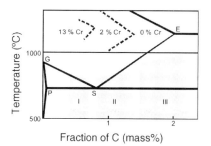

Group	State	Example		Fraction of carbides (%)
		Unalloyed	Alloyed	
I	Hypoeutectoid	C60U	45NiCrMo16	<1
II	Hypereutectiod	C105U	90MnCrV8	5–10
III	Hypoeutectic	200CrMn8	X210CrW12	>15

Figure 3.10 Division of cold working steels in three groups by means of a schematic Fe–C diagram (a) and the resulting structural constitution (b). [46].

following hot deformation. Despite the comparatively high content of retained austenite of these steels, degrees of hardness of 63–68 HRC are reached [46].

The multiphase structures of tool steels consist of a highly alloyed matrix, in which carbides of different chemical compositions and of different types are included. The carbides have a significant influence on the mechanical properties. Their impact is determined by their constitution, amount, size, shape and arrangement. In general, the chromium-rich carbide M_7C_3 is considerably harder than the pure iron carbide Fe_3C, whereby the chemical composition of the carbides can fluctuate over a wide range [45]. During annealing at temperatures from about 100 °C, first low-alloyed carbides precipitate in several steps, and with increasing annealing temperature also high-alloyed carbides precipitate, because of which the martensite hardness decreases continuously [45].

3.3.1
Tool Steel X210CrW12

The tool steel X210CrW12 (1.2434) belongs to group 2 of the alloyed steels for cold working. This ledeburitic cold working steel is in its 'usual' structural constitution a low-dimension changing, air-hardening, martensitic chromium steel with high wear resistance. Its application area comprises cutting tools and punching tools, tools for chipless forming and generally wear-resistant tools and components. Figure 3.11 shows a pseudo-binary section of the four-component system Fe–Cr–W–C. The steel X210CrW12 precipitates bar-shaped M_7C_3 primary carbides from the liquid already during solidification due to its high chromium content of 12%, whereas close to equilibrium no carbides of the type $M_{23}C_6$ or M_3C appear. The great hardness of the chromium-rich M_7C_3 carbide leads to a hardness of 63–68 HRC even with retained austenite contents of >20% and, therefore, to an excellent wear resistance which,

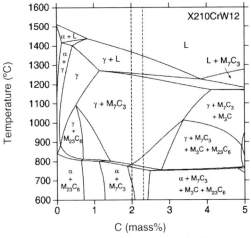

Figure 3.11 Pseudo-binary phase diagram for the Fe–Cr-W–C system (after [4]).

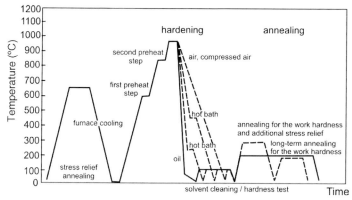

Figure 3.12 Heat treatment scheme for steel X210CrW12 [51].

however, concurs with a relatively low toughness [46]. Due to its high carbon content, the steel grade 1.2434 exhibits a wide melting range with a distinctive plateau, which is developed due to the eutectic solidification and the overlaid carbide precipitation. For the steel X210CrW12, the content of eutectic M_7C_3 carbides amounts to about 14%, where during soft annealing an additional 10% of annealing carbides can be precipitated, which are also of the type M_7C_3 [47].

In Figure 3.12, typical heat treatments for the material X210CrW12 are depicted. In terms of hardness and wear resistance in the hardened state, it is aspired to conduct the complete finishing in the soft-annealed state up to the net shape. The dimensional stability of the material, which is influenced by the different volumes of the different phases, the thermal stress during cooling between border and core and also the transformation stress as a result of the temporally displaced structural transformation between border and core, is of essential importance [48]. Depending on the thickness of the carbide bands, the dimensional change takes place mainly parallel to the carbide band direction, where in the transverse direction a plastic shortening may occur. This is irreversible and cannot be adjusted by subsequent annealing [48]. Annealing treatments after the hardening procedure also have a great influence on the tool's dimensional stability, as pointed out in the double illustration of hardness and dimensional stability subject to the annealing temperature in Figure 3.13 [46]. With increasing temperature, a volume contraction takes place due to the transformation of the tetragonal martensite into its cubic modification and then a volume increase occurs caused by the transformation of the retained austenite into martensite. With a further temperature increase, another volume contraction takes place due to the compression of the newly formed martensite by stress relief. The location of the temperature ranges for the volume changes is influenced by the alloy composition and the associated retained austenite concentration [49].

Important process-relevant material data such as heat transmission coefficients, coefficients of thermal expansion, heat capacities and thermal conductivities from room temperature up to liquidus were determined by means of basic experiments. Furthermore, flow curves in the partial liquid state were established with case-compression experiments [4, 50].

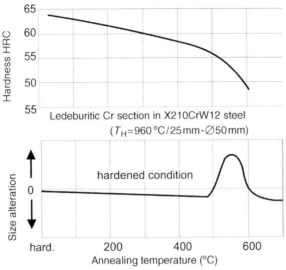

Figure 3.13 Hardness and dimensional change of steel X210CrW12 subject to annealing temperature [46].

3.3.2
Roller-Bearing Steel 100Cr6

The roller-bearing steel 100Cr6 (1.3505) is an hypereutectoidic chromium- and carbon-rich steel. In contrast to the melting behaviour of X210CrW12, the steel 100Cr6 exhibits no distinctive plateau within the melting range, where, however, the chemical composition of the specific batch has to be considered in the examination [18, 19, 52].

Due to the characteristic requirements (high hardness, rolling resistance, wear resistance, no structural changes and adequate toughness) for this steel, its development led to a continuous improvement of metallurgical processes for the manufacture of homogeneous steels with high purity. The reason for this is that mainly non-metallic inclusions generate local stress peaks in bearing material and thus serve as the starting position of fatigue cracks [53]. The roller-bearing components made from the steel 100Cr6 require a minimal hardness of 58 HRC, which can be reached by means of classical heat treatment by quenching from about $850\,°C$ in oil, water or salt baths to about $200\,°C$ and subsequent annealing at $150–180\,°C$. The hardness constitution is essentially composed of martensite, retained austenite and carbides. Parts of the retained austenite can also be transformed through annealing with a second annealing step or by means of freezing. The retained austenite concentration within the martensitic microstructure amounts to between 7 and 15% and increases with rising austenitization temperature, which is due to the higher fraction of dissolved carbon and chromium within the austenite. An overview and discussion of the transformation structures of 100Cr6 after one- and two-step austenitization and with different bath temperatures during annealing can be found in the literature [54, 55].

To reduce the danger of cracking in martensitically hardened components, an isothermal transformation can be conducted in the lower bainite level which leads to a tougher and more dimensionally stable structure without such a high hardness [53, 56]. During the martensite hardening and the transformation into the bainite level, different internal stress conditions develop [55]. A transformation in the lowest bainite level has the advantage of no retained austenite, which delays transformation that results in an undesired distortion of the components. Therefore, bainitization has gained strongly increasing importance in relation to the improvement of the dimensional stability of components [56]. The choice of the optimal transformation conditions depends strongly on the alloy composition, where the major disadvantage of bainitization is the longer processing time in comparison with martensite tempering and thereby increases the costs.

Titanium in steel 100Cr6 results in a decrease in the fatigue life due to the development of relatively small (<10 μm) hard and sharp-edged carbides. Fatigue experiments with two martensitically hardened and two bainitically transformed states showed that the crack formation in all cases started at second-phase particles, mostly titanium carbonitrides. Inclusions with diameters of less than 10 μm caused fatigue crack formation. The crack development in the low cycle fatigue domain occurred much earlier in bainite structures than in martensite structures. The subsequent stable crack development is mostly determined by the hardness and begins with minor cyclical stress intensities that increase the harder the steel is. A higher retained austenite fraction acts under cyclical stress, delaying the introduction of critical crack dispersion [57]. Due to the described phenomena, the titanium level in the steel 100Cr6 is usually limited to 30 ppm [53].

Examinations regarding the formation of carbide segregations in tool steels with about 1% C revealed the development of quasi-ledeburitic carbides, the surface proportion increasing with rising content of carbide-developing elements. Concerning the arrangement of the quasi-ledeburitic carbides within the structure, two especially characteristic forms arise: in steels with less than 2% alloying elements they exist as accumulations and with alloying elements above 6% they are precipitated in the form of a network. Steels with alloying element contents between 2 and 6% exhibit a hybrid of carbide accumulations and partial network. The form of quasi-ledeburitic carbides could be divided into five characteristic basic forms: lamellar-eutectic, splintery, island-like, agglomerated and compact. In unalloyed and low-alloyed steels, agglomerated and compact forms were found. The splintery and agglomerated forms occurred in steels with about 3% W, whereas the lamellar and the island-like forms occurred in steels with over 5% Cr [58].

In steel 100Cr6, mainly compact carbide accumulations formed, where considerable concentration differences can occur due to pronounced segregations so that also the chromium-specific carbide $(Cr, Fe)_7C_3$ stabilizes. If the coarse carbides that developed during casting are not dissolved by means of a suitable heat treatment, these carbide segregations lead to tiered carbide accumulations in the hot-rolled state. For the dissolution of ledeburitic carbides and existing carbide bands after the warm transformation, diffusion annealing at temperatures beneath the solidus (1150–1200 °C) is necessary. The holding times here are strongly dependent on the component dimensions [58, 59].

Calculations concerning the disintegration, segregation and element contents of steels with 1%C and 1.5% Cr during solidification show segregation fractions of $I_s(Cr) = 3–4$. The segregation fraction is calculated according to Equation 3.10 from the maximum and minimum carbon contents:

$$I_s(Cr) = c_{max}/c_{min} \tag{3.10}$$

The segregation fraction increases with rising C content up to maximum 1.5% carbon and decreases again above this critical value, due to the beginning of carbide formation. As soon as the carbon content at which carbides can form has been reached, further addition leads to premature development of eutectic with less chromium content during solidification. Therefore, the segregation fraction of the element chromium decreases, so that no development of eutectic carbides results for steel 100Cr6 [60].

For steel 100Cr6, furthermore, the risk exists that at temperatures from 1140 to 1160 °C the material becomes overheated and the grain boundary areas start to melt [59]. The reason for this is the local enrichment with chromium and carbon in the surroundings of dissolved, ledeburitic carbide grains, because relatively high chromium and carbon contents remain in their surroundings after their dissolution (Figure 3.14) [61].

Important process-relevant material characteristics of the steel 100Cr6 such as heat-transmission coefficient, dilation coefficient, heat capacity and thermal conductivity from room temperature up to fusion heat are currently determined by means of basic experiments. Furthermore, flow curves in the partial liquid state were established with case-compression experiments [4, 50].

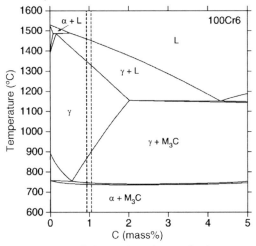

Figure 3.14 Pseudo-binary phase diagram for the Fe–Cr–C system.

3.4
Structural Parameter Development and Material Selection

In general, the characterization of the microstructure in the partial liquid state is often carried out on quenched samples. However, in most methods for the examination of partial liquid materials such as holding experiments, compression tests, primary material examinations and also rheology tests, the microstructure prevailing in the partial liquid state is of interest. Therefore, the samples have to be quenched so quickly that the microstructure at room temperature reflects the partial liquid state. Due to diffusion processes and phase transformations, this is often not successful so that the microstructural changes caused by quenching have to be considered. If the structure was changed much during quenching, the determined structural parameters have to be corrected or can possibly not be determined at all. Therefore, exact knowledge and examinations of the mechanisms that occur and their effects on the structural parameters are of special importance. For structures for which an exact determination of the structural parameters is not possible due to extremely large changes during quenching, error-tolerance ranges have to be prepared. Within the scope of the examinations, therefore, the steels 100Cr6 and X210CrW12 already used for thixoforming were observed in detail to evaluate the magnitudes of error in the determination of structural parameters of samples quenched from the partial liquid state. These two steels were chosen to examine the behaviour of a complexly melting steel with eutectic (X210CrW12) and a simply melting steel without eutectic (100Cr6). The austenite in the carbon-rich steel X210CrW12 exists in a metastable state at room temperature after quenching, whereas the steel 100Cr6 transforms martensitically. The results of DTA examinations, thermodynamic calculations in equilibrium (Thermo-Calc) and in consideration of diffusion-controlled processes (DICTRA) are discussed and compared.

3.4.1
Development of the Structural Parameters of Steel X210CrW12

The examination of the structure development and the determination of the structural parameters of steel X210CrW12 in the partial liquid state was carried out by means of cylindrical samples ($d = 15$ mm, $h = 25$ mm) with a chemical composition according to Table 3.2 at temperatures from 1200 to 1400 °C. For this purpose, they were held isothermally in an aluminium oxide pan inside a convectively heated furnace for 30 min and subsequently quenched with water. The temperature measurement throughout the experiments is achieved continuously by means of a

Table 3.2 Chemical composition of the examined steel X210CrW12 (all data in mass%, except N in ppm).

C	Si	Mn	Cr	Mo	Ni	Cu	W	Fe	N
2.08	0.41	0.30	11.57	0.07	0.21	0.04	0.74	Rest	251

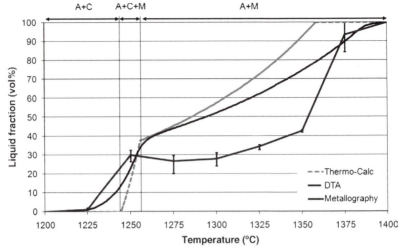

Figure 3.15 By means of Thermo-Calc and DTA (10 K min^{-1}) calculated and metallographically determined liquid-phase contents with fluctuation margin (with A = austenite, K = carbide, S = cast, according to Thermo-Calc calculation).

thermo-element placed within the sample's centre. A temperature of 200 °C after a cooling time of 7 s was reached, so that cooling rates of 100–200 K s^{-1} were realized.

Figure 3.15 shows the metallographically established liquid-phase contents of the quenched samples, denoted by the error bars. Above the three-phase area (above about 1250 °C), the samples exhibit a constant liquid-phase content of about 30%, where the metallographic determination was carried out by measuring the secondary austenite and the eutectic areas. Figure 3.16 shows exemplarily the approach and the problems in the determination of the phase contents. The results in Figure 3.15

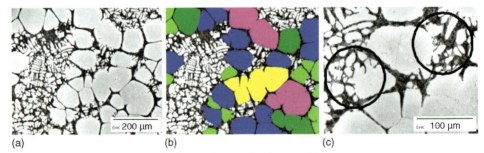

(a) (b) (c)

Figure 3.16 Micrograph of a sample quenched from 1350 °C. In the partial image (a) the completely light phase exhibits a volume content of 72.5% while the detected areas in the partial image (b) show a content of 55%. The different colours symbolize the different grain-size grades. The enlarged partial image (c) clarifies the difficulty in distinguishing between primary and secondary austenite.

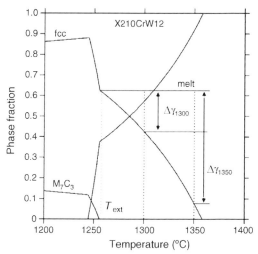

Figure 3.17 Schematic clarification of the development of
pre-eutectic, secondary austenite by means of the calculated
phase contents of steel X210CrW12 for its standard composition.

suggest that with quenching from higher temperatures metallographically lower
liquid-phase contents are being determined than expected on the basis of thermodynamic calculations and DTA. Therefore, the following examinations are concerned
with the question of whether the solid and liquid phase contents or rather the
structural parameters above the three-phase area can be determined sufficiently
accurately with conventional analysis methods. The focus is the question of how far
the secondary austenite generated during quenching is discernible from the primary
austenite being in equilibrium during the liquid phase. Here, the regular grain
growth of the globulites is distinguishable from the irregular grain growth, where the
secondary austenite builds up in the form of dendrites on the primary grains and
forms protuberances [71].

Figure 3.17 illustrates the problem of secondary austenite formation for temperatures of 1300 and 1350 °C by means of a diagram showing the equilibrium phases of
steel X210CrW12. At the eutectic temperature, the structure exhibits an austenite
content of about 60%. If the temperature is increased, the austenite content decreases
to about 10% at 1300 °C, so that theoretically $\Delta\gamma_{1300} = 20\%$ is obtained, and increases
up to a temperature of 1350 °C to $\Delta\gamma_{1350} = 50\%$. As depicted in Figure 3.16, this phase
content of secondary austenite is very hard to discern after quenching.

3.4.1.1 Metallographic Analysis of Quenched Specimens from the Three-Phase Area

Based on the thermodynamic calculations and the preliminary assumptions, it is
expected that for the material X210CrW12 the determination of the liquid-phase
fraction poses no great difficulties up to the temperature at which the material still
remains in the three-phase area (austenite–carbide–liquid). The solid phase should
occur in form of metastable retained austenite due to its high carbon content and
should be easily distinguishable from the formerly liquid phase, which is shaped as a

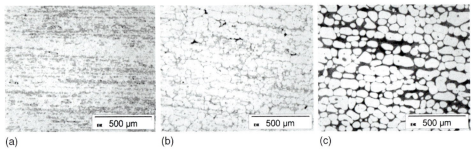

(a)   (b)   (c)

Figure 3.18 Quenched structure after a holding time of 30 min at
1200 °C (a), 1225 °C (b) and 1250 °C (c).

fine eutectic and also easily measurable. Due to the high temperature dependence up
to liquid-phase contents of about 40%, this area represents only a small temperature
interval. By means of quenching experiments, these assumptions could be confirmed
experimentally up to a temperature of about 1250 °C. At 1200 °C, a completely
austenitic matrix with finely distributed carbides occurs (Figure 3.18). Above about
1225 °C, the material starts to melt locally at a few grain boundary triple points (b),
which fits well with the determined solidus temperature of 1230 °C from the
DTA. Because of the inhomogeneous primary material, first areas with liquid phase
form already locally at temperatures below the expected solidus temperature. In
Figure 3.18c, the structure of an sample quenched from 1250 °C is exhibited. At this
temperature, the material is close to the upper boundary of the three-phase area. In
the quenched structure, a retained austenite content (light) of about 70% can be
determined, which is in good agreement with the value from DTA of 66%. The in part
strong dependence with the use of different thresholds for the same micrograph is
depicted in Figure 3.19, where the solid phase particles are assigned to different
grain-size grades, that is underlaid darker. At a lower threshold of 160, many particles

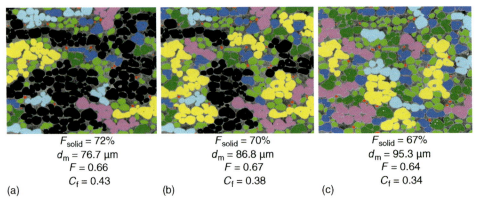

$F_{solid} = 72\%$ $F_{solid} = 70\%$ $F_{solid} = 67\%$
$d_m = 76.7\ \mu m$ $d_m = 86.8\ \mu m$ $d_m = 95.3\ \mu m$
$F = 0.66$ $F = 0.67$ $F = 0.64$
$C_f = 0.43$ $C_f = 0.38$ $C_f = 0.34$

(a)   (b)   (c)

Figure 3.19 Structural evaluation of a sample at 1250 °C at
different lower grey-value thresholds of 160 (a), 180 (b) and 200
(c) for the solid phase.

are still connected to each other (black areas), whereas they exist isolated at a lower threshold of 200 (light green and blue areas). The difference of the measured solid-phase fraction in the depicted image is about 5%, whereas the influence on the average grain size ($\Delta d_m = 20\,\mu m$) and the contiguity of the solid phase ($\Delta C_f = 0.1$) is considerably larger. Even at a threshold of 200, globulites are still connected to each other (yellow areas), which are regarded as separate particles by the automatic image recognition and have to be corrected 'by hand' according to the experiences of the metallographer.

It can be noted that the structural parameters of X210CrW12 samples quenched from the three-phase area (austenite–carbide–cast) can be determined allowing for diverse factors of influence, because the austenite can be distinguished relatively clearly from the eutectic. It cannot be ruled out that the globulites grow marginally during quenching and that, therefore, higher solid-phase contents are measured. However, this effect is overridden by the systematic errors during structure detection. A metallographic determination of the carbide content is not possible because the carbides occur mostly at the phase boundaries and cannot be differentiated from the quenched eutectic. In the determination of the structural parameters, the most important factors of influence are the thresholds set by the user for the detection of the different phases and the minimal particle size (number of pixels) for the consideration of a particle. The minimum number of pixels for the consideration of a particle can be discounted if the existing eutectic is very fine. In addition to the evaluation-specific inaccuracy, the determined liquid phase content is also dependent on the particular observed sample area. Therefore, the deviation of different images of the same sample can vary widely, which is also depicted by the dispersion bars in Figure 3.15.

3.4.1.2 Metallographic Analysis of Quenched Specimens from the Two-Phase Area

Representative structural images of the quenched samples and the determined liquid-phase contents of the quenching experiments above the three-phase area are summarized in Figure 3.20. The structural images show that above a temperature of 1300 °C secondary austenite accumulates at some austenite particles and protuberance formation occurs. Furthermore, at this temperature the development of fine, secondary austenite (highlighted by arrows) starts in large liquid-phase areas. At a temperature of 1325 °C, coarse, secondary austenite is increasingly formed. If the primary austenite grains lie close to each, other the secondary austenite adheres to the large primary grains. If the diffusion paths become long enough in larger liquid-phase areas, the coarse secondary austenite exists isolated in the dark matrix and only adheres to the large austenite grains in the boundary areas. This phenomenon intensifies with rising temperature, so that a sharp boundary between primary and secondary austenite can no longer be drawn, which becomes apparent from the quenched sample of 1350 °C. At temperatures of up to about 1325 °C, the protuberance-formed accumulating austenite and the fine secondary austenite isolated in the liquid can be considered without large errors and disposed of during image re-editing. At higher temperatures, the distinction is no longer possible, because on the one hand the eutectic coarsens with increasing temperature and on the other the

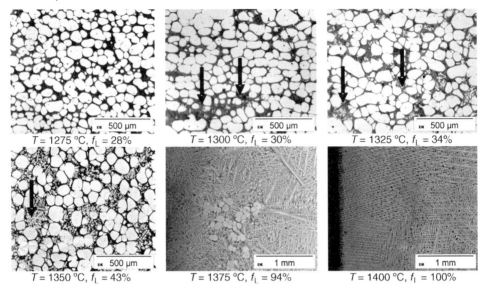

Figure 3.20 Structures and determined liquid-phase contents of quenched samples above the three-phase area, where fine, secondary austenite is highlighted by the arrows.

globulites become increasingly smaller, which makes the differentiation more complicated.

Due to the large difference (>20 vol.%) between the metallographically determined and the DTA and Thermo-Calc designated phase contents, it has to be assumed that in addition to the described local, irregular increase in secondary austenite, a smooth globulite growth occurs in all directions in space. This mechanism is more pronounced at higher temperatures, so that it cannot be neglected as with lower liquid-phase contents. An examination of the differently formed austenite is impossible by means of conventional metallography.

Because conventional examination methods meet their limits for materials with fine, low-contrast structures, where a differentiation by means of grey-scale areas is impossible, texture analysis and analysis of the element distribution provide methods to solve many problems [62]. Therefore, microprobe examinations and EBSD (Electron Beam Scattering Detection) measurements were executed to achieve a differentiation of primary austenite and regularly growing secondary austenite.

3.4.1.3 Microprobe Examinations of the Quenched X210CrW12 Specimens

In addition to the metallographic examination, two samples quenched from 1290 and 1330 °C were examined with a microprobe to differentiate between primary and secondary austenite by means of element distribution. X-ray intensity distribution images with a step range of 2 μm show that the austenite grains for these two temperatures exhibit nearly identical element content intensities (Figure 3.20). In Figure 3.21, the intensities of the elements Si, W, Cr and C are displayed. The examination shows that no gradients giving information about the former phase

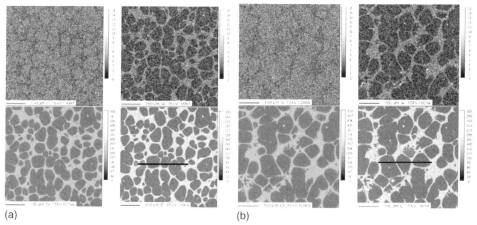

(a)   (b)

Figure 3.21 Element-distribution images (Si, W, Cr, C) of dimensions 500 × 500 μm of specimens quenched from 1290 °C (a) and 1330 °C (b) of the material X210CrW12. Markings of the line-scan distances are displayed in Figure 3.23.

boundaries could be measured in the element distribution. The measurements prove the already determined, by means of calculation, accumulation of chromium, tungsten and carbon and the depletion of silicon in the eutectic. The element distribution images in Figure 3.22 also show constant alloying element contents in the austenite for a continuous measurement, where no significant difference between the primary grain and the accumulated secondary grain can be seen.

The associated line-scan measurements for specimens quenched from 1290 °C exhibit an erratic change of the carbon and tungsten content around the phase

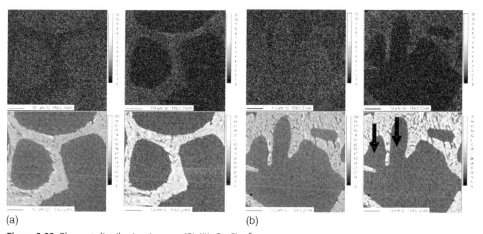

(a)   (b)

Figure 3.22 Element-distribution images (Si, W, Cr, C) of dimensions 60 × 60 μm of specimens quenched from 1290 °C (a) and 1330 °C (b) of the material X210CrW12. The secondary austenite probably accumulated during quenching is identified with arrows.

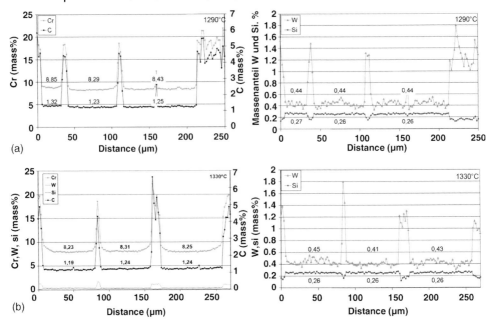

Figure 3.23 Carbon, chromium, tungsten and silicon contents of the line-scan measurements of specimens quenched from 1290 °C (a) and 1330 °C (b).

boundaries (Figure 3.23). The calculated depletion of silicon within the liquid phase is also confirmed. In contrast to this, the transition of chromium is somewhat flattened. However, the grains exhibit diverse average element contents. Thus, the chromium and carbon contents in the left grain are about 7% (8.85–8.29%) and (1.32–1.23%) higher, respectively, than in the middle grain of the line scan, whereas the silicon and tungsten contents are approximately the same in the different grains. One reason for the differences in the chromium and carbon contents can be found in the local geometric circumstances. Another reason could be that even after a 30 min holding time the equilibrium state is still not reached due to some inhomogeneity of the rolled primary material and locally higher element contents still exist [71].

In the sample quenched from 1330 °C, the difference in element content of different grains is smaller, which can be ascribed to the higher diffusion rate at higher temperatures. Whereas the carbon content changes erratically at the grain boundary, the chromium content exhibits a more flattened transition from austenite to eutectic than for the sample quenched from 1290 °C. The difference within one grain is to some extent smaller than the difference between different grains, so that a differentiation of the accumulated austenite is no longer possible by means of microprobe examinations. The tendency decreases with increase in temperature because of increasing diffusion.

Figure 3.24 shows enlarged the chromium and carbon contents in the neighbourhood of a liquid-phase area. It can be assumed that in the area primary austenite

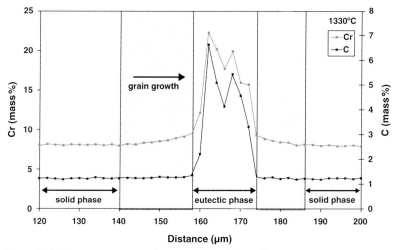

Figure 3.24 Element contents at a former solid–liquid boundary surface, where the grey underlaid areas show the fields in which the former phase boundary surfaces could have been located.

existed, whereas in the areas of the eutectic with carbon and chromium contents above 2% or 8% (peak), liquid phase was present. In the grey underlaid transition area, the chromium content increases slowly, which fits well with the diffusion calculations. However, no conclusion can be drawn about where the austenite–liquid phase boundary was located within the transition area at the time of quenching. Due to the shallow transition of the element concentrations, the grain growth cannot be determined with this analysis method and the determination of the liquid-phase content or the structural morphology in the partial liquid state is, therefore, also not possible by means of a microprobe.

Table 3.3 shows the thermodynamically calculated and measured element contents within the austenite. Whereas the carbon contents are in very good agreement, the deviations are higher for tungsten. Also for chromium and silicon, slightly higher element contents are calculated than measured experimentally. However, the tendency concerning the depletion and the enhancement of the elements in the different phases could not be confirmed.

In conclusion, it can be noted that an estimation of the former grain boundary positions in the partial liquid state seems to be very difficult or even impossible due to

Table 3.3 Comparison of the calculated and measured element contents (line-scan) within the austenite.

	C (%)	Cr (%)	Si (%)	W (%)
1305 °C (calculated)	1.13	8.95	0.31	0.31
1290 °C (line-scan)	1.23–1.32	8.29–8.85	0.26–0.27	0.44
1285 °C (calculated)	1.23	9.16	0.32	0.34

the described conditions. As no nucleation is required, the present globulites will immediately start to grow after quenching, hence a differentiation of primary and secondary austenite seems to be impossible. The thermodynamic equilibrium calculations provide good reference values concerning the composition of the particular stable phases as the experimental and calculated values agree fairly well. The results of the thermodynamic diffusion calculations could also be confirmed, because no differentiation between primary and secondary austenite is possible by means of a differentiation of the local element concentrations at technically realizable cooling rate.

3.4.1.4 EBSD-Measurements of Quenched X210CrW12 Specimens

In addition to the microprobe experiments, EBSD measurements were also executed to obtain conclusions about the structure in the partial liquid state via the crystal orientation. It became apparent that also with this method no differentiation of primary and secondary austenite is possible (Figure 3.25). The image in (a) shows austenitic globulites in a matrix, which consists of fine, two-phase structure of fcc (light) and bcc or carbide components of the eutectic (dark). A differentiation of the carbides from the body-centred components is not possible by means of the applied measurement so that both phases are equally coloured. The image in (c)shows the orientation distribution and the image in (b) depicts the corresponding inverse pole figure. The grain marked with an arrow exhibits a twin boundary in the middle. It shows in its upper part a very uneven grain boundary, presumably due to the secondary austenite accumulated with the same orientation as the primary grain. Figure 3.26 shows the backscattered electron (BSE) image of a sample quenched from 1330 °C (a). The orientation distribution in the image in (b) shows that even the secondary austenite of the eutectic possesses the same orientation as the neighbouring large, primary austenite particles. Even at the high realized quenching rates of more than $100 \, K \, s^{-1}$, the growth rate of the phase boundary surfaces seems to be so high that the same orientation is propagated within a certain sphere of influence

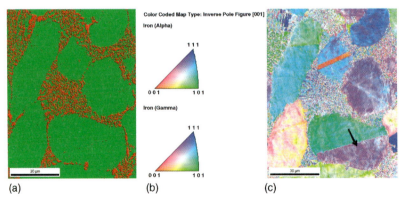

(a) (b) (c)

Figure 3.25 Austenite (light) and ferrite or carbide distribution (dark) (a) and inverse pole figure with associated orientation distribution [(b) and (c)] of a specimen quenched from about 1275 °C.

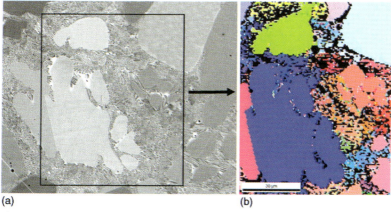

(a)   (b)

Figure 3.26 BSE image (a) and orientation distribution (b) of a specimen quenched from 1330 °C.

around the primary particle. The image in (b) exhibits the areas of the eutectic, which are distinctly dominated by the primary particle. Therefore, an accurate distinction between the primary and secondary austenite of this material is not possible with this examination method [71].

In summary, it can be noted that for steel X210CrW12 even at very high cooling rates no differentiation of primary and secondary austenite is possible with the different analysis methods. As shown by the results of the EBSD measurements, the growth rate of the austenite is apparently so rapid and the mobility of the phase boundary surface is so high that the orientation of the primary austenite grains even continues into the finest branches of the eutectic. Therefore, it has to be assumed that also the even grain growth of the austenite grains of steel X210CrW12 happens so fast and results in a major error in the liquid-phase determination at high temperatures above the three-phase area. Therefore, a dependable structural parameter determination is not viable.

3.4.2
Determination of the Structural Parameters of Steel 100Cr6

Analogous to the quenching experiments with steel X210CrW12, quenching experiments with steel 100Cr6 were carried out and were also examined metallographically concerning the structural parameters by means of microprobe and EBSD measurements. To reduce the influence of the banded carbide structure on the melting behaviour, a laboratory cast instead of a rolled primary material was examined (Table 3.4).

Table 3.4 Chemical composition of the laboratory cast (all data in mass%, except N in ppm).

C	Si	Mn	Cr	Mo	Ni	Al	Cu	Fe	N
0.93	0.32	0.27	1.44	0.02	0.06	0.04	0.07	Rest	62

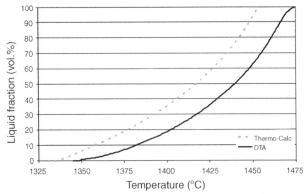

Figure 3.27 By means of Thermo-Calc and DTA (10 K min^{-1}) calculated liquid-phase contents of steel 100Cr6.

Figure 3.27 shows the liquid-phase contents calculated using Thermo-Calc and DTA. The difference between the two curves is continuously increasing due to the influence of the heating rate, the difficulty in evaluating the continuously increasing DTA signal and the extensive scattering of several measurements of the same material due to local inhomogeneities (carbon enrichments).

3.4.2.1 Metallographic Analysis of Quenched Specimens of Steel 100Cr6

For the examination of the structure of steel 100Cr6 in the partial liquid state, quenching experiments were conducted for which the samples were embedded in glass to avoid oxidation. As an example, the water-quenched structures from 1350, 1375, 1400 and 1425 °C after a holding time of 20 min are depicted in Figure 3.28. The samples quenched from 1350 °C exhibit some light areas, at which the material already starts to melt locally due to the enriched alloying elements at the grain boundaries (a). The quenched formerly partial liquid structures show that no conclusion can be drawn about the structural conditions in the semi-solid state, because the structures look approximately the same over the complete semi-solid interval. Independent of the quenching temperature, a martensitic formation with lighter grain boundary areas is discernible with minor amounts of retained austenite between the needles. The martensite needles are somewhat coarser at lower temperatures, where at the retained austenite content seems to increase due to the higher carbon

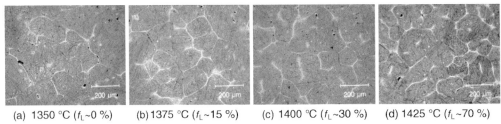

(a) 1350 °C (f_L~0 %)   (b) 1375 °C (f_L~15 %)   (c) 1400 °C (f_L~30 %)   (d) 1425 °C (f_L~70 %)

Figure 3.28 Microstructures of 100Cr6 specimens quenched from 1350, 1375, 1400 and 1425 °C.

content with decreasing temperature. Microhardness measurements of the light grain boundary areas indicate that it is mainly a matter of a layer of retained austenite with fine carbide precipitations, the development of which is strongly dependent on the cooling conditions.

In conclusion, it can be noted that an identification of the phase components is impossible due to the structural morphology and the low contrast of the metallographic images. Furthermore, it can be assumed based on the thermodynamic calculations and due to the higher process temperature and the, therefore, higher diffusion ability that as with steel X210CrW12, a considerable amount of secondary austenite is still formed during quenching and is subsequently martensitically transformed, so that it is metallographically not distinguishable.

3.4.2.2 Microprobe Examination of the Quenched 100Cr6 Samples

In Figure 3.29, a BSE image and the corresponding element distribution images of the elements silicon, manganese, chromium and carbon of a sample quenched from 1425 °C are depicted. Only chromium exhibits significant concentration differences, the other element being distributed fairly homogeneously. As in steel X210CrW12, the formerly liquid phase is chromium enriched, although, by means of line-scan measurements, no conclusion about the former solid–liquid phase boundary can be drawn. Thermodynamic calculations support the high segregation aptness of the chromium around the end of the solidification of the liquid phase (Figure 3.30). Because of the faster diffusion due to the higher process temperature of steel 100Cr6, the concentration differences of the two phases are considerably less pronounced than for steel X210CrW12, so that no determination of the liquid-phase contents is possible by means of element distribution.

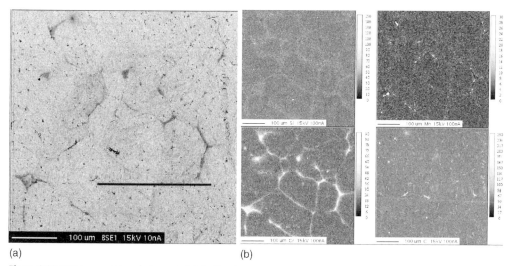

(a)   (b)

Figure 3.29 BSE image (a) and element distribution images (Si, Mn, Cr, C) with dimensions 500 × 500 μm of a specimen quenched from 1425 °C (b).

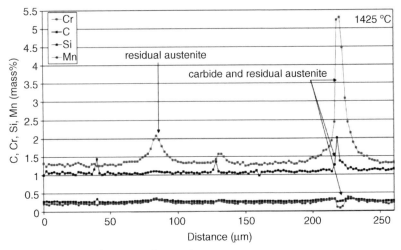

Figure 3.30 Carbon, chromium, silicon and manganese contents of the line-scan measurement of a sample quenched from 1425 °C.

A comparison of the chromium and carbon distributions (Figure 3.29) and a close inspection of the line scan (Figure 3.30) indicate that carbides precipitated in the segregated grain boundary areas whereas most of the segregated zones consist of retained austenite. In the areas in which higher chromium contents and marginally higher manganese contents without simultaneously raised carbon concentrations exist, retained austenite should be present in a metastable form due to a concurrent reduction of the M_s temperature. In the right part of the line scan, detection of carbide can be assumed, due to higher chromium and carbon concentrations, with simultaneous depletion of manganese. It is clear from the manganese distribution that the manganese concentration in the neighbourhood of the carbide is slightly raised and the carbide is, therefore, surrounded by austenite with higher stability. A detailed discussion of the structural development is presented with regard to the resulting mechanical properties in Section 3.6.

3.4.2.3 EBSD Measurements of Quenched 100Cr6 Specimens

Figure 3.31a shows the BSE image of a specimen quenched from 1425 °C. The black areas are fine cracks formed due to the high cooling rate. The image in (b) shows that retained austenite is present in the probably former liquid-phase areas. These can be taken as the lower boundary of the liquid-phase content and can provide a rough indicator of the former structure in some samples. The orientation measurement in (c) shows a random distribution of the martensitic structure. An accurate differentiation of the formerly solid and liquid sample areas is, however, not possible.

The quenching experiments on steel 100Cr6 confirm the tendency of the thermodynamic calculations that in comparison with the partial liquid X210CrW12 no liquid-phase determination is possible due to the diffusion of the elements in the

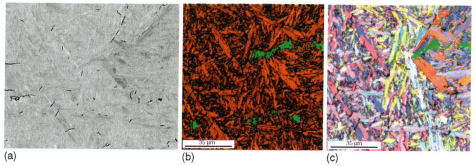

Figure 3.31 BSE image (a), austenite (light) and ferrite or carbide distribution (dark), (b) and orientation distribution of a specimen quenched from 1425 °C (c).

partial liquid 100Cr6. Furthermore, the liquid-phase determination is complicated by the superimposed martensite development during quenching.

3.4.3
Concluding Assessment of the Microstructure Parameter Determination

For the determination of the structural parameters and the phase contents in the partial liquid state, thermodynamic calculations and quenching experiments were conducted on steels X210CrW12 and 100Cr6. For steel X210CrW12, the characterization of the microstructure was considered to be difficult above the eutectic temperature, because liquid phases transform into secondary austenite and are difficult to discern from the primary phase. The same limitation holds even more strongly for systems without eutectic components.

For steel X210CrW12, a clear distinction between primary and secondary austenite seems to be possible only with reservations. It cannot be discerned whether the globulites grew in a homogeneous way or merely formed protuberances. At a quenching temperature of 1350 °C, the fraction of secondary austenite increases markedly, whereas in comparison with the situation at 1300 °C considerably more secondary austenite with fine, dendritic formation in the eutectic is observable. This dendritic austenite and also the austenite protuberances can be definitely attributed to a former liquid phase whereas it is not evident whether an even globulite growth took place during quenching. It can be assumed that during quenching of steel X210CrW12, three different growth types of the secondary austenite can be discerned. During structural evaluation, the uneven growth (formation of protuberances) and the dendritic growth can be considered metallographically. In contrast, the homogeneous grain growth cannot be quantified either with conventional image analysis or by means of the microprobe and EBSD analysis methods.

In comparison with the results for steel X210CrW12, no conclusion can be drawn concerning the structure in the partial liquid state due to the higher process temperatures and the, therefore, higher permeability of the elements and the transformation behaviour during the solidification of steel 100Cr6. Furthermore, the low austenite stability complicates the structural examination due to the solved carbon and chromium contents in steel 100Cr6 and the associated martensitic transformation during quenching.

Concerning the liquid-phase determination, it can finally be noted that it should be resorted to liquid-phase contents calculated thermodynamically and detected by means of DTA measurements, because metallographically determined values cannot be used. A determination of the structural parameters such as the grain form and the contiguity on quenched samples is not possible especially with higher liquid phase contents, but is already problematic with medium or low liquid-phase contents.

3.5
Melting Behaviour

In this section, the results of investigations concerning the influence of carbide distribution and the effect of titanium nitride on the melting behaviour of steel 100Cr6 are described, because examinations of thixoforged damper brackets within the EU project 'Thixocomp' exhibited very high segregation of solid and liquid phases for long flow lengths during thixoforging. The reason for this was a coarse-grained, cloudy melting of the rolled primary material with high intra-globular liquid-phase contents. Damper brackets generated from the fine-grained, globular melting material HS6-5-3 exhibited a considerably lower segregation tendency. To examine the primary material influence and the specific adjustment of the fine-grained, globular structure, the rolled state was, therefore, compared with the cast states (with and without titanium doping).

3.5.1
Thermodynamic Preliminary Considerations and Microstructural Examinations Concerning the Structural Regulation of Steel 100Cr6 by Means of TiN Particles

To determine the necessary amount of titanium for the formation of a sufficient amount of TiN particles, thermodynamic calculations were executed with a standard composition of 1.00% C, 0.35% Si, 0.30% Mn, 1.50% Cr and 50 ppm N. Figure 3.32 shows that with a liquid-phase content of 50% about 300 ppm of Ti can be solved, whereas at the solidus temperature only 50 ppm of Ti can be solved. Because 1000 ppm (0.1%) Ti would be necessary to form TiN in the pure melt, titanium is unsuitable as a grain development addition for rheo-processes. The calculations show, furthermore, that addition of 50–200 ppm of Ti should be sufficient for the microstructural adjustment. Higher contents would lead to an early development of TiN particles in the cast or in the partial liquid state and would, therefore, have no growth-hampering influence on the austenite grain size. The calculations show, furthermore, that under the given circumstances no unwanted formation of CrN is to be expected.

Based on the thermodynamic calculations, three different laboratory melts were generated in a vacuum-induction furnace. As input -materials, rolled primary material and sponge titanium were used. A laboratory melt without alloying of

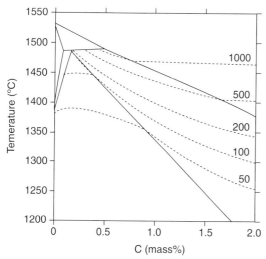

Figure 3.32 Solubility limit of titanium nitride with addition of 50, 100, 200, 500 and 1000 ppm to steel 100Cr6.

titanium was produced to compare rolled and cast primary material, which does not exhibit carbide bands. In addition, melts with 100 and 250 ppm of Ti were produced to examine the influence of TiN particles on the structural development. Because the laboratory melt with 250 ppm Ti showed the same results as the melt with 100 ppm, in the following only the states 'rolled', 'cast' and 'cast + 100 ppm Ti' will be discussed. The chemical compositions are given in Table 3.5 and the primary structures are shown in Figure 3.33. The element distribution images in Figure 3.34 show the Ti (C, N) particles that were formed, which exhibit enrichment with carbon and nitrogen. In the following, these will be called TiN particles for simplification.

The experimental setup and the temperature–time curve of the quenching experiments are displayed in Figure 3.35. The temperature of the samples heated in aluminium oxide pans was measured during the whole experiments by means of a thermal element in the sample centre. They reached the target temperature after about 10 min and were held isothermally for 20 min and subsequently quenched with water and examined metallographically.

The structural images of the quenched samples in Figure 3.36 show an ASTM grain size of −2 for the rolled material at 1250 °C vertical to the rolling direction, whereas the cast states exhibited considerably smaller ASTM grain sizes of 2 and 4.

Table 3.5 Chemical compositions of the as rolled and as cast state (all data in mass%, except Ti and N in ppm).

State	C	Si	Mn	Cr	Mo	Ni	Al	Cu	Ti	N
Roll	1.08	0.33	0.28	1.47	0.02	0.05	0.03	0.06	<10	65
Cast	0.93	0.32	0.27	1.44	0.02	0.06	0.04	0.07	<10	62
Cast + Ti	1.00	0.33	0.28	1.45	0.02	0.05	0.04	0.07	102	57

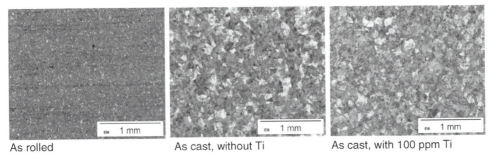

As rolled   As cast, without Ti   As cast, with 100 ppm Ti

Figure 3.33 Primary structure of the examined 100Cr6 states.

The alloying of titanium results in a reduction in the austenite grain growth in the solid state because of the formation of TiN particles. The experiments in the partial liquid state at 1400 °C showed that this effect disappeared because the cast alloyed with titanium and the cast without titanium possess very similar grain sizes. Therefore, TiN cannot be used for the reduction of the unwanted grain growth during holding in the partial liquid state. Only the indirect effect of the reduction of the austenite grain growth during heating can be used in the solid state. In this case, the grain size of the titanium-alloyed cast is lower than that of the cast without titanium, as the examinations of the third test series show. For short holding times in the partial liquid state, the growth constraint of the TiN particles in the solid state still has an influence on the microstructure in the partial liquid state.

The metallographic image parallel to the rolling direction in the third test series shows that (bottom left) the carbide bands continue within the austenite grains and

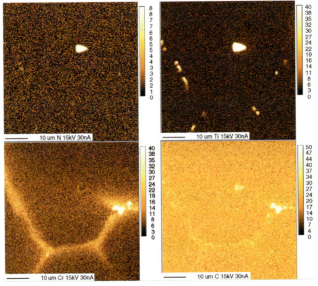

Figure 3.34 Element distribution images (N, Ti, Cr, C) of titanium-doped 100Cr6 specimens quenched from the partial liquid state.

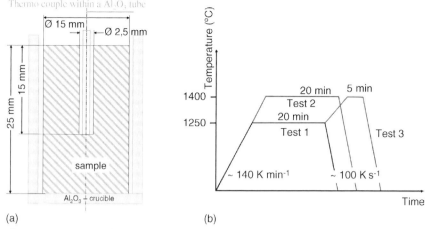

(a)   (b)

Figure 3.35 Experimental assembly (a) and temperature-time curve of quenching experiments (b).

thus form a high fraction of intra-globular liquid phase and result in long-stretched austenite grains parallel to the rolling direction. The 20 min hold at 1250 °C, the subsequent heating to the partial liquid state and the further hold of 5 min at 1400 °C are not sufficient to eliminate the negative influence of the rolled primary material.

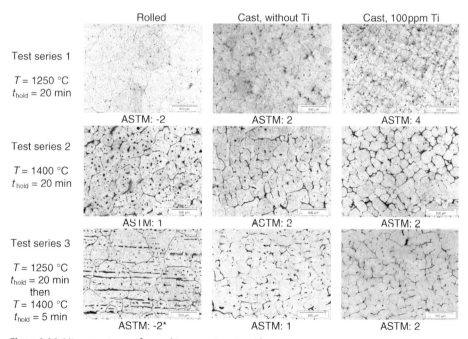

Figure 3.36 Microstructures of quenching experiments with different primary material conditions (*parallel to rolling direction).

In contrast, the two cast exhibit a fine-grained, globular structure without noteworthy intra-globular liquid phase content.

3.5.2
Concluding Appraisal of TiN Doping

In conclusion, it can be noted that the rolled primary material, in which the carbides exist long-stretched and chain-like, should not be used for thixoforming, due to the disadvantageous melting behaviour and the increased segregation tendency. Whereas the hot-rolled material exhibits coarse-grained and cloudy melting behaviour with a high content of intra-globular liquid phase, the two cast states show fine-grained, globular melting behaviour without noteworthy intra-globular phase contents. In the solid state, an influence of the TiN particles on the austenite grains could be detected. Because the grain-fining effect of the TiN particles is lost in the partial liquid state and the TiN particles have a negative influence on the dynamic mechanical component properties due to their angular morphology, TiN doping should not be applied. The examinations of the laboratory melts show that a structural regulation is unnecessary by means of TiN particles, because the as-cast material exhibits a rather fine-grained, globular structure. Concerning the primary material selection, it can be generally noted that primary material for the production of low-segregation components should not contain pronounced carbide banding because these (and also very homogeneous material due to the self-blocking–remelting effects) lead to a complex, non-globular melting behaviour with high intra-globular liquid-phase contents. Similar results for examinations of steel HP9/4/30 can be found in the literature [63].

3.6
Microstructure Analysis and Material Properties

In the partial liquid state, steels exhibit a multiphase structure. Due to their equilibrium concentrations, sometimes highly different element contents between the liquid and solid phases can occur, which lead to different phase transformation behaviours. Resulting from these conditions, an isolated observation of both structural areas and their interaction concerning the possible mechanical properties is necessary. If the steels are continuously cooled or quenched from the partial liquid state and subsequently aged, this can lead to new, complex structures. In this section, the results of the examination and development of adaptive heat-treatment strategies exploiting the former partial liquid state are presented and the potential is assessed by means of the two steels X210CrW12 and 100Cr6.

3.6.1
Structural Changes and Properties of the Tool Steel X210CrW12

For the experimental examination of the transformation behaviour, cold-rolled rods of steel X210CrW12 with a diameter of 56 mm and a composition of 2.12% C, 12.12%

Cr, 0.81% W, 0.33% Mn and 0.21% Si were used. Cuts of 30 mm length underwent different heat treatments and forming experiments. As reference state, a first test series was austenitized on the basis of the norm at a temperature of 970 °C and a holding time of 30 min and subsequently quenched in oil. In the following, some of the specimens were annealed at 300 °C for 2 h. To examine the influence of a transformation on the structural development, the cold-rolled primary material was annealed at 1270 °C for 20 min and heated to 900 °C, then compressed by 100% and subsequently cooled with air. For the comparison with conventionally hardened material, a third test series was executed, in which steel X210CrW12 was held at 1270 °C for 20 min in the partial liquid state and subsequently cooled in air. Some of the specimens of this test series were subsequently isothermally annealed for 1–168 h at 490 °C, for 1–24 h at 540 °C and for 30–75 min at 595 °C. Additionally, small dilatometer samples with a length of 10 mm and a diameter of 5 mm were examined concerning their transformation behaviour and their microstructure at a temperature of 1210 °C and at continuous cooling rates of 4.4, 6.7, 13.3, 20, 24, 40 and 120 K min^{-1}. The experiments were carried out slightly beneath the solidus temperature at 1210 °C, because the specimens would have been distorted by the spring force of the clamping device at higher temperatures. The length changes were established by means of an inductive position transducer, which offers a precision of about 10 nm. The macro-hardness (HV10) and the microhardness (HV0,1) were determined.

3.6.1.1 Thermodynamic Preliminary Considerations and Microstructural Examinations

Preliminary examinations of the structural development of steel X210CrW12 rapidly cooled from the solidification interval show that the structure existing at room temperature consists mainly of an interpenetrating two-phase network formed by an austenitic primary phase and a $\gamma + Cr_7C_3$ eutectic. The primary phase equals the solid phase located in the solidification interval, whereas the eutectic phase content reflects the original liquid phase. In contrast, when applying a conventional heat treatment the austenitic phase transforms to a considerable extent into martensite, whereas an annealing treatment at significantly higher temperatures stabilizes the austenite so much that it remains the dominant phase at room temperature. In the following, the material scientific aspects of the structural development are, therefore, examined more closely also in respect of the kinetic phenomena during cooling from high annealing temperatures.

The pseudo-binary phase diagram shows a carbon content of 2.1 mass% for a recommended hardening temperature of 970 °C (Figure 3.11). For these conditions, Thermo-Calc calculations show a phase content of 79 mol% austenite and 21 mol% Cr_7C_3 carbide. The chemical composition of the austenitic phase is made up of 4.6% Cr, 0.76% C, 0.45% W, 0.30% Si and 0.28% Mn. According to the empirical equation $M_s = 635 - 435\%C - 17\%(°C)$ the calculated M_s is about 196 °C [64], so that with sufficiently rapid cooling the austenite transforms into martensite and the structure of the hardened steel is characterized by a martensitic matrix with embedded Cr_7C_3 carbides at room temperature. Here, carbides in the sub-micrometer range and up to a size of 50 μm are formed (Figure 3.37). Due to the relatively high carbon content in

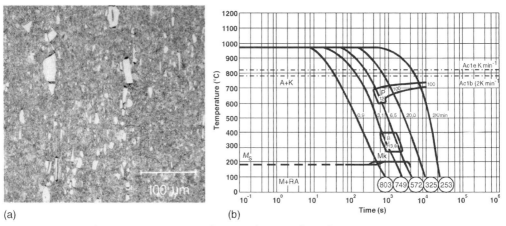

(a)

(b)

Figure 3.37 Microstructure after air cooling (a) and CCT diagram of steel X210CrW12 for a hardening temperature of 980 °C (b) [51].

the martensitc matrix and the high content of Cr_7C_3 carbides, a hardness of about 760 HV10 or 63 HRC is reached. The CCT diagram for steel X210CrW12 with 11.5% Cr, 0.78% W, 2.03% C, 0.32% Mn, 0.39% Si and 0.11% V in (b) shows that the macro-hardness and the M_s for a hardening temperature of 980 °C could be well confirmed experimentally (hardness $\approx$ 800 HV10; $M_s \approx$ 190 °C). Furthermore, inert transfor-mation behaviour is exhibited at a critical cooling rate of about $2 \, K s^{-1}$. As a char-acteristic orientation factor, the value for the transformation A_{c1b} (beginning of the austenitization at heating) is at about 790 °C. Therefore, already cooling rates of about $2 \, K s^{-1}$ are sufficient to reach a martensitic transformation of the austenite without the formation of bainite and pearlite phases.

In the second test series, the influence of forming on the structural development was examined. Samples that were held isothermally and then quenched in water exhibit a primary content of globular, former solid phase of about 60% and an average grain size of 44 μm (±21 μm). The liquid phase initially existing at 1270 °C is dark and exists at room temperature as finely structured $\gamma + Cr_7C_3$ eutectic, whereas the light, austenitic primary phase exhibits typical twin contrasts (highlighted by black arrows). Furthermore, sporadic coalescence develops (highlighted by white arrows). The examinations show no differences in the structural development of compressed and uncompressed samples (Figure 3.38). In both cases, a homogeneous structure exists, which no longer provides any indication of the rolled primary material condition (no more influence of carbide bands is recognizable). Hardness measurements on the compressed samples exhibit marginally lower hardness values (Table 3.6).

Figure 3.39 shows a sector of the pseudo-binary cut through the six-component system with 11.5% Cr, 0.7% W, 0.3% Mn and 0.25% Si. It is readily evident that the alloying condition point exists in the two-phase area $L + \gamma$ (liquid + austenite) at 1270 °C and the liquid undergoes a eutectic transformation during cooling. Thermo-Calc calculations show a phase content of 60 mol% (61.4 vol.%) austenite, or 40 mol% (38.6 vol.%) for the liquid at 1270 °C. At that condition, the chemical composition of the

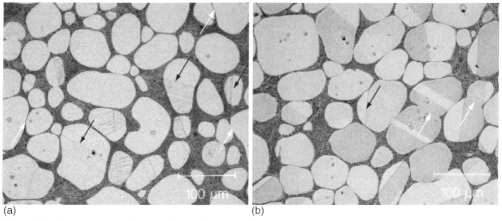

(a)   (b)

Figure 3.38 Microstructures after annealing within the solidification interval at 1270 °C for 20 min without (a) and with subsequent transformation of 100% and air cooling (b). The light arrows highlight the coalescing globulites, and the twin boundaries are marked by dark arrows.

austenitic solid phase exhibits, with 9.4% Cr, 1.31% C, 0.39% W, 0.28% Si and 0.24% Mn, significantly higher contents of the austenite-stabilizing elements chromium and carbon than at conventional hardening temperatures, so that a martensite start temperature of -147 °C results. Therefore, the appearance of the metastable austenitic phase is confirmed by thermodynamic calculations and rules of experience.

Figure 3.40 allows a detailed appraisal of the stability of the austenitic high-temperature phase. It shows the dependence of the carbon and chromium contents on the annealing temperature (a). According to these values, the martensite start temperature can be calculated depending on the annealing temperature (b). Therefore, the maximum solubility of carbon and chromium is reached at a process temperature of the semi-solid, forming above the solidus temperature, that is in the three-phase area $L + \gamma + Cr_7C_3$, so that the austenite stability is the highest there and M_s is reduced to values below -150 °C. With a temperature window suitable for thixoforming between 1260 and 1290 °C, the development of metastable austenitic structural areas is therefore expected.

Thermo-Calc calculations of the transformation temperatures in equilibrium show that during cooling at the A_{c1} temperature (~780 °C) which is $\gamma \rightarrow \alpha$, transformations

Table 3.6 Measured macro- and microhardnesses of the different structures.

Heat treatment	Hardness (HV10)	Primary grain (HV0.1)	Eutectic (HV0.1)
1270 °C, 20 min, air	370 ± 9	310 ± 6	620 ± 150
1270 °C, 20 min, air, Deformation	366 ± 11	295 ± 8	534 ± 109
970 °C, 12 min, oil	774 ± 20	—	—

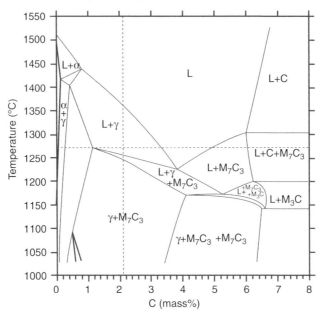

Figure 3.39 Thermo-Calc calculated pseudo-binary phase diagram in equilibrium for the system Fe–Cr–W–C–Mn–Si with 11.5% Cr, 0.7% W, 0.3% Mn and 0.25% Si.

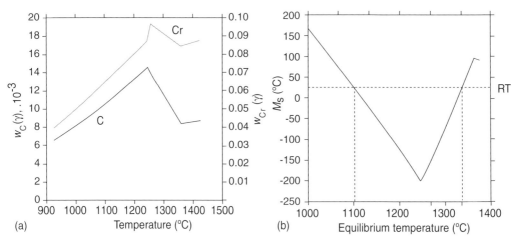

Figure 3.40 Thermodynamic calculations for the carbon and chromium contents within austenite under equilibrium conditions (a) and from this calculated M_s temperatures as a function of the annealing temperature using the empirical equation $M_s = 635 - 475[\text{mass\% C}(\gamma)] - 17[\text{mass\% Cr}(\gamma)]$ (b) [64].

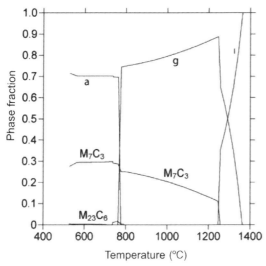

Figure 3.41 By means of Thermo-Calc calculated phase contents of steel X210CrW12 in equilibrium.

with concurrent formation of M_7C_3 and $M_{23}C$ carbides take place (Figure 3.41). This diffusion-controlled cooling rate is, however, only observed at an adequately slow cooling rate, as depicted in Figure 3.37b for cooling from a temperature of 980 °C by the 'pearlite nose'. If, however, the austenite is stabilized at a higher annealing temperature through the high solubility of carbon and chromium, not only does the M_s temperature change but also the transformation kinetics differ, as shown in Figure 3.42 in form of a

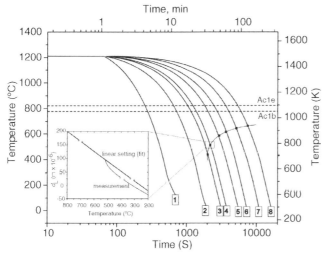

Figure 3.42 CCT diagram for steel X210CrW12 for different cooling rates from 1210 °C after an annealing of 30 min at 1270 °C. The enlarged partial image illustrates the transformation induced change in length of sample 4.4. The numbers used correspond to the experimental conditions in Table 3.7.

CCT diagram for cooling from 1210 °C. Figure 3.42 shows significantly more inert transformation behaviour for cooling from very high temperatures in comparison with 'usual' hardening treatments. The pearlite nose dislocates in the direction of longer times by about half a dimension, that is, in the more advantageous direction. For interpreting the CCT diagram, it has to be considered that under certain circumstances carbide precipitation can already occur at temperatures above A_{c1} or A_{r1}. It has to be considered here that in equilibrium the carbide content amounts to about 12.8 mol% at 1210 °C according to Thermo-Calc calculations, whereas it amounts to about 21 mol% at 980 °C. A pre-eutectic carbide precipitation could, however, not be proven in the dilatometer. The transformation delay indicates, however, an increased C or Cr content in the austenite.

The results for specimens annealed at 1270 °C and subsequently held in the dilatometer for 60 min at 1210 °C and cooled at different cooling rates are depicted in Table 3.7 and Figure 3.42. The curves numbered from 1 to 8 represent the different cooling times from 1210 °C to room temperature. Specimens cooled with a cooling rate lower than 30 K min^{-1} exhibited a considerable change in length related to the transformation that occurred, as highlighted by the inset. The measured hardness values show that the hardness for the quenched state increases from 366 HV10 to the maximum hardness of 614 HV10 at a cooling rate of about 10 K min^{-1}, whereas it drops again monotonously at lower cooling rates.

The metallographic images of the dilatometer experiments in Figure 3.43 show a continuous transformation of the primary, austenitic phase and a coarsening of the carbides subject to the cooling time. For cooling times of 10 or 30 min from 1210 °C to room temperature, the matrix remains austenitic. With increasing cooling time, the eutectic phase content ($\gamma + Cr_7C_3$) becomes coarser and singular Cr_7C_3 carbides form within the grain core of the austenite. At a cooling time of 50 min, a dark phase forms which was identified as fine needle-shaped pearlite and is treated in detail in the course of the discussion of the isothermal ageing experiments of the third test series. The content of this fine pearlite increases with increase in cooling time, and the phase content of Cr_7C_3 carbides in the grain core of the retained austenite also increases. In experiment 8, nearly the whole austenitic matrix structure is transformed.

Table 3.7 Cooling conditions of the dilatometer trials and the corresponding hardness values.

Test	Cooling time, 1210 °C–RT (min)	dT/dt (K min^{-1})	t8/5 (min)	Hardness (HV10)
1	10	120	2.5	366
2	30	40	7.5	354
3	50	24	12.5	420
4	60	20	15	498
5	90	13.3	22.5	537
6	120	10	30	614
7	180	6.7	50	584
8	270	4.4	67.5	501

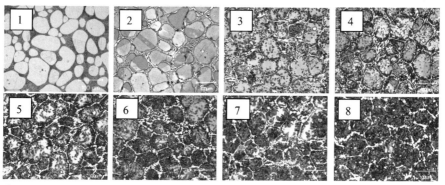

Figure 3.43 Microstructure development of the steel X210CrW12 after differently paced coolings from 1210 °C after prior annealing in the partial liquid state (1270 °C/30 min).

In the following, the experiments of the third test series are discussed. The samples were annealed at 1270 °C, quenched and subsequently held isothermally for 1–168 h at 490 °C, for 1–24 h at 540 °C and for 30–75 min at 595 °C. Figure 3.44 shows that at an ageing temperature of 490 °C a maximum hardness of 800 HV10 is reached after an annealing time of 100 h, at 540 °C a maximum hardness of 745 HV10 is reached after 10 h and at the highest annealing temperature of 595 °C a maximum hardness of 720 HV10 is reached after 30 min. In conformity with previously published research on high-chromium content steels [51, 65], an examination of the microstructure

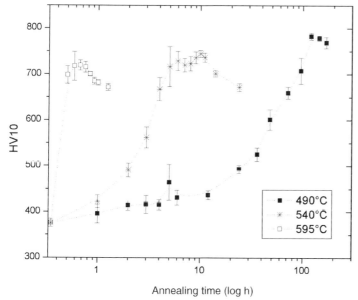

Figure 3.44 Hardness as a function of the ageing time for annealing temperatures of 490, 540 and 595 °C.

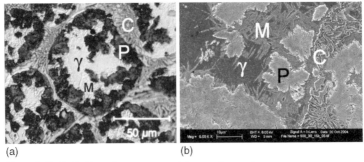

Figure 3.45 Optical interference–contrast image of a specimen annealed at 595 °C for 30 min after quenching from 1270 °C (a) and scanning electron microscope image (b) [with enlarged eutectic carbides (C), needle-shaped pearlite (P) and partially in martensite (M) transformed austenite (γ)].

shows the formation of needle-shaped, acicular pearlite and martensite needles within the not yet transformed austenite, as depicted for an annealing temperature of 595 °C and an annealing time of 30 min (Figure 3.45) or 75 min (Figure 3.46). After a holding time of 30 min, when the maximum hardness is reached, about half of the retained austenite is transformed into fine, needle-shaped pearlite whereas the rest of the austenite partially transforms into martensite. After a holding time of 75 min, that is, in the over-aged state, a high pearlite content and a significantly lower retained austenite content exist, which transforms completely into martensite. At ageing temperatures of 490 and 540 °C, a similar behaviour can be observed. The only difference is the formation of very low amounts of Widmanstätten carbides within the austenite. The development of this form of carbide can be found increasingly at 490 °C, at which it always concurs with pronounced martensite formation, as depicted in Figure 3.46b. For all ageing temperatures, the morphology of the acicular pearlite is very fine and

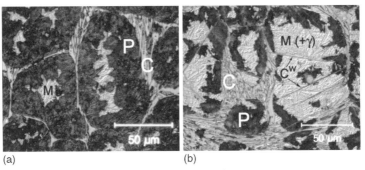

Figure 3.46 Optical interference–contrast images of specimens annealed at 595 °C for 75 min after quenching from 1270 °C (a) and annealed at 490 °C for 96 h (b) [with coarsened eutectic carbides (C), Widmanstätten carbides (CW), needle-shaped pearlite (P) and partially in martensite (M) transformed austenite (γ)].

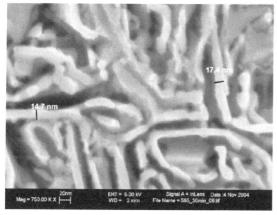

Figure 3.47 High-solution scanning electron microscope image of a deeply etched specimen, annealed at 595 °C for 30 min, which shows acicular pearlite.

only observable at high magnifications. SEM images show that the lamellae exhibit a thickness of up to 20 nm (Figure 3.47).

Examinations concerning the wear characteristics of steel X210CrW12 were executed by means of 'rubber wheel abrasion' tests. The condition of a conventional standard heat treatment (970 °C/30 min/oil + 300 °C/2 h) was compared with the condition of a formerly partial liquid X210CrW12 (1270 °C/30 min) and subsequently aged at 490 and 595 °C to its maximum hardness. The determined volume loss reveals excellent wear resistance, in comparison with literature values of nitrided steels (Table 3.8). In the annealed state, the formerly partial liquid material contain retained austenite contents, as already established in the literature concerning extraordinary wear characteristics. Examinations of high-chromium content cast iron

Table 3.8 Hardness values (HV10) and volume loss during the wear experiments (dV) of the conventionally hardened and the thixo-heat-treated steel X210CrW12 in comparison with wear-resistant, nitrided chromium steels.

		X210CrW12			Cr-alloyed tool steels					
	970 °C 30 min and 300 °C 2 h	1270 °C/20 min and subsequently								
		Quenched	595 °C 40 min	490 °C 120 min						
Hardness (HV10)	722	370	719	784	490	545	615	745	800	940
dV (mm³)	4.2	10	3.6	3.8	13.3	9.9	8.0	11.8	3.3	2.1

consisting of a austenitic–martensitic matrix with high contents of chromium carbides exhibit unusual wear qualities. The best wear resistance was established in different examinations to be at a retained austenite content of about 20% [66, 67]. Other examinations show that the smashing of carbides leads to a reduction in the wear resistance [68]. Because X210CrW12 processed in the partial liquid state exhibits significantly lower carbide sizes than the conventionally processed material, this leads to a reduction in carbide smashing and thus a further improvement in the material properties.

It can be noted that the austenite stability due to the carbon and chromium contents of the austenite in the partial liquid state (1.23 mass%) is, in comparison with the austenite using a conventional hardening temperature (0.76 mass%), very much increased and the phase transformation is displaced towards longer times. Therefore, SSM tool steels are very transformation inert and even useable for very high component thicknesses. The hardness increase at lower cooling rates can be explained by the precipitation of secondary carbides and the concurrent transformation of austenite into bainite and pearlite. A transformation into martensite could not be observed.

Isothermal holding at high temperatures after quenching from the semi-solid interval shows a considerable hardening potential, which, at up to 800 HV10 (64 HRC), is higher than that of conventional hardening (Table 3.8). Due to the destabilization of the austenite, a microstructure of very fine austenite is formed, as already known in the literature [65]. In this way, a multiphase structure of transformed eutectic, pearlite, martensite, retained austenite and sporadically also Widmanstätten carbides is formed, which exhibits excellent hardness and wear properties. The measured hardness changes are, therefore, a result of the complex interplay of different resulting structures of the austenite transformation: low pearlite and martensite content at the beginning of the heat treatment (low hardness), optimum pearlite/martensite composition at the beginning of the heat treatment (maximum hardness) and high pearlite and low martensite content with over-ageing of the steel and a softer product structure. The determined ferrite contents correlate very well with the measured hardness values and support the exhibited results (Figure 3.48).

To verify the transformation kinetics and to determine the rate-determined mechanism, the activation energy was evaluated using an Arrhenius relationship. From the temperature–time data of the maximum hardness (595 °C/0.6 h, 540 °C/10 h, 490 °C/120 h), a value of 285 kJ mol^{-1} K^{-1} is obtained, which agrees very well with published data for the chromium diffusion within austenite ($Q = 280$ kJ mol^{-1} K^{-1}) of chromium contents of 10–15 mass% [69].

3.6.1.2 Examinations Concerning Long-term Heat Resistance

Long-term annealing at fictional service temperatures of 400 and 450 °C was executed for steel X210CrW12 to carry out a benchmarking of the new heat treatment strategy in comparison with the industrially used reference state. For this, the conventionally hardened state (980 °C/30 min, H_2O or air) was compared with different thixo-states (1270 °C or 1300 °C/30 min, air or H_2O, 540 °C/10 h, air). Because no significant differences concerning the annealing temperature (1270 or 1300 °C) or the quenching medium could be found, in the following only the results based on the state

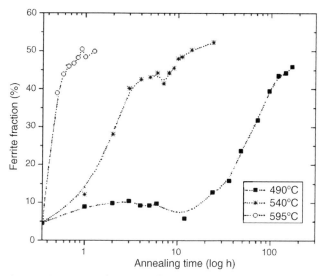

Figure 3.48 Determined contents of magnetic phase (ferrite) as a function of the annealing time for ageing temperatures of 490, 540 and 595 °C.

1270 °C/30 min/air are considered. Figure 3.49 depicts the hardness curve of the samples held isothermally at 400 and 450 °C. It is apparent that the hardness values decrease with increasing temperature, the conventionally hardened material exhibiting a considerably steeper decrease than the thixo-material. Due to the seemingly very stable structure, this material exhibits constant values of more than 58 HRC at a

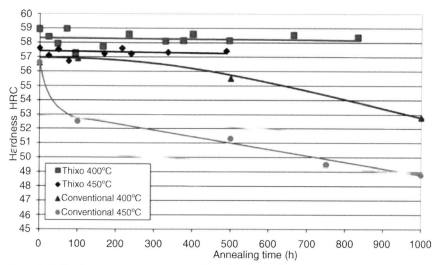

Figure 3.49 Hardness curves for conventionally hardened and of thixoformed steel X210CrW12 under long-term annealing.

temperature of 400 °C and a holding time of 1000 h, whereas the conventionally hardened material decreases to values below 53 HRC. At a fictional service temperature of 450 °C, the difference between the two states becomes even more pronounced, because the conventionally hardened material exhibits at this temperature already a very steep decline below 53 HRC after 100 h.

In conclusion, it can be noted that by means of the application of an adjusted heat-treatment strategy, the insertion temperature of the cold working steel X210CrW12 could be increased considerably and could even be inserted at temperatures at which usually only warm-working steel is used. Depending on the particular application, service temperatures of up to 400–450 °C seem realistic.

3.6.2
Structural Changes and the Properties of the Hypo-eutectic Steel 100Cr6

Due to the chromium and carbon concentrations, the austenite has an M_s temperature of about 350 °C, so that it transforms into martensite during quenching. The structural images in Figure 3.50 confirm the calculations and reveal a martensitic structure, at which a light seam is located at the former austenite–grain boundaries. By close examination of the element distribution images and the line-scan of the microprobe examinations, it can be assumed that the white phase is a two-phase structure of retained austenite with finely ingrained carbides, which could be confirmed by microhardness measurements [70].

Future research should deal with the systematic observation of the structural development of steel 100Cr6 without a connected carbide network. The transformation behaviours of both the solid and liquid phases have to be considered to be able finally to derive a heat-treatment strategy that delivers useful mechanical properties based on the partial liquid state.

In addition, the volume shrinkage and the concurring pore formation for steel 100Cr6 will be an important factor for the mechanical properties.

3.6.3
Final Evaluation of the Heat-treatment Strategies

The SSM potential of steel concerning the metallic properties was examined by means of systematic heat treatments on steels X210CrW12 and 100Cr6. After processing in the partial liquid state, new structures could be produced by means of isothermal

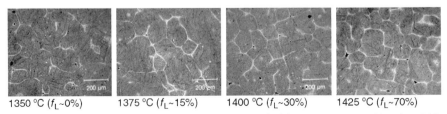

1350 °C (f_L~0%)   1375 °C (f_L~15%)   1400 °C (f_L~30%)   1425 °C (f_L~70%)

Figure 3.50 Microstructure of 100Cr6 specimens quenched from 1350, 1375, 1400 and 1425 °C.

ageing or controlled, continuous cooling due to the element distribution in the liquid and solid phases. Due to these, mechanical properties can be achieved that are superior to those of conventional hardening. For example, the application temperature of steel X210CrW12 can be raised considerably by means of the described heat treatments.

To consolidate the industrial application of steel 100Cr6 in the thixoformed state, further systematic examinations have to be conducted and applicable heat-treatment strategies have to be developed. Important for this is the avoidance of a connected carbide network on the grain boundaries.

Based on the examinations up to this point, a transformation of the segregated grain-boundary areas has to take place in such a way that a structural compound of austenite, martensite and finely distributed carbides can form. The necessary transformations can be reached either by means of specific cooling directly following the transformation process or by means of a special heat treatment by heating from the cooled state, so that no degradation of the 'global' mechanical properties of the material occurs.

Table 3.9 shows the element contents of the separate phases with a fraction of 50% solid and 50% liquid phase. The M_s temperatures calculated from this are given in the last column. For the purpose of comparison, the respective values for the conventional hardening of the materials are also given. In Table 3.10, the summarized measured microhardness of the separate structural components of steels X210CrW12 and 100Cr6 are given. The values in these two tables clarify that due to the strong reallocation of the elements chromium and carbon in both materials, a multiphase structure is formed which exhibits potential for new material properties due to its different transformation behaviour. For steel X210CrW12, this was already

Table 3.9 Thermo-Calc calculated carbon and chromium contents of the solid and liquid phases at equilibrium with phase contents of 50 mol% and of the austenite at conventional hardening temperatures; calculated M_s temperatures as a function of the annealing temperature using the empirical equation
$M_s = 635 - 475[\text{mass}\% \ C(\gamma)] - 17[\text{mass}\% \ Cr(\gamma)] - 33[\text{mass}\% \ Mn(\gamma)]$ [64].

		C (mass%)	Cr (mass%)	Mn (mass%)	M_s (°C)
X210CrW12	Solid (austenite)	1.23	9.16	0.23	−112.6
1285 °C	Liquid	3.07	14.85	0.37	(eutectic transformation)
$f_s = 50\%$					
Near 970 °C	Austenite	0.76	4.62	0.30	185.5
100Cr6	Solid (austenite)	0.55	1.36	0.24	342.7
1418 °C	Liquid	1.46	1.65	0.36	−98.5
$f_s = 50\%$					
850 °C	Austenite	1.00	1.50	0.30	124.6

Table 3.10 Measured microhardness of the structural components of different material conditions of the steels X210CrW12 and 100Cr6.

Heat treatment	Phase	Hardness
X210CrW12, after quenching from 1270 °C	Solid (austenite)	300 HV0.1
	Liquid (eutectic)	450–700 HV0.1
X210CrW12, microstructure after adapted heat treatment	Different carbides, acicular pearlite, martensite and retained austenite	750–800 HV10
X210CrW12, after conventional hardening (970 °C)	Austenite	760 HV10
100Cr6, after quenching from 1400 °C	Martensite, carbides, retained austenite	750 HV30
100Cr6, after conventional hardening (850 °C/oil)	Martensite, carbide, retained austenite	830 HV30

examined in detail in the scope of this research, whereas for the steel 100Cr6 further test series have to be conducted. Only after completing these examinations can a final statement about the thixoformability of steel 100Cr6 be made.

3.7
Conclusions

Thixoforming of steel is in general possible, but still remains in an early and far from mature development stage. Not least because of the high process temperatures and the oxidation aptness of partial liquid steels, this procedure is still faced by great challenges and much optimization is needed before it can be applied industrially.

The preconditions for selecting suitable alloys for thixoforming are well understood: it is mainly the temperature sensitivity and the particular liquid-phase development during heating and soaking. High-alloyed steels with precipitates that stabilize the microstructure at high process temperatures seem to be more suitable than lean-alloyed steel concepts.

Thixoforming requires a thorough understanding of the microstructure development during semi-solid processing. It has been shown that this is not possible in a highly accurate manner by metallographic examination of quenched samples alone. Proper process control will therefore need instantaneous microstructure state monitoring by online numerical modelling.

Characteristic features are observed in thixoformed specimens which are inherent to the thixoforming process; for example the alloy decomposition due to significant different solubilities of liquid and solid phases and due to the sponge effect in the liquid–solid suspension or the occurrence of pores and inhomogeneous grain size distributions.

On the other hand, these features can be used to develop unique properties, as has been shown by the improvements in hardness/wear and temperature stability as a consequence of novel microstructures or microstructure gradients. In general, heat treatments after thixoforming should be adapted to this process in order to make full use of its potential.

Thixoforming will not become a low-cost alternative to forging or to casting. There is a chance that, in combination with difficult geometric requirements, with specific mechanical needs or with simultaneous fusion/bonding of different materials, thixoforming will become an alternative to established process routes.

References

1 Joly, P. and Mehrabian, R. (1976) The rheology of a partially solid alloy. *Journal of Materials Science*, **11**, 1393–1418.

2 Kumar, P., Martin, C. and Brown, S. (1993) Shear rate thickening flow behaviour of semi-solid slurries. *Metallurgical and Materials Transactions A*, **24** (5), 1107–1115.

3 Kumar, P., Martin, C. and Brown, S. (1994) Constitutive modelling and characterization of the flow behaviour of semi-solid metal alloy slurries: – I. The flow response. *Acta Metallurgia et Materialia*, **42** (11), 3595–3602; Martin, C., Kumar, P. and Brown, S. (1994) Constitutive modelling and characterization of the flow behaviour of semi-solid metal alloy slurries – II. Structural evolution under shear deformation. *Acta Metallurgia et Materialia*, **42**, (11), 3603–3614.

4 Kopp, R. (ed.) (2004) *Formgebung metallischer Werkstoffe im teilerstarren Zustand und deren Eigenschaften, Arbeitsbericht des 3. Antragszeitraumes des Sonderforschungsbereichs 289*, Deutsche Formungsgemeinschaft (DFG).

5 Kiuchi, M. and Kopp, R. (2002) Mushy/semi-solid metal forming technology – present and future. *Annals of the CIRP*, **51** (2), 653–670.

6 Fan, Z. (2002) Semisolid metal processing. *International Materials Reviews*, **47** (2), 49–85.

7 Suéry, M., Matin, C. and Salvo, L. (1996) Overview of the rheological behaviour of globular and dendritic slurries. Proceedings of the 4th International Conference on Semi-solid Processing of Alloys and Composites, Sheffield, pp. 21–29.

8 Frohberg, M. (1984) Viskositätsmessungen an flüssigen Metallen und Metalllegierungen. *Metall*, **38**, (12), 1152–1160.

9 Barnes, H. (1997) Thixotropy – a review. *Journal of Non-Newtonian Fluid Mechanics*, **70** (1), 1–33.

10 Chiarmetta, G. and Rosso, M. (2000) Proceedings of the 6th International Conference on Semi-solid Processing of Alloys and Composites, Turin.

11 Tsutsui, Y., Kiuchi, M. and Ichikawa, K. (2002) Proceedings of the 7th International Conference on Semi-solid Processing of Alloys and Composites, Tsukuba.

12 Alexandrou, A. *et al.* (2004) Proceedings of the 8th International Conference on Semi-solid Processing of Alloys and Composites, Limassol.

13 Wu, S., Wu, X. and Xiao, Z. (2004) A model of growth morphology for semi-solid metals. *Acta Materialia*, **52**, 3519–3524.

14 Flemings, M. (1991) Behavior of metal alloys in the semisolid state. *Metallurgical Transactions A*, **22A** (5), 957–981.

15 Gullo, G. (2001) *Thixotrope Formgebung von Leichtmetallen – Neue Legierungen und Konzepte*, Dissertation der

Eidgenössischen technischen Hochschule Zürich, ETH-Dissertation Nr. 14154.

16 Ichikawa, K., Murakami, T., Miyamoto, S., Nakayama, Y. and Tokita, M. (2002) Production and properties of nano-stuctured bulk pure irons by spark plasma sintering process. Proceedings of the 7th International Conference on Semi-solid Processing of Alloys and Composites Tsukuba, pp. 379–384.

17 Uggowitzer, P., Gullo, G. and Wahlen, A. (2000) Metallkundliche Aspekte bei der semi-solid Formgebung von Leichtmetallen, in *Vom Werkstoff zum Bauteilsystem* (eds H. Kaufmann and P.J. Uggowizer), LKR-Verlag Ranshofen, pp. 95–107.

18 Balitchev, E., Hallstedt, B. and Neuschütz, D. (2005) Thermodynamic criteria for the selection of alloys suitable for semi-solid processing. *Steel Research*, **76** (2/3), 92–98.

19 Püttgen, W. and Bleck, W. (2004) DTA-measurements to determine the thixoformability of steels. *Steel Research*, **75** (8/9), 531–536.

20 Kaufmann, H., Fragner, W., Galovsky, U. and Uggowitzer, P. (2005) Fluctuations of alloy composition and their influence on sponge effect and fluidity of A356-NRC. Proceedings of the 2nd International Light Metals Technology Conference.

21 Bleck, W. (ed.) (2004) *Werkstoffkunde Stahl für Studium und Praxis*, Verlag Mainz, Aachen.

22 Zhu, M., Kim, J. and Hong, C. (2001) Modelling of globular and dendritic structure evolution in solidification of an Al–7 mass% Si alloy. *ISIJ International*, **41**, 992–998.

23 Kurz, W., Giovanola, B. and Trivedi, R. (1986) Theory of microstructural development during rapid solidification. *Acta Metallurgica*, **34** (5), 823–830.

24 Rappaz, M. and Gandin, Ch. (1993) Probabilistic modelling of microstructure formation in solidification processes. *Acta Metallurgia et Materialia*, **41** (2), 345–360.

25 Uggowitzer, P. and Kaufmann, H. (2004) Evolution of globular microstructure in new rheocasting and super rheocasting

semi-solid slurries. *Steel Research*, **75** (8/9), 525–530.

26 Kraft, T. and Exner, H. (1996) Numerische Simulation der Erstarrung, Teil 2: Mikroseigerungen in ternären und höherkomponentigen Legierungen. *Zeitschrift für Metallkunde*, **87** (8), 652–660.

27 Kraft, T. and Exner, H. (1997) Numerische Simulation der Erstarrung, Teil 3: Vorhersage der Gefügestruktur. *Zeitschrift für Metallkunde*, **88** (4), 278–290.

28 Kraft, T. and Exner, H. (1997) Numerische Simulation der Erstarrung, Teil 4: Unterkühlungseffekte. *Zeitschrift für Metallkunde*, **88** (6), 455–468.

29 Wang, H., Davidson, C. and StJohn, D. (2004) Semisolid microstructural evolution of AlSiMg alloy during partial remelting. *Materials Science and Engineering A*, **368**, 159–167.

30 Tzimas, E. and Zavaliangos, A. (2000) Evaluation of volume fraction of solid in alloys formed by semisolid processing. *Journal of Materials Science*, **35**, 5319–5329.

31 Suery, M. (2004) Microstructure of semi-solid alloys and properties. Proceedings of the 8th International Conference on Semi-solid Processing of Alloys and Composites, Limassol.

32 Ludwig, O., DiMichiel, M., Falus, P., Salvo, L. and Suery, M. (2004) *In-situ* 3D microstructural investigation by fast X-Ray, microtomogaphy of Al–Cu alloys during partial remelting. Proceedings of the 8th International Conference on Semi-solid Processing of Alloys and Composites, Limassol.

33 Loué, W. and Suéry, M. (1995) Microstructural evolution during partial remelting of Al–Si7Mg alloys. *Materials Science and Engineering*, **A203**, (1), 1–13.

34 Salvo, L., Cloetens, P., Maire, E., Zabler, S., Blanding, J., Buffière, J., Ludwig, W., Boller, E., Bellet, D. and Josserond, C. (2003) X-ray micro-tomography an attractive characterization technique in materials science. *Nuclear Instruments and Methods in Physics Research B*, **200**, 273–286.

35 Kleiner, S., Beffort, O. and Uggowitzer, P. (2004) Microstructure evolution during reheating of an extruded Mg–Al–Zn alloy into the semi-solid state. *Scripta Materialia*, **51**, 405–410.

36 Püttgen, W., Bleck, W., Seidl, I., Kopp, R. and Bertrand, C. (2005) Thixoforged damper brackets made of the steel grades HS6-5-3 and 100Cr6. *Advanced Engineering Materials*, **7** (8), 726–735.

37 Salvo, L., Suéry, M., De Charentenay, Y. and Loué, W. (1996) Microstructural evolution and rheological behaviour in the semi-solid state of a new Al-Si based alloy. Proceedings of the 4th International Conference on Semi-solid Processing of Alloys and Composites, Sheffield, pp. 10–15.

38 Maire, E., Buffiere, J., Salvo, L., Blandin, J., Ludwig, W. and Letang, J. (2001) On the application of X-ray microtomography in the field of materials science. *Advanced Engineering Materials*, **3** (8), 539–546.

39 Gurland, J. (1958) The measurement of grain contiguity in two-phase alloys. *Transactions of the Metallurgical Society of Aime*, **8**, 452–455.

40 Underwood, E. (1970) *Quantitative Stereology*, Addison-Wesley, Reading, MA.

41 Gurland, J. (1979) A structural approach to the yield strength of two-phase alloys with coarse microstructures. *Materials Science and Engineering*, **40** (9), 59–71.

42 Wolfsdorf, T., Bender, W. and Voorhees, P. (1997) The morphology of high volume fraction solid-liquid mixtures: an application of microstructural tomography. *Acta Materialia*, **45**, 2279–2295.

43 Gullo, G., Steinhoff, K. and Uggowitzer, P.J. (2000) Microstructural changes during reheating of semi-solid alloy AA6082 modified with barium. Proceedings of the 6th International Conference on Semi-solid Processing of Alloys and Composites, Turin, pp. 367–372.

44 Poech, M. and Ruhr, D. (1994) Die quantitative Charakterisierung der Gefügeanordnung. *Praktische Metallographie*, **31** (2), 70–77.

45 Kulmburg, W. (1998) Das Gefüge der Werkzeugstähle – Ein Überblick für den Praktiker. Teil 1: Einteilung, Systematik und Wärmebehandlung der Werkzeugstähle. *Praktische Metallographie*, **35** (4), 181–202; (1998) Teil 2: Besonderheiten des Gefüges der einzelnen Stahlgruppen. *Praktische Metallographie*, **35**, (5), 267–279.

46 Kulmburg, A. (1982) *VDI-Berichte*, **432**, 31–43.

47 Berns, H. and Trojahn, W. (1985) Einfluss der Wärmebehandlung auf Restaustenit, Verschleißwiderstand und mechanische Eigenschaften ledeburitischer Chromstähle. *HTM*, **40** (2), 62–72.

48 Haberling, E. and Wiegand, H. (1983) Thyssen Edelstahl Technische Berichte, **9**, 89–95.

49 Berns, H. (1974) Restaustenit in ledeburitischen Chromstählen und seine Umwandlung durch Kaltumformen, Tiefkühlen und Anlassen. *HTM*, **29** (4), 236–247.

50 Shimahara, H. and Kopp, R. (2004) Investigations of basic data for the semi-solid forming of steels. Proceedings of the 8th International Conference on Semi-solid Processing of Alloys and Composites, Limassol.

51 Werkstoffdatenblatt K107 http://www.boehler.de, (2006).

52 Lecomte-Beckers, J., Rassili, A., Carbon, M. and Robelet, M. (2004) Characterization of thermophysical properties of semi-solid steel for thixoforming. Proceedings of the 8th International Conference on Semi-solid Processing of Alloys and Composites, Limassol.

53 Hengerer, F. (2002) Wälzlagerstahl 100Cr6 – Ein Jahrhundert Werkstoffentwicklung. *HTM*, **57** (3), 144–155.

54 Vetters, H. (2002) Umwandlungsgefüge des Wälzlagerstahls 100Cr6 nach ein- und zweifacher Austenitisierung. *HTM*, **57** (3), 168–173.

55 Hengerer, F., Lucas, G. and Nyberg, B. (1974) Zwischenstufenumwandlung von Wälzlagerstählen. *HTM*, **29** (2), 71–79.

56 Vetters, H. (2002) Wälzelemente aus 100Cr6, bainitisch umwandeln oder martensitisch härten? *HTM*, **57** (6), 403–408.

57 Berns, H., Trojahn, W. and Wicke, D. (1987) Umlaufbiege- und Zugschwellermüdung des Stahles 100Cr6. *HTM*, **42** (4), 211–216.

58 Peter, W., Finkler, H. and Kohlhaas, E. (1970) Entstehung und Beseitigung besonderer Karbidseigerungen in Werkzeugstählen mit rd. 1% C. *Archiv für das Eisenhüttenwesen*, **41** (8), 749–756.

59 Hildebrand, M., Göhler, H., Rabe, W., Hödel, M. and Pohl, B. (1973) Karbidausscheidungen im Stahl 100Cr6 und ihre Beeinflussung durch Lösungsglühen. *Neue Hütte*, **18** (12), 727–730.

60 Kato, T., Jones, H. and Kirkwood, D. (2003) Segregation and eutectic formation in solidification of Fe–1C–1.5Cr steel. *Materials Science and Technology*, **19** (8), 1070–1076.

61 Vodopivec, F. (1971) Auflösung ledeburitischer Carbide und Ausscheidung von Korngrenzenzementit in Wälzlagerstahl. *Archiv für das Eisenhüttenwesen*, **42** (4), 283–286.

62 Fuchs, A., Bernthaler, T., Stahl, B., Klauck, U., Reinsch, B. and Schneider, G. (2001) Bildanalyse komplexer Werkstoffgefüge durch Texturanalyse und Korrelation mit den Eigenschaften durch neuronale Netze. *Zeitschrift für Metallkunde*, **92** (8), 979–985.

63 Omar, M., Atkinson, H., Palmiere, E., Howe, A. and Kapranos, P. (2004) Microstructural development of HP9/4/30 steel during partial remelting. *Steel Research*, **75** (8/9), 552–560.

64 Finkler, H. and Schirra, M. (1996) Transformation behaviour of the high temperature martensitic steels with 8–14% chromium. *Steel Research*, **67** (8), 328–342.

65 Kaya, A. and Edmonds, D. (1998) Nonclassical decomposition products of austenite in Fe–C–Cr alloys. *Metallurgical and Materials Transactions A*, **29A** (12), 2913–2924.

66 Matsuo, T., Kiminami, C., Botta Fo, W. and Bolfarini, C. (2005) *Wear*, **259**, 445–452.

67 Tong, J., Zhou, Y., Shen, T. and Deng, H. (1989) The influence of retained austenite in high chromium cast iron on impact abrasive wear. International Conference on Wear of Materials 1989, Denver, CO, pp. 65–70.

68 Hanlon, D., Rainforth, W. and Sellars, S. (1999) *Wear*, **225–229**, 587–599.

69 Brandes, E. and Brook, G. (eds) (1998) *Smithells Metals Reference Book, 7*, Elsevier, Amsterdam.

70 Püttgen, W., Hallstedt, B., Bleck, W., Löffler, J.F. and Uggowitzer, P.J. (2007) On the microstructure an properties of 100Cr6 steel processed in the semi-solid state. *Acta Materialia*, **55**, 6553–6560.

71 Püttgen, W., Hallstedt, B., Bleck and Uggowitzer, P.J. (2007) On the microstructure formation in chromium steels rapidly cooled from the semi-solid state. *Acta Materialia*, **55**, 1033–1042.

4

Design of Al and Al–Li Alloys for Thixoforming

Bernd Friedrich, Alexander Arnold, Roger Sauermann, and Tony Noll

4.1
Production of Raw Material for Thixoforming Processes

A summary of the most important process alternatives for the forming of partially liquid metals is shown in Figure 4.1. In the left-hand column, 'conventional' thixoforming – which is currently most widely used for industrial applications – is schematically represented and in the right-hand column, the so-called 'slug on demand' process.

The stress induced and melt activated (SIMA) process requires conventional continuous-cast or extruded billets as feed material. Cold deformation of the billets induces high residual plastic strain. On reheating, recrystallization leads to a fine globulitic granular structure. The SIMA material exhibits very good properties, although it is only economical for special applications due to its high cost [2]. Mechanical agitation is one of the principle processes for the production of raw material [3], but was never really successful. It was not possible to counter the corrosion and erosion of the agitating mechanism within the system [1], and on the process side the continuous casting procedure could not be synchronized with the agitation process.

The magnetohydrodynamic (MHD) process is to date the most widely used industrial process for the production of raw material [1]. Grain refinement is achieved by motion of the bath induced by a magnetic field. This involves installing a solenoid coil on the permanent mould, fed with an alternating current. This measure can also be used to keep the permanent mould temperature homogeneous at a desired value. In the semi-solid processing (SSP) process, just as in the MHD process, magnetic agitation is used to achieve grain refinement. The difference is that each billet is individually produced in its own chamber [4]. This offers the advantage of flexible alloy adjustment. Furthermore, the hot billets can be reprocessed straight from the chamber, so that advantages can be achieved from the viewpoint of energy considerations. On the other hand, there are economic advantages in the alternative of semi-finished production.

Thixoforming: Semi-solid Metal Processing. Edited by G. Hirt and R. Kopp
Copyright © 2009 WILEY-VCH Verlag GmbH & Co. KGaA, Weinheim
ISBN: 978-3-527-32204-6

Thixoforming

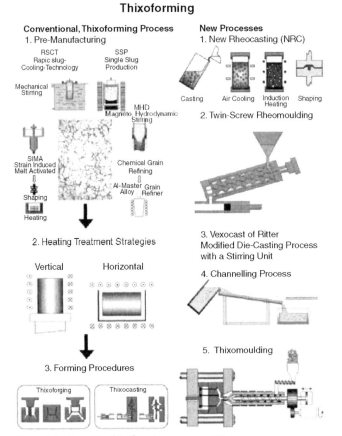

Conventional, Thixoforming Process
1. Pre-Manufacturing

RSCT
Rapid slug-
Cooling-Technology

SSP
Single Slug
Production

Mechanical
Stirring

MHD
Magneto Hydrodynamic
Stirring

SIMA
Strain Induced
Melt Activated

Shaping

Heating

Chemical Grain
Refining

Al-Master Grain
Alloy Refiner

2. Heating Treatment Strategies

Vertical   Horizontal

3. Forming Procedures

Thixoforging   Thixocasting

New Processes
1. New Rheocasting (NRC)

Casting   Air Cooling   Induction   Shaping
                          Heating

2. Twin-Screw Rheomoulding

3. Vexocast of Ritter
Modified Die-Casting Process
with a Stirring Unit

4. Channelling Process

5. Thixomoulding

Figure 4.1 Alternative thixoforming processes [1].

Chemical grain refinement is a process that offers lower processing costs and is from an economic point of view considered as an alternative to the other processes mentioned above [5]. The fine grain structure is achieved with the aid of alloying elements (e.g. AlTi5B1) that act as heterogeneous nucleating agents [5]. The grain refiners are in the form of a wire injected directly into the melt. Other elements can also be added to enhance the grain refining properties of these elements (so-called modification) [4]. It was furthermore verified that the chemical grain refinement of aluminium alloys has an advantageous effect on the 'castability' and other technologically important properties. The further targeted development of an A356–AlSi7Mg0,3 alloy by means of grain refinement and modification using Ti and Sr was investigated in 2003 [6]. One possibility for further refinement of the grain structure is the combination of chemical refinement with the MHD process [7].

The raw material has to be reheated precisely to a target temperature in the solidus–liquidus range [8]. Rapid heating prevents undesired grain growth and makes higher cycling rates possible. In this respect, inductive heating offers advantages over other heating processes due to its inherent controllability and its frequency-dependent, relatively high penetration depth. During reheating, the eutectic phase melts first. The liquid portion rises steeply with minimal heating at temperatures above 560 °C. From about 570 °C, the entire eutectic phase is molten. At temperatures above 570 °C, the aluminium dendrites begin to melt. The liquid portion rises only slowly in this temperature range. The working temperature for reheating is precisely within the transformation range between fully molten eutectics and commencement of dendrite melting. In addition to its low temperature sensitivity to the phase relationship, A356 has the characteristic that, in its partially liquid state, ~50% primary phase and ~50% eutectic exist. Due to this composition, there are material-specific advantages in terms of processing. Materials for which it is not possible to adjust the solid–liquid mixture so accurately require considerably higher processing costs [7].

In the 'slug on demand' or rheocasting processes, the production of a suitable thixoformed raw material is done straight from the fluid melt with direct forming of the partially liquid metal [1]. Various new rheocasting processes are briefly explained in the following (see Figure 4.1, right-hand column).

In new rheocasting (NRC), the melt is cast in a controlled manner into a crucible and thus cooled to the desired temperature between the solidus and liquidus states. This creates a fine-grained, globulitic solid phase. After a short homogenization phase, the partially solidified material is fed directly into the forging or casting process. In twin-screw rheomoulding, the melt is fed into the machine via a casting gate. There are two interacting screw feeders, with a special profile to achieve high shear rates, to convey the material axially into the machine and where it is simultaneous cooled to the partially liquid state. The high shear rate creates a fine-grained, globulitic grain structure. The partially liquid metal is dosed and pressed out of an opening into the filling chamber for forming in a modified die-casting machine. To date this process has been used only for modelling alloys of the types PbSn15 and MgZn30. One difficulty with this process is the choice of a suitable material for the screw feeders, in order to avoid adherence of the metal and to minimize wear [1].

The Vexocast process was developed by the company Ritter in Germany [1]. The metallic suspension is produced directly from the melt in a separate treatment chamber which is in combination with a holding furnace and with the casting chamber. The melt is inductively agitated in the treatment chamber and in the freezing range a small quantity of metal powder is added to cool it. One essential feature of the process is its short time range of only about 15 s for production of the metallic suspension. The channelling process represents a further application method of the 'slug on demand' process. The two essential constituents of the trial setup for the channelling process are the casting channel and the forming basin. The channel has to provide for rapid cooling of the melt, down to a temperature below the liquidus temperature. This creates a very large number of active nuclei suspended in the melt. The melt flows from there into a ceramic forming basin installed directly underneath the channel. On cooling to the partially liquid range, the high nucleus density created results in the formation of a globulitic primary phase.

The thixomoulding process entails injection die casting of partially liquid light metal alloys, predominantly Mg alloys [9, 10]. The original material, in the form of granulate or turnings, is transferred by powerful shear movements in a heated steel screw feeder into a liquid metallic suspension. Subsequently, the screw feeder performs a pushing motion to inject the material into the actual forming tool [11]. The proportion of solid material can be up to 65% [12].

4.2
Chemical Grain Refinement of Commercial Thixoalloys

If a metal alloy has to be processed in a partially liquid state, it must conform to many requirements. The following criteria are among the most important:

- In the semi-solid range, the alloy constitutes a globulitic solid phase with grain sizes of <100 μm.
- Within the semi-solid range, the alloy exhibits low temperature sensitivity of the liquid-phase portion.
- The solid-phase skeleton fractures down under shear load, transferring the metallic suspension to low viscosity in the range of 1 Pa s.

Furthermore, it is important that there is minimal grain growth in the material during the holding period in the partially liquid state, because the refinement of the solid phase (α-aluminium) has a positive effect on its viscosity, the tendency for separation of the mixture and the component wall thicknesses achievable. A summary of the relevant requirements in relation to grain structure morphology, rheology and thermochemistry is given in Table 4.1. Many of these criteria are best met by the casting alloys already used on an industrial scale, such as A356 and AlSi7Mg0.3 – which explains their wide use in the thixoforming process. The most important non-technical requirements for thixoforming materials are low material costs, which has to be realized in terms of raw material prices, simple production procedures, minimum number of processing operations and minimum wastage of residual materials. The choice of a suitable material inevitably determines all further processing operations. Systematic development of raw materials for thixoforming with the aid of alloying elements and grain refiners has been carried out to only a limited extent to date [6, 13].

4.2.1
Methodology for Tuning of Commercial Alloys for Semi-solid Processing by Means of Grain Refinement

Because of its range of applications in the automotive, machinery and aerospace industries, and also in shipbuilding and electrical engineering, the material A356 (EN AC-AlSi7Mg0.3) is by volume one of the most important aluminium casting materials. This is primarily due to its low density and good mechanical properties, such as high strength and toughness, and also its high corrosion resistance and

Table 4.1 Essential requirements for thixoforming materials.

Requirement (determination method)	Parameter	Formula[a]	Target value[a]
Small grain size (metallography)	Mean grain diameter	$d_{\mathrm{m}} = \mathrm{f}\left(\dot{T}, c\right)$	$d_{\mathrm{m}} < 100\,\mu\mathrm{m}$
Globulitic structure (metallography)	Form factor	$f = \frac{4\pi A}{U^2}$	$f > 0.6$
Small cross-linking of dendrites (metallography)	Contiguity	$C^{tS} = f^S \dfrac{2S^{SS}}{2S^{SS}+S^{SL}}$	$0.4 < C^{tS} < 0.6$
Small cross-linking volume of dendrites (metallography)	Contiguity volume	$Vc = VsCs$ $V_c = V_s C_s$	$0.1 < V_c < 0.3$
Low viscosity of the melt under shear load (rheometer tests)	Viscosity under shear load	$\eta = k\dot{\gamma}^{n-1}$	$\eta \approx 1\text{--}10\,\mathrm{cP}$
Defined solid–liquid phase ratio (DTA, simulation, metallography)	F_{liq}	$f_S = 1 - \left(\dfrac{T_M - T_L}{T_M - T}\right)^{\frac{1}{1-p}}$	$40\text{--}60\,\mathrm{mass\%}$
Large melting interval (DTA, simulation)	$dT = (T_{\mathrm{liq}} - T_{\mathrm{sol}})$	$dT = \mathrm{f}\left(c, \dot{T}, p\right)$	max. 130 K, min. 70 K
Temperature sensitivity (DTA, simulation)	Stop at 50% liquid portion	df_l/dT	<0.015
Temperature sensitivity	Temperature sensitivity	$Df_S = 0.01(df_s/dT)T_{ss}$	≤ 0.06
Temperature sensitivity (DTA)	Temperature sensitivity	dT_{40-60}	>10 K

[a] η = viscosity; k = constant $\dot{\gamma}$ = shear velocity; n = Oswald–de Waele exponent; d_{m} = mean grain diameter; f = forming factor; A = grain surface area; U = circumference of a grain; S^{SS} = the grain boundary surface area between the solid phase, that is, the surface area between the cohering grains not separated melt; S^{SL} = phase boundary surface area between solid phase and melt; dT = melt range; $f(c; T; p)$ = function of (alloying element content; rate of cooling; pressure); f_s = proportion of solid material; T_M = melt temperature of a component; T_L = liquidus temperature; p = exponent defined by the phase state equilibria.

suitability for welding and heat treatment. Due to these material characteristics, the alloy has a good potential for lightweight construction. Furthermore, the material is the most important raw material for the production of components by thixoforming. The solidification range of ~60 °C is within a range making it readily workable in its partially liquid state. The material reacts particularly well in the range of temperature-sensitivity of the phase portion provided that the liquid-phase proportion is set at >50%. On the basis of this material characteristic profile, the alloy A356 (EN AC-AlSi7Mg0.3) was exemplarily prepared to meet the requirements making it workable in its partially liquid state. This objective should in the first instance be followed up independently of any other processing route. Process technology for raw material production is primarily focused on the input of a fine-grained basis grain structure. A series of different technical solutions are available for this purpose. The necessary fine grain structure should hence be created by the addition of a chemical grain refinement agent in combination with modification of the eutectic grain structure. Further, the interaction of the employed chemical grain refiners with the modifying agents and other alloying elements needs to be studied (Figure 4.2).

Additionally, the adaptation of the alloys has to be managed such that during reheating in the partially liquid state, grain growth is minimized [14].

4.2.2
Adaptation of the Chemical Composition

The objective of this task is to identify the effect and the interaction of the standard chemical grain refiner AlTi5B1 with both the recognized modifying agents, sodium and strontium, and other elements (manganese and lead). To date little is known

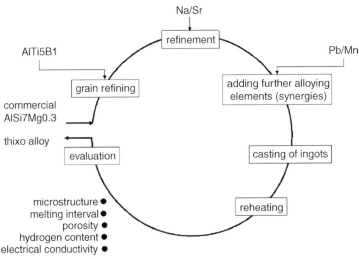

Figure 4.2 Procedure for adaptation of the alloy composition A356 (EN AC-AlSi7Mg0.3) to the requirements of thixoforming.

about the mechanisms involved in the interactive effect between grain refinement and modifying agents. Within the framework of melt testing, using the model-supported testing plan, various chemical compositions were produced and then evaluated in respect of the effect variables important for thixoforming. Particular attention was devoted to the grain size in the primary phase, modification of the eutectic phase, the melt range and the rheological characteristics.

For these parameters, a mathematically linear approach was chosen for determination of the efficiency. For the parameters titanium concentration and reheating time, the focus is on the question of how the parameters can be set in order to achieve optimum results for the effect variables. The standard grain refiner AlTi5B1 was used for chemical grain refinement. Lead was selected to test its possible positive effect on grain refinement. In minimal concentrations Pb already reduces the surface tension of aluminium and should therefore have a positive effect on grain refinement and/or modification. For strontium and sodium, 200 ppm was defined as the upper step value, in order to ensure its efficiency but at the same time to avoid over-modification. For the reheating time, periods between 5 and 15 min are technically relevant. Particularly long holding times may be unavoidable, however, in the event of process malfunctions during reheating, so the suitability of the raw material was also tested in this respect.

The titanium content and the reheating time can be identified as significant influences on the grain size. As the titanium content increases, the grain size can be expected to decrease in the α-aluminium phase. With longer reheating time, as a result of maturing processes, grain growth can be expected to occur. In addition, a significant interaction between titanium and strontium can be identified, although the effect of strontium alone is only minimal and can be identified as being statistically insignificant. The grain refining effect of titanium is evidently weakened by the addition of strontium.

To indicate the affectability of the grain size, in the following detailed evaluations the titanium and strontium contents and the reheating time were considered in particular (Figure 4.3). By the addition of 0.4% titanium (in the form of AlTi5B1), the grain size can be decreased from about 85 to 65 μm. As the reheating time is extended. grain growth can be expected to occur; however, this is inhibited by the addition of titanium. This effect of the grain refiner is also dependent on the strontium content. A systematic positive effect of manganese and lead addition cannot be ruled out. The change in the grain size as a result of grain refining treatment, and also the effect of the duration of the reheating time with simultaneous modification using strontium, are illustrated in Figure 4.3a.

4.2.3
Influence on the Form Factor of the Primary Phase

A further criterion for evaluation in the assessment of thixoforming materials is the form factor. The rounder the grain or the greater the form factor is, the more pronounced the thixotropic material characteristic in a forming operation will become. The duration of reheating is recognized as an essential parameter affecting

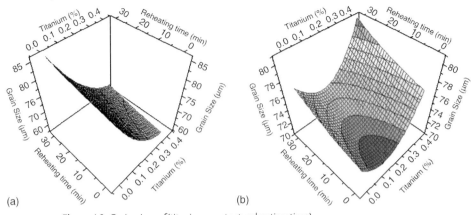

Figure 4.3 Grain size = f(titanium content, reheating time),
(a) without Sr and (b) with Sr. (b) Modification of the eutectic grain
structure depending on the content of Sr (ppm) and Ti.

the form factor. The effect of alloying elements on forming is, however, largely unexplained. For the material A356 (EN AC-AlSi7Mg0.3), the main parameter affecting the form factor of the primary phase (α-aluminium) surprisingly appears to be the addition of strontium, which seems to have an even greater effect in comparison with the reheating time. The other elements, on the other hand, cannot be identified as significant parameters. The dependence of the form factor on the titanium added and on the reheating time with manganese added and also modification using strontium is shown in Figure 4.4.

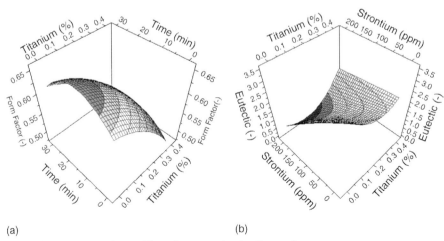

Figure 4.4 Dependence of form factor on reheating time and
titanium content (0.3% Mn, 200 ppm Sr, 0 ppm Na, 0% Pb).

With increasing titanium content, the form factor is clearly impaired, that is, the titanium added during solidification evidently promotes dendritic growth. At the same time, the form factor passes through a maximum with extended reheating, lasting between 10 and 20 min. After that, uneven grain growth occurs, during which smaller α-aluminium dendrites agglomerate to larger grains, so that the form factor 'apparently' becomes worse. Under favourable conditions ($t = 15–20$ min, Ti $= 0\%$, Pb $= 0\%$, Na $= 0\%$, Sr $= 200$ ppm), a form factor of ~0.7 can be achieved. This also approximately corresponds to the form factor achieved when reheating conventional raw material, for instance when using the MHD process. The result appears to be plausible because during reheating the states of dendrite formation, grain growth due to Oswald ripening and strengthened formation of the networking structure as a result of grain contact and coalescence can be differentiated [15].

The optimum grain structure adaptation is achieved when the dendrites can be transferred into individual globulites. The prospects for success of this treatment are essentially defined by the basis grain structure and the chemical composition [14]. By adding 0.3 mass% Mn and modification using 200 ppm of strontium, the form factor after 30 min of reheating can be favourably affected, because formation of the dendrites into a globulitic form during the same period leads to a better form factor than without addition of manganese and strontium (Figure 4.5).

4.2.4
Modification of the Eutectic Structure

The involvement of the eutectic structure in the evaluation of a material's suitability for thixoforming represents a new development step. Modification treatment and division of the modification into different classes are on the other hand standard practice for processing of conventional aluminium castings. A grain structure having a very fine, uniformly distributed eutectic represents Class 1. Coarse, needle-shaped, unevenly distributed silicon precipitations represent Class 3. Silicon precipitations in

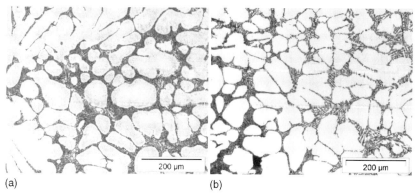

(a)　　　　　　　　　　　　　　　　(b)

Figure 4.5 Microstructure of A356, (a) without Sr and Mn refining ($f = 0.45$) and (b) with Sr and Mn refining ($f = 0.55$).

the intermediate group represent Class 2. Successful modification (Class 1) enhances internal feeding, minimizes the porosity and the tendency for heat cracking of a material, positively affects the flow and form-filling characteristics and improves the mechanical properties of the component.

The formation of the eutectic grain structure in testing of thixoalloys can be expected to be primarily affected by strontium. On the other hand, titanium and manganese have very little effect. Due to the resulting interactions, alloying elements such as titanium also have a significant worsening effect on the refinement and uniformity of the eutectic if strontium is also present. If strontium is added, titanium should be limited to between 0.15 and 0.25 mass% (Figure 4.4b). If 0.3 mass% of manganese is added, the negative effect of titanium on modification of the eutectic can be reduced to some extent.

4.2.5
Influencing the Semi-solid Temperature Interval

Determination of the freezing range for alloys is carried out by means of differential thermal analysis (DTA). Both the grain refinement and modification should have a favourable effect on the semi solid temperature interval. The liquidus temperature becomes higher with increasing titanium content. In addition, it could be demonstrated that the silicon and magnesium contents, and also the lead content in particular, should also be taken into consideration because the liquidus temperature reacts significantly to the alloying content.

The grain refinement element is primarily suitable in favourably affecting the freezing range overall. All other alloying elements are of little significance in comparison. Lead has a generally negative effect on the freezing range, so this element should be avoided. The study confirmed the suitability of the thermal analysis principle for assessment and control of the melt range. The hoped-for evaluation of the modifying treatment and grain refinement by means of thermal analysis was not achieved, however, due to the interfering influence of the other alloying components (primarily Mg and Si). The grain-refined raw material differs from commercial raw materials in respect of its smaller grain size and a more uniform grain size distribution in the border and middle areas of the cast. At the same time, it was noticeable that the grain-refined raw material had a somewhat more dendritic grain structure composition than commercial raw material.

4.2.6
Characteristics of the Raw Material

Because the commercially available MHD grain structures and grain-refined raw material can vary significantly from one another after reheating, both materials were treated with the same reheating programme. The raw materials were heated to the target temperature (580 °C) horizontally in an induction unit. The chemically grain-refined raw material was thereby transformed into a fine globulitic grain structure similarly to the conventional MHD material. The average grain diameter and the

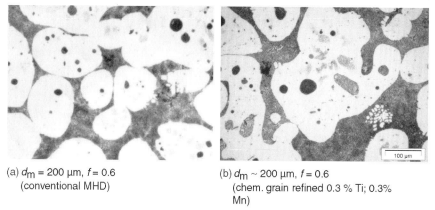

(a) d_m = 200 μm, f = 0.6
(conventional MHD)

(b) d_m ~ 200 μm, f = 0.6
(chem. grain refined 0.3 % Ti; 0.3% Mn)

Figure 4.6 Grain structure of the billet centre, A356 after reheating.

form factor are comparable (Figure 4.6). During reheating, however, it was observed that for the chemically grain-refined material less melt ran out of the billet than with conventional raw material. This presents the possibility that when using grain-refined raw material the dripping losses that are normally incurred during vertical reheating may be reduced.

The reheated raw material is processed by die casting into the principle component (Figure 4.7). A step form is used with which the form filling of a component with decreasing wall thickness can be examined. In contrast to commercial raw material, using chemically grain-refined raw material, with the same form filling rate (v_{piston} = 0.05 m s^{-1}), even the overflow of the casting chamber could be filled. This implies that an improved form filling capacity can be achieved.

Samples were taken from the component from steps of 15 and 25 mm height and then metallographic tests were performed (Figure 4.8).

For both of the raw materials no significant variation in grain structure with the step height was determined. The chemically grain-refined material had a somewhat finer grain structure than the commercial raw material. The advantages that were already apparent in respect of the aluminium phase and eutectic after reheating of the

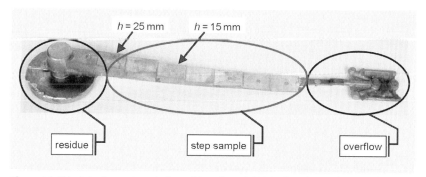

Figure 4.7 Principle step-part made from thixo-adapted A356 (after chemical grain refinement).

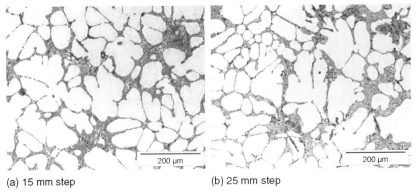

(a) 15 mm step   (b) 25 mm step

Figure 4.8 Grain structure of thixo-cast step samples.

different raw materials were also retained after forming. Examination of the chemical homogeneity over the length of the component indicated that for both raw materials no significant mix separation had occurred.

4.3
Fundamentals of Aluminium–Lithium Alloys

4.3.1
State of the Art

Aluminium–lithium alloys have been in use since 1950. In particular, wrought alloys such as AA8090 and AA2091 are used in the form of continuous-pressed profiles or rolled sheets in aerospace constructions. Also, forged components such as Al–Li brake callipers are used in motor racing [16]. Their low density, high specific solidity and high modulus of elasticity make these materials highly suitable for a multitude of high-load applications. Because of the high raw material cost of lithium and high expenditure on processing, however, they have not been very widely used to date. The value of $5 \in \mathrm{kg}^{-1}$ of tolerable extra costs in transportation systems for reduction in weight per kilogram indicates that their introduction in automotive mass production would only occur if significant cost reductions were feasible.

However, aluminium–lithium alloys are subject to low toughness values, stress-cracking corrosion and heavy anisotropy of mechanical properties [18, 19]. Aluminium–lithium alloys can be produced by melt technologies and powder-based metallurgical methods. The latter require more complex operating lines (cost issue) but permit a higher lithium content to be applied. However powder-based metallurgy can be subject to high contamination of the powder during the jet-spraying process (e.g. oxide formation – unfavourable for ductility) and also the high reactivity of the lithium itself. Melt-based metallurgical production is more common but the achievable lithium content is limited to ~10%. During continuous casting of Al–Li semi-finished products, an aluminium melt (99.5%) with the slightest Li content will become scratched far more quickly than an Li-free melt. Li contents of >5 ppm can already give rise to problems during casting, when jet-spray/float systems are used. According to Bick and Markwoth [20], between lithium in the aluminium melt and silicon in refractory material (wall components, jets), an $LiAlSi_3O_8$ phase should form, which is deposited on the walls of the jet along with the above-mentioned oxides formed in the melt ($LiAlO_2$ and Al_2O_3), and this can lead to blockage of the casting jets. Studies have been conducted [21] into the development of suitable aluminium–lithium alloys for fine casting processes. The casting tests using conventional Al–Li forging alloys identified deficiencies in the casting properties. Furthermore, the tendency for thermal cracking remained – and that should be eliminated by means of suitable grain refinement. A cost-effective casting technique may in the future be possible by adaptation of the low-pressure or reverse-pressure die-casting process using an inert gas.

The main motivation is to evaluate the potential for processing of Al–Li–X alloys in the partially liquid state, because the special processing conditions involved can be expected to produce a significant improvement in the material's properties and at the same time greater scope of formability. The main challenges in the production and processing of aluminium–lithium alloys include:

- Hot-tearing susceptibility during the casting process.
- Because of its high reactivity, lithium oxidizes very rapidly and needs to be handled very carefully.
- High gravitational separation during casting.
- High rate of waste/high proportion of turnings.
- Corrosion susceptibility (particularly stress-cracking corrosion).
- Recycling issues.

The acceptance of Al–Li alloys is currently limited, which is especially due to doubts in respect of fracture toughness and the fact that the costs are always three times higher than for existing high-strength alloys, probably due to the special requirements during melting and casting.

The alloy systems studied are derived on the one hand from the precipitation-hardened system Al–Li–Cu and on the other from the mixed-crystal strengthened system Al–Li–Mg. For comparison, the commercial alloys AA8090 (Al–Li–Cu–Mg), AA2090 (Al–Li–Cu) and AA1420 (Al–Li–Mg) were included in the study. In solid aluminium at 610 °C lithium has a maximum solubility of ~4% (16 at.%). Due to the low density of this element ($0.54 \, g \, cm^{-3}$), every 1% addition of lithium gives rise to a 3% reduction in density. Amongst the soluble elements, lithium is also noteworthy in that it causes the modulus of elasticity of aluminium to increase markedly (6% increase for every 1% added). By formation of the regulated metastable δ' (Al_3Li) phase, binary and more complex alloys containing lithium can be hardened, which is coherent with the Al matrix and has a particularly low mismatch with the Al matrix. Because of these properties, alloys containing lithium are under development as a new generation of low-density, high-rigidity materials for application in airframes, whereby improvements of up to 25% in the specific rigidity are possible, thus offering the opportunity for the production of a large range of higher performance aluminium alloys by conventional melt-based metallurgical means. In the aluminium industry, these may also contribute to their resistance to the growing trend toward the application of non-metallic composite materials as structural materials in the aerospace industry [22].

Hardened binary Al–Li alloys primarily suffer from low elasticity and toughness under the massive stress peaks that can occur, because coherent δ' precipitations are sheared by moving displacements. This can give rise to fractures along the grain boundaries. Correspondingly, this work has focused very heavily on the study of elements forming suitable precipitations or dispersoids allowing the displacements to be dispersed more homogeneously. The alloy systems developed to date are compared in Table 4.2.

4.3.2
The System Al–Li–Cu

The first commercially applied alloy of this system was A2020. Copper on the one hand reduces the solubility of lithium in its solid state, whereby the precipitation of δ' is promoted, and on the other hand gives rise to the precipitation of phases, such as the GP zones and θ', as familiar from binary Al–Cu alloys. The alloy A2020 combines properties of high strength at room temperature ($R_{p0.2} \approx 520$ MPa) with good creep behaviour at temperatures up to 175 °C and has a modulus of elasticity 10% higher than that of other aluminium alloys. Newer Al–Li–Cu alloys have a lower Cu content and higher Li:Cu ratios, of which the best-known example is the alloy A2090. In addition to the δ' phase, this alloy is hardened by means of the hexagonal T1 phase (Al_2CuLi). Zirconium additions ($\sim$0.12%) are also applied in order to form fine dispersoids of the type Al_3Zr, which control recrystallization and grain size. Another effect of this element is to promote grain formation of the δ' precipitations on these dispersoids. Overall, zirconium increases the properties of strength and toughness [13].

Table 4.2 Selection of Al–Li alloys developed to date for the aerospace industry.

Alloy	Li	Cu	Mg	Zr	Others
A2020	1.3	4.5	—	—	0.5 Mn; 0.25 Cd
A2090	1.9–2.6	2.4–3.0	<0.25	0.08–0.15	—
A2091	1.7–2.3	1.8–2.5	1.1–1.9	0.04–0.16	—
A8090	2.2–2.7	1.0–1.6	0.6–1.3	0.04–0.16	—
A8190	1.9–2.6	1.0–1.6	0.9–1.6	0.04–0.14	—
A8091	2.4–2.8	1.6–2.2	0.5–1.2	0.08–0.16	—
A8092	2.1–2.7	0.5–0.5	0.9–1.6	0.08–0.15	—
A8192	2.3–2.9	0.4–0.7	0.9–1.4	0.08–0.15	—
X2094	1.3	4.5	0.4	0.14	0.4 Ag
1420	2.0	—	5.2	0.11	—

4.3.3
The System Al–Li–Mg

Magnesium reduces the solubility of lithium, although its effect appears to be to strengthen the mixed-crystal stability. In alloys containing more than 2% of magnesium and that are heat-treated at relatively high temperatures, a non-coherent, cubic Al_2LiMg phase is formed. Unfortunately, this phase generally does not form at the grain boundaries and has an unfavourable effect on toughness. Therefore, the addition of magnesium does not essentially improve the strength of the binary Al–Li alloys. The alloy 1420 has a relatively low yield limit ($R_{p0.2} \approx 280$ MPa), but high corrosion resistance [23]. Higher yield limits ($R_{p0.2} \approx 360$ MPa) are achieved in the modified alloy 1421. This alloy contains 0.18% of the expensive element scandium, which forms fine, stable dispersoids of the phase Al_3Sc. This is isomorphous with an Al_3Zr structure and it promotes increased strength.

4.3.4
The System Al–Li–Cu–Mg

Examples in this group are A2091, A8090 and A8091. Ageing at temperatures of around 200 °C leads to coprecipitation of the semi-coherent phases T1 and S′ in addition to the coherent phase δ′. The S′ phase (Al_2CuMg) cannot be sheared by displacements and therefore enhances a homogeneous dispersal of stresses. Because the nucleus formation of this phase is similarly difficult, Al–Li–Cu–Mg alloys are normally used in the T8 state (helicopter support arms made from A8090).

4.3.5
The System Al–Li–Cu–Mg–Ag

The costs of delivering freight into low Earth orbit are estimated at $\sim$US$8000 kg^{-1}, so that there is considerable encouragement to reduce the weight of space vehicles by the use of lighter or stronger materials. Therefore, alloys containing lithium come into consideration for welded fuel tanks instead of the conventional aluminium alloys 2219 and 2014. Minimal addition of silver and magnesium significantly improves the strength properties. For Al–Cu–Li-Mg-Ag alloys with lithium contents of around 1–1.3% the material's strength results from adding small quantities of Mg and Ag, which stimulates nucleus formation for a fine, hardened T1 phase which coexists alongside the S′ and θ′ in a maximally hardened state. Each of these precipitations is created in another Al–matrix plane and is resistant to shearing due to displacements. These observations have led to the development of an alloy known as Weldalite 049 (Al–6.3Cu–1.3Li–0.4Mg–0.4Ag–0.17Zr). This material can have strength characteristics in excess of an R_m of 700 MPa in the T6 and T8 states. On the basis of the strength/density ratio, the corresponding steel would need a strength in excess of 2100 MPa, so Weldalite 049 is the first aluminium alloy produced from conventional casting blocks to be designated as having ultra-high strength. Modifications to this alloy using a low copper content (X2094 and X2095) develop values of $R_{p0.2}$ of

~600 MPa in the T6 state, which is double that of the conventional aluminium alloy 2219-T6 [22].

The original objective for the development of Al–Li–Cu–(Mg) alloys was as a substitute for the conventional aluminium airframe alloys, with an expected density reduction of about 10% and increase in rigidity of 10–15%. These alloys are classified as high-strength (A2090, A8091), medium-strength (A8090) and damage-tolerant (A2091, A8090) in respect of their mechanical properties. Their densities range from 2.53 to 2.60 g cm^{-3} and modulus of elasticity from 78 to 82 GPa. Nevertheless, problems do exist in respect of their low lateral toughness in continuously cast profiles. These problems appear to relate to their microstructure, primarily in the grain boundary areas. Under continual loading of these alloys at increased temperatures (>50 °C), the incidence of creep fracture can also increase to a relatively high degree [24].

4.3.6
Effects of Alkali Metal Contaminants

One known factor that reduces the ductility and fracture toughness of alloys containing lithium is, along with hydrogen, the presence of alkali metal contaminants (in particular sodium and potassium) and particularly at the grain boundaries. In conventional aluminium alloys, these elements are combined with elements such as silicon in harmless, permanent compounds. However, they react readily with lithium. Commercial alloys containing lithium are produced by conventional melting and casting techniques and usually contain 3–10 ppm of alkali metal contaminants. This level can be reduced to below 1 ppm by means of vacuum melting and refining [25]. It can be demonstrated that this leads to a considerable improvement in fracture toughness at room temperature (Figure 4.9).

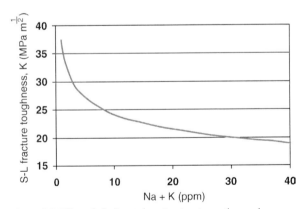

Figure 4.9 Effect of alkali metal contaminants on the crack toughness on A2090 extruded profiles [26].

4.4
Development of Aluminium–Lithium-Alloys for Semi-solid Processing

The concept of adaptation/optimization of familiar Al–Li alloys for processing in a partially liquid state is essentially based on the following points:

- influencing the micro-grain structure and the texture properties by suitable selection of the composition and the process sequence;
- optimization of the degree of recrystallization and the grain size in relation to the applicable component geometry;
- adaptation of the casting grain structure by addition of elements such as Ce, Sc, Ti, B and Ag;
- minimization of the proportions of balancing phases T2 and δ, of Fe and Si phases, of Na and K and of hydrogen in solution;
- variations in the main alloying elements (Li, Mg, Cu, Sc, etc.).

The semiempirical methodology and sub-operations employed are shown in Figure 4.10. After selecting suitable Al–Li–X basis systems, first an evaluation was made of the important key parameters for processing in a partially liquid state, with the aid of thermochemical calculations. After tuning of the alloys by means of microalloying element additives, melt-based metallurgical synthesis of trial alloys as well the characterization was performed. Finally, thixocast and rheocast principle parts and real components were jointly produced and evaluated. Where necessary, an adaptation of the selected alloy composition was carried out in addition to repeating the development operation for reproducibility check.

4.4.1
Thermochemical Modelling

The thermochemical calculations in the selected Al–Li–X system models [27–29] produced the following results, compared with experimental data [30]:

- In the Al–Li–Cu system, the alloys AlLi4Cu4 (A3) and AlLi2.5Cu4 (A2) display the lowest temperature sensitivity in the liquid-phase portion. Both are below the critical value of $df_l/dT < 0.015$ for a liquid-phase proportion of 50% and therefore should be suitable for processing in the semi-solid range. The alloys AlLi2.5Cu2.5 (A1) and AlLi4Cu1.5 (A4) display a smaller processing window in relation to temperature sensitivity of the liquid-phase portion, but should prove of interest for processing with low liquid-phase proportions (<30%) for thixoforging

- In the Al–Li–Cu system, account must be taken of the occurrence of brittle intermetallic phases T2–Al5CuLi3 and T1–Al6(Cu,Zn)Li3. This could in some cases lead to worsening of the mechanical properties, in particular elongation.

- In the Al–Li–Mg system, the alloys AlLi4Mg8 (A7) and AlLi2.1Mg5.5 (A8) display the lowest temperature sensitivity in the liquid-phase portion. Both are below the critical value of $df_l/dT < 0.015$ for a liquid-phase proportion of 50% and therefore

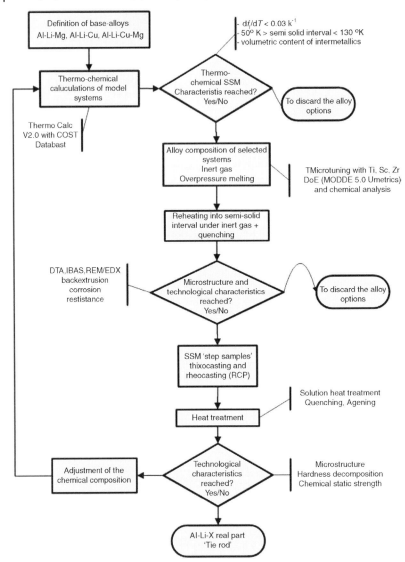

Figure 4.10 Flow diagram of the methodology applied for adaptation of Al–Li–X alloys to thixoforming.

should be suitable for processing in the semi-solid range. Similarly to the argument for A1 and A4, the alloy AlLi2.5Mg4 (A6) should be suitable for processing with low liquid-phase proportions (<30%) for thixoforging.

- In the Al–Li–Mg system, account must be taken of the occurrence of the brittle intermetallic phases σ-AlLi and T-Al₂LiMg, in particular for the alloy A7.

- The complex alloy in the system Al–Li–Cu–Mg (A5–AlLi2.1Cu1.2Mg0.7) is excluded due to an insufficient process window for semi-solid forming.

Table 4.3 Properties of selected Al–Li–X alloys relevant to SSM processing; calculations made by Scheil–Gulliver approximation.

Alloy	Cu (wt%)	Li (wt%)	Mg (wt%)	T_{50} (°C)	df_l/dT	$T_{50} - T_s$ (°C)	Intermetallic phases
A1 (AA2090)	2.5	2.5	—	637	0.023	105	4.7% T_2, 1% T_1
A2	4.0	2.5	—	629	0.015	97	4.8% T_2, 2.7% T_1, 0.3% T_B
A3	4.0	4.0	—	605	0.008	39	12.5% T_2, 0.6% AlLi
A4	4.0	1.3	—	639	0.024	107	2.5% T_1, 2.1% T_B
A5 (AA8090)	1.2	2.5	0.7	638	0.026	197	3.3% T_2, 0.5% AlLi, 0.4% τ
A6	—	2.5	4.0	613	0.012	155	5.2% τ, 2.7% γ
A7	—	4.0	8.0	542	0.004	84	15.5% τ, 5.9% γ, 1.1% AlLi
A8 (AA1420)	—	2.1	5.5	603	0.010	145	3.4% τ, 5.2% γ

The important key parameters on the basis of the thermochemical calculations are summarized in Table 4.3.

4.4.2
Choice of a Suitable Basis System

Since for semi-solid workability the formation of a globulitic microstructure is a prerequisite quench tests were carried out from the partially liquid range. The average solid phase particle diameter should be less than 100 μm in this respect. The grain structure formations indicate that the grain-refined alloys A1, A2, A4 and A5 formed the larger globulitic particles with virtually identical solid phase contents. The alloys A3, A6, A7 and A8 on the other hand display the desired grain sizes (<100 μm). The alloys A3 and A7 from this group were taken as samples for detailed examination, because these displayed the largest semi-solid ranges and a finer globulitic grain structure.

With the AlLi4Cu4(A3) alloy, in the following quench tests were carried out based on 540–620 °C after holding times of 2, 5, 10 and 20 min. For each quenched sample, four recorded grain structure images (Figure 4.11) were evaluated. The liquid-phase

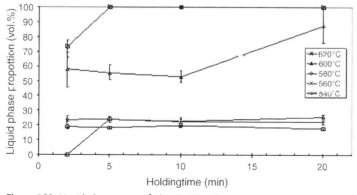

Figure 4.11 Liquid-phase curve of AlLi4Cu4Ti with respect to holding time and temperature in the semi-solid interval.

Table 4.4 Process window for the grain-refined alloy AlLi4Cu4 + TiZr (A3).

$T_{\mathrm{s}} \approx 550\text{–}560\,^{\circ}\mathrm{C}$	$T_{\mathrm{L}} \approx 660\text{–}679\,^{\circ}\mathrm{C}$
$T^{40-60} \approx 590\text{–}610\,^{\circ}\mathrm{C}$	$\Delta T^{40-60} = 20\,^{\circ}\mathrm{C}$
$T_{\mathrm{process}} \approx 590\text{–}600\,^{\circ}\mathrm{C}$	$t_{\mathrm{hold}} = 2\text{–}5\,\mathrm{min}$

portion at above 540 °C is independent of the holding time in this respect. The liquid-phase progressions for a holding time of 2 min, determined on the basis of the quench tests, were very closely consistent with the DTA results. The mean grain sizes (600 °C) amount to 60–80 µm and remain very constant up to a holding time of 20 min. The form factor at 600 °C, with values of 0.5–0.6, is more favourable than at 590 °C (0.35–0.45), because more solid phase is released and therefore a higher form factor is created. This means that there is a process window as indicated in Table 4.4. The stated temperatures vary because of the non-homogeneous alloy distribution within a billet.

With the AlLi4Mg8 (A7) alloy, analogous quench tests were performed based on 460–580 °C. As for AlLi4Cu4 (A3), there is no indication of any significant interdependence between the liquid-phase portion and the holding time. Again there is a good correlation of the liquid-phase contents determined by both DTA and quench tests. The mean grain sizes vary at both temperatures, between 35 and 65 µm, and the form factor between 0.37 and 0.48. Because of its grain structure, the examined material is therefore suitable for processing in the semi-solid state. Because of the large fluctuations in relation to the chemical compositions, it is difficult to identify a definite process window for the material on the basis of the DTA and grain structure tests (assumed: $T_{\mathrm{process}} \approx 540 \pm 15\,^{\circ}\mathrm{C}$, $t_{\mathrm{hold}} \approx 2\text{–}5\,\mathrm{min}$). A higher homogeneity would be necessary for reproducible billet heating.

4.4.3
Fine Tuning of the Grain Refinement for Al–Li Alloys

For statistically confirmed determination of the parameters and effect variables during micro-tuning of the Al–Li–X alloys by means of adding Sc and Zr, the parameters for Sc and Zr content and also the casting temperature were defined for the system AlLi2.1Mg5.5. The indicator (effect variable) was the desired average grain size in the cast billets, which should be minimized.

As shown in Figure 4.12, the effects of added Sc and Zr on the resulting grain size are significant. The variation of the casting temperature, on the other hand, has no effect on the grain size. The grain size can be statistically confirmed and reproducibly set at the desired value via the Sc and Zr contents. Figure 4.13 shows in this respect the contents of Sc and Zr that have to be added in order to set a defined grain size in the alloy. The mathematical model makes a prognosis of contents of 0.25% Sc and 0.25% Zr for a grain size of 30 µm.

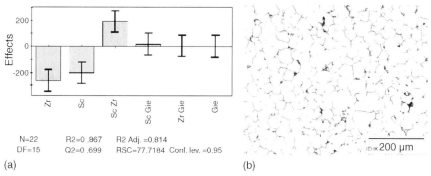

(a)

N=22   R2=0 ,867   R2 Adj. =0,814
DF=15   Q2=0 ,699   RSC=77,7184 Conf. lev. =0,95

(b)

Figure 4.12 Effects of Sc and Zr and casting temperature on the grain size of AlLi2.1Mg5.5ScZr (a) and microstructure of a billet (b).

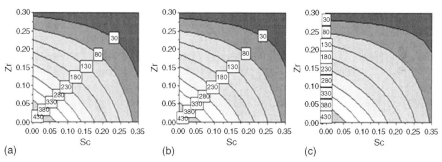

(a)   (b)   (c)

Figure 4.13 Dependence of the average grain size in μm (white boxes) on the Sc and Zr contents of AlLi2.1Mg5.5ZrSc at three casting temperatures [(a) 750, (b) 800 and (c) 850 °C].

4.4.4
Assessment of the Raw Material Development Programme

The results of the optimization programme can be summarized as follows:

- With the aid of the overpressure melt process, Al–Li–X alloys with Li contents of 1.5 4 wt% can be reproducibly produced. In order to prevent contamination by Na and K, it is necessary to employ master alloys and high-grade metals.

- In spite of the complex permanent mould system, chemical non-homogeneity of the raw material billet cannot be entirely eliminated. In particular, the elements Cu and Mg are concentrated at the edge of the billet and are below the specified desired values in the centre of the billet.

- DTA testing confirmed the results of the thermochemical calculations, while for the AlLi4Cu4, AlLi4Mg8 and AlLi2.1Mg5.5 alloy systems the largest process window for processing in the semi-solid range can be expected.

- The evaluations of the microstructure in the semi-solid range indicate that in particular the Ti and Zr grain-refined alloys AlLi4Cu4 and AlLi4Mg8 and also the Sc and Zr microalloyed alloys AlLi2.1Mg5.5 achieve very good results with globular and fine-grained structures of ~60 µm. In addition, these alloys have a high proportion of pre-melting eutectic phase, which should support thixotropic flow during forming.

4.5
Consideration of the Forming Pressure on Thixoalloy Development

For full consideration of the process window during forming in the partially liquid state, solidification under pressure has to be taken into consideration, because at the end of the forming process pressures of up to 1000 bar can be reached. This effect calls for changes in the phase transformation temperatures and considerable shifting of phase state ranges, even the occurrence of new phases and modified solubilities for the alloying elements. Preliminary metallurgical studies regarding the effect of the pressure on the properties of metals and alloys solidified under pressure indicate that alloys solidified under these conditions possess greatly enhanced mechanical properties. This effect must be taken into account for the solidification of alloys in the partially liquid state, in particular for aluminium alloys because the elevated pressure has a great influence on the solidification temperature of the residual melt and so, due to the sudden solidification (pressure-induced undercooling), non-homogeneity of the component can be the result. The pressures building up in the partially liquid state during the forming process can, however, have a positive effect, for example considerable improvement of mechanical properties. Therefore, pressure, forming rate and temperature control must be adapted to each other in order to produce homogeneous components with enhanced mechanical properties.

When continuously casting Al–Si raw material for thixoforming by means of MHD agitation, in near-surface regions of the rod an increase in the silicon content occurs in comparison with the core area [36]. For the most widely used thixoalloy AlSi7Mg the silicon content close to surface can even increase to an over-eutectic level if the agitation effect is very strong. The cause may be, for instance, a shifting of the eutectic point due to the overpressure [37]. This leads to non-homogeneity in subsequently produced thixocast components. It is recognized from the literature and also from our own experience that in the case of remelting of the surface regions, when a shrinkage gap is created a transfer of eutectic residual melt in the edge area is possible through the framework of the primary crystal structure, if an overpressure is acting internally [38, 39]. However, these studies only give information about the effect of pressures of up to ~10 bar. For a comprehensive understanding of the effects of pressure, in particular for thixoforming, it is not sufficient to limit oneself to such a low level of pressure.

During component production by thixoforming, as a rule final pressures of about 1000 bar are created, which influences the still liquid-phase portion of the suspension. As mentioned above, this effect calls for changing of the phase transformation temperatures and also considerable shifting of phase state ranges, even the occurrence of new phases and modified solubilities of the alloying elements. Cernov and Schinyaev [40] have already published a pressure dependence of this nature, although this was purely theoretical and derived from the Clausius–Clapeyron equation under the assumption that the heat of the melt and the volume change during melting are constant, that is, independent of pressure and temperature. Experimental validation of this is insufficiently covered. It has been postulated [37] that for a shift of the eutectic composition, of for instance the Al–Si eutectic, from 12.6 to 13.6%, an absolute pressure of well over 1000 bar would be necessary. However, even at low pressures, an extension of the melt range also occurs, similarly to the solubility range of the aluminium solid solution. Therefore, particularly during forming under pressure in the partially liquid state, large effects on the grain structure composition can be expected.

For some Al alloy systems, the effects of solidification under pressure on mechanical properties were studied [41]. For the system Al–Mg10, for example, considerable increases in the yield limit, tensile strength and elongation were indicated with increasing solidification pressure at pressures up to 1000 bar. For the conventional casting alloys A356–AlSi7Mg, on the other hand, the solidification pressure had little effect on mechanical properties. The work to date studying the influence of pressure on the properties of metals is related to solidification from the fully liquid state. No studies are known that have dealt with the solidification of partially liquid suspensions under pressure. This question is, however, of particular importance for this task, being one of the possible causes of chemical non-homogeneity in thixo-components, and this underlines the need for this research.

4.5.1
Experimental

The alloys to be studied for pressure dependence were selected from the binary Al–Mg system (samples 1–4), from the binary Al–Li system (samples 5–8) and from the ternary Al–Li–Mg system (samples 9–17); this allowed for a good coverage of the relevant Al domain (Figure 4.14).

For the production of the selected alloys, the proven inert gas overpressure method was used. Synthesis of the various Al–Li alloys was carried out by the melt-based metallurgical process directly, from highest purity metals (Al 99.999%, Li 99.8% and Mg 99.8%). The castings were produced with an overpressure in the furnace in a circular steel permanent mould ($D = 80$ mm, $H = 220$ mm). The weight of the Al–Li–X trial billet produced was · 2.8 3 kg. In order to check the homogeneity of the billets produced, selected billets were sampled at four different levels (in total 12 samples per billet). The deviations in the central regions of the billets were minimal. For this reason, the samples for the DTA testing (six samples: diameter $= 10$ mm, length $= 60$ mm per billet) were taken from the central regions, from the two billets containing 1% Li and also containing 2% Mg/1% Li. Also, it was not possible to take the samples 10% Mg/4% Li, 16% Mg/0.5% Li, 16% Mg/0.7% Li, 16% Mg/2% Li and 16% Mg/4% Li as they were so brittle that they broke during handling of the samples.

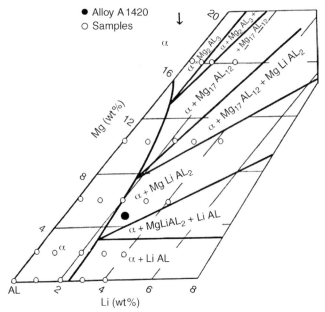

Figure 4.14 Alloy compositions in the Al–Li–Mg system selected for determination of pressure dependence.

The samples were investigated in a specialized DTA system at pressures from 1 to 2000 bar. The samples contained a hole of size ~1.2 mm for a thermocouple. Three other thermocouples assured the homogeneity of the temperature within the furnace surrounding the sample. The chamber was evacuated and an argon pressure of 15–20 bar was applied in order to improve heat transfer. The samples were heated to ~780 °C and for measurements under 'normal conditions' they were cooled (at 3–5 °C min^{-1}) for solidification.

For recording of the DTA curves under high pressure, a pressure of 2000 bar was applied on the hot sample solidified again at the same rate of cooling. Two tests were run for each billet. After DTA analysis, the samples and two untreated samples per billet were used for tensile strength testing.

4.5.2
Results of High-pressure DTA

The results generally indicate that for almost all samples solidified under a pressure of 2000 bar (except sample 6), a *substantial* increase in the liquidus temperature was identified (Table 4.5). Figure 4.15 shows examples of DTA recordings for the Al3–Li sample under normal pressure and under 2000 bar; all results are summarized in Table 4.5.

The effect of the pressure on the increase in the liquidus temperature is most significant in the binary systems Al–Li ($\Delta T = 16.2$–47.4 °C). In the binary systems Al–Mg the pressure effect is somewhat less ($\Delta T = 16.2$–24.2 °C), and similarly for the

Table 4.5 Liquidus temperatures of alloys solidified under normal pressure and under 2000 bar, determined by means of DTA.

Composition (wt%/at.%)			$T_{liquid,}$ normal pressure (°C)	$T_{liquid,}$ 2000 bar (°C)	ΔT ($T_{liquid, 2000 bar}$ − $T_{liquid, 1 bar}$) (°C)
Al	Mg	Li			
98/97.8	2/2.2	0	625.4	649.6	24.2
92/93.23	6/6.75	0	615.2	626.9	11.7
90/89	10/11	0	588.1	600.9	12.8
84/82.5	16/17.5	0	541.1	562.7	21.6
99/96.2	0	1/3.8	635.7	651.9	16.2
98/92.6	0	2/7.4	627.3	620.8	−6.5
97/89.3	0	3/10.7	596.9	644.3	47.4
96/86.1	0	4/13.9	590.6	628.3	37.7
97/94.1	2/2.2	1/3.8	631.3	646.1	14.8
96/90	2/2.0	2/7.3	623.5	632.5	9.0
95/87.3	2/2	3/10.7	612.4	626.5	14.1
94/84.1	2/2	4/13.9	600	623.2	23.2
93/89.8	6/6.4	1/3.8	568.7	587.2	18.5
92/86.4	6/6.3	2/7.3	601.3	611.0	9.7
91/83.2	6/6.1	3/10.7	585.6	613.5	27.9
90/80.2	6/5.9	4/13.9	573.1	598.5	25.4
89/85.6	10/10.7	1/3.7	584.9	592.1	7.2
96.4/92.3	2.1/2.2	1.46/5.4	610.4	635.7	25.3

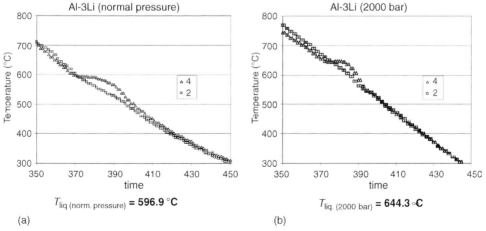

$T_{liq. (norm. pressure)}$ = **596.9 °C**

$T_{liq. (2000 bar)}$ = **644.3 °C**

(a)

(b)

Figure 4.15 Examples of DTA records under normal and high pressure showing differences in the thermal effect (liquidus temperature at 597 °C/644 °C).

ternary system Al–Li–Mg ($\Delta T = 7.2$–$23.2\,^\circ$C). A systematic correlation between the pressure effect on the increase in the liquidus temperature and the structure of the corresponding binary and ternary systems could not so far be identified.

4.5.3
Effect of the Solidification Pressure on the Microstructure

All samples that were solidified under a pressure of 2000 bar were very brittle, which is due to a phase modification and/or dissolution of a brittle phase on the phase boundaries. An analysis of the microstructure indicates that the morphology of the phase changes and that non-metallic contamination has accumulated at the phase boundaries of the sample that was solidified at 5–8 K s^{-1} under pressure (Figures 4.16 and 4.17).

In the samples solidified under normal conditions, the intermetallic inclusions are uniformly finely distributed within the alloy matrix, whereas in the samples solidified under a pressure of 20 bar, these intermetallics are separated irregularly in a coarser form (Figure 4.16). In the samples solidified under a pressure of 2000 bar, the separation of the intermetallic is somewhat less again and at the phase boundaries as well as between the mix crystals the non-metallic contamination is eliminated

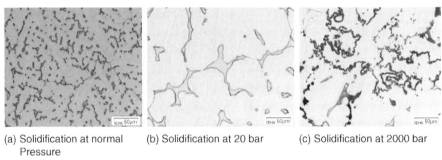

(a) Solidification at normal   (b) Solidification at 20 bar   (c) Solidification at 2000 bar
    Pressure

Figure 4.16 Microstructures of Al–10Mg–1Li solidified under different pressures.

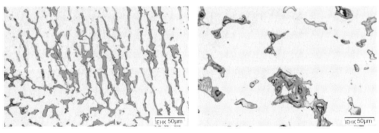

(a) Solidification at normal pressure   (b) Solidification at 2000 bar

Figure 4.17 Microstructures of Al–6Mg–4Li solidified under different pressures.

(Figures 4.16 and 4.17, both on the right). As the high-purity argon (6.0) used still contains 0.2 vol.-ppm oxygen and the fact that during compression to 2000 bar this content increases to 400 vol.-ppm, the oxide phase formation in the highly reactive Al–Li–Mg alloys is understandable. The solubilities of Li and Mg in the aluminium mixed crystal that were increased under pressure and the formation of non-metallic contamination are responsible for the reduced mechanical properties of the samples solidified under pressure. By using oxygen-free argon and applying heat treatment this negative effect could possibly be reduced.

4.5.4
Possible Impacts of High Pressure for the Thixoforming Process

The cooling curves for Al–Li–Mg alloys under 2000 bar indicate that the pressure has a very great influence on the melting temperature of these alloys. Increases in the liquidus temperature of up to 47 °C were identified. This has a direct effect on the distribution of the alloying elements and on the liquid–solid phase proportions. This effect is shown exemplarily for an Al-4% Mg–2.1% Li–0–8% sample/alloy (Figure 4.18).

Under normal pressure, this alloy has a freezing range of 83 °C ($T_{liquidus}$ 639 °C, $T_{solidus}$ 556 °C). A 50% liquid-phase fraction is achieved at 619 °C, corresponding to an undercooling of the melt by 20 °C. During forming of the suspension, the pressure increases suddenly and a pressure of, for instance, 2000 bar increases the liquidus temperature by 25.3 °C. Merely due to the rise in pressure the undercooling changes from 20 to 45.3 °C. This high value of undercooling leads to the formation of an additional solid-phase portion. Such a process requires a considerable displacement in the process window. This phenomenon, which also takes place to a slight extent at low pressures [37], can cause increases in alloying elements in the component and also changes in viscosity.

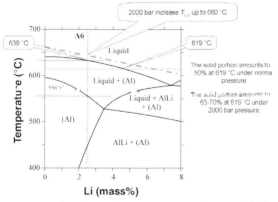

Figure 4.18 Al–Li phase diagram showing the change of the fluid/solid phase distribution.

4.6
Preparation of Principle Components from Al–Li Thixoalloys

4.6.1
Thixocasting

On the basis of the thermochemical calculations, DTA analyses and grain structure characterization in the partially liquid state, two compositions of the Al–Li system were defined as having the greatest potential for thixocasting. Both have a lithium content of ~4.0 wt%. One alloy contains ~4.0 wt% of magnesium and the other ~4.0 wt% of copper, each with low amounts of added titanium and zirconium. The objective of this part of the study was also to prepare a suitable heating strategy for inductive heating of the reactive aluminium–lithium alloys. Pyrometallurgically produced Al–Li billets (diameter 80 mm, length 200 mm) were mechanically cleansed of external oxide layers and macro-cavities. With reference to the thermochemical calculations and the DTA testing completed with the heating experiments, a suitable process temperature could be specified for both alloys (Table 4.6)

In respect of the high reactivity of the alloys, the raw material billets were packed in a foil of high-purity aluminium for inductive heating, in order to reduce any contact with the atmosphere as far as possible. When using the protective aluminium foil, there is no need for additional purging of the basin with inert gas during heating. The 76 mm billets were heated to the partially liquid state within a period of 420 s, with a solid-phase proportion of 50%. A billet format (diameter 76 mm, length 160 mm) suitable for the horizontal heating system was used. The state of the partially liquid billets was first estimated subjectively by manual cutting. For the alloy AlLi4Mg8Ti a very good processing state was achieved at 562 °C, whereas the alloy AlLi4Cu4Ti first displayed a suitable cutting characteristic at ~601 °C. At the end of the homogenizing phase, both materials had a temperature difference between the edge and the centre of ~1.5 °C.

During the practical examinations, it could be demonstrated the metal alloyed with magnesium (AlLi4Mg8Ti) can even be safely heated in the partially liquid state without the protective foil. The alloyed magnesium gives rise to the formation of an oxidation layer on the surface of the billet, thus greatly inhibiting any further ingress of oxygen. After inductive reheating, the partially liquid billet was manually

Table 4.6 Process window for the alloys A3 and A7.

A3	$T_s \approx 550$–$560\,°C$	$T_L \approx 660$–$679\,°C$
	$T^{40-60} \approx 590$–$610\,°C$	$\Delta T^{40-60} = 20\,°C$
	$T_{process} \approx 590$–$600\,°C$	$t_{hold} = 2$–$5\,min$
A7	$T_s \approx 460$–$470\,°C$	$T_L \approx 650$–$660\,°C$
	$T^{40-60} \approx 530$–$560\,°C$	$\Delta T^{40-60} = 30\,K$
	$T_{process} \approx 530$–$550\,°C$	$t_{hold} = 2$–$5\,min$

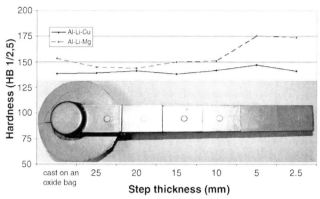

Figure 4.19 Hardness development of step samples derived from an Al–Li–X principle component.

transferred into a ceramic carrier basin and placed in the preheated casting chamber (Foundry Institute of RWTH Aachen). The effect of the solidification rate (wall thickness 1–25 mm) on the casting grain structure was studied by casting into a step form (Figure 4.19). The form was preheated to ~250 °C using three heater–cooler units. When the component was removed from the form too early, ignition of the casting residue took place that was still in a partially liquid state. During removal from the form, some of the thinner component walls cracked. In general, the alloy AlLi4Mg8Ti displays better flow characteristics and visual surface quality than the alloy Ali4Cu4Ti.

Figure 4.20 shows the grain structure of the 25 and 2.5 mm steps for both materials. Both materials display a virtually perfect thixo-grain structure with a pronounced globulitic primary phase. The magnesium-alloyed variant has a significantly smaller particle size of 35 µm than the Al–Li–Cu alloy with 48 µm. The form factor of the globulitic primary phase for both alloys is ~0.74. The 'image analysis' evaluation of the samples could not detect any notable signs of mix separation across the step thickness. The solid-phase proportion for the copper-alloyed type is ~75% and for the magnesium-alloyed version ~65% in both step thicknesses/cooling rates.

Selected thixocasting step samples of the two aluminium–lithium alloys underwent T6 heat treatment (AlLi4Mg8, 2 h at 430 °C solution annealed, quenched in water, 15 h at 120 °C precipitation heat treatment; AlLi4Cu4, 2 h at 510 °C solution annealed, quenched in water, 15 h at 120 °C precipitation heat treatment). Even after the heat treatment, the globulitic grain structure was retained. In the case of the alloy AlLi4Cu4Ti, several large globulites collected near the surface, which, with 100–150 µm, had double the average grain size than that in the centre of the steps (50–80 µm). In contrast, the alloy AlLi4Mg8Ti displayed a uniform grain size distribution between the centre of the step and the step surface in all steps (Figure 4.21).

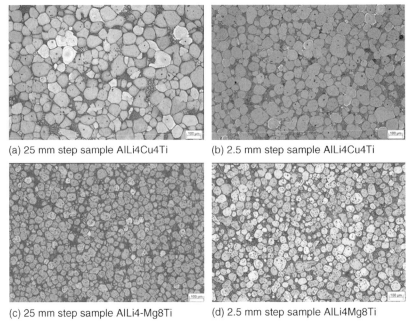

(a) 25 mm step sample AlLi4Cu4Ti

(b) 2.5 mm step sample AlLi4Cu4Ti

(c) 25 mm step sample AlLi4-Mg8Ti

(d) 2.5 mm step sample AlLi4Mg8Ti

Figure 4.20 Microstructures of thixocast step samples (AlLi4Cu4Ti and AlLi4Mg8Ti) [33].

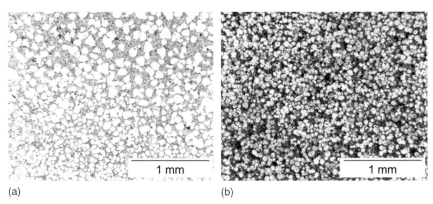

(a)

(b)

Figure 4.21 Phase separation near the surface of an Al–Li4Cu4 (Al–Li4Cu4Ti) step sample (a) and near the surface of an AlLiMg8 (AlLi4Mg8Ti) step sample (b).

4.6.2
Rheocasting

On the basis of the experience with thixocasting, the 0.15 wt% Sc and 0.15 wt% Zr modified alloy AlLi2.1Mg5.5 (A8) was selected, because due to the lower Li and Mg contents the embrittlement is reduced due to minimized intermetallic phase

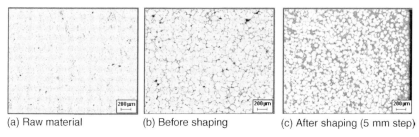

(a) Raw material   (b) Before shaping   (c) After shaping (5 mm step)

Figure 4.22 Development of the microstructure of AlLi2.1Mg5.5ZrSc in the course of the RCP [32].

formation. Because, first, mechanical properties were to be determined, for the melt-based metallurgical raw material production refined aluminium (99.999%) and high-purity lithium (99.9%) were used, so that the Na content would be kept as low as possible. Implementation of the rheo-container process (RCP) for AlLi alloys required the modification of the process window, in particular that of the melt treatment. Due to the expected reactivity of the alloys, the SiC melting crucible was closed in and flushed with argon gas; see Section 4.4.2 for details of the process. Figure 4.22 shows the development of the microstructure commencing with the raw material and continuing through to forming the billets synthesized by means of the overpressure melting process display a very fine granular structure of $d_m \sim 40\,\mu m$ average diameters. In the course of the RCP, the primary phase is formed with a more globular but also more coarse-grained aspect ($d_m \approx 65\,\mu m$).

The very homogeneous distribution of the primary alpha-solid solution across all wall thicknesses in the step sample is shown in Figure 4.23. The average grain size of

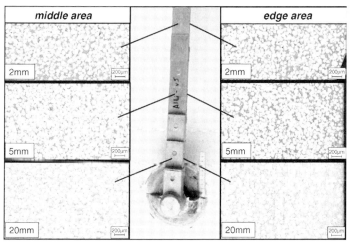

Figure 4.23 Microstructure of an AlLi2.1Mg5.5ZrSc alloy processed in the partially liquid state (at $T = 601\,°C$) by means of the RCP [32].

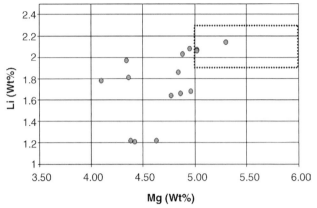

Figure 4.24 Li and Mg contents of the Al–Li–X step samples (for specification of A1420, see the rectangle).

the AlLi2.1Mg5.5ZrSc alloy was determined as 65 μm, as mentioned above. A significant tendency for mix-separation of the solid and liquid phases was not observed. Only in the immediate border area of the component did the liquid phase become concentrated. In statistically distributed samples for determining the alloy composition, there were indications that the target composition could largely not be set in relation to the main alloying elements Li and Mg (Figure 4.24). This is due to the already demonstrated non-homogeneity in the raw material or to burning off of the alloying elements during the forming process. This needs to be improved for the extensive forming of real components.

The modification in the microstructure during heat treatment is shown in Figure 4.25. Two different T6 heat-treatment strategies were tested. Already after a short solution annealing for 3 h, dissolution of the eutectic and intermetallic phase portions commences, enhancing the aluminium-mix crystal. This is accompanied by slight grain growth. If the solution annealing time is extended to 24 h, the modifications in the microstructure are drastic in contrast.

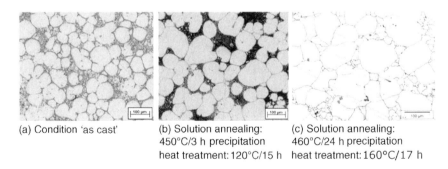

(a) Condition 'as cast'

(b) Solution annealing: 450°C/3 h precipitation heat treatment: 120°C/15 h

(c) Solution annealing: 460°C/24 h precipitation heat treatment: 160°C/17 h

Figure 4.25 Microstructure of AlLi2.1Mg5.5ZrSc alloy processed by means of the RCP in its 'as cast' state and after two T6 heat treatment states (5 mm step) [32].

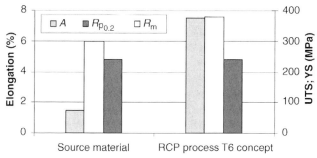

Figure 4.26 Mechanical properties of step samples made from Al–Li2.1Mg5.5Zr0.15Sc0.15 (as cast and after annealing) [34].

Almost only the aluminium-mix crystal remains with a slight residual phase of Al_2LiMg on the grain boundaries. In this state, the microstructure resembles that of an aluminium moulding alloy. The addition of the elements Sc and Zr allowed good grain refinement to be achieved and undesired grain growth and recrystallization to be successfully inhibited during the solution annealing.

The mechanical properties of various step samples were analysed in their 'as cast' and T6 states. The 'as cast' properties revealed no technically noteworthy results, either for static strength or for elongation values (Figure 4.26). After a 24 h solution annealing at 460 °C, quenching in water and subsequently precipitation heat treatment at 160 °C for 17 h, the average tensile strength was increased from 160 to above 380 MPa, the average yield limit from 100 to above 230 MPa and the average elongation from 2 to over 7%. Some samples even achieved a tensile strength of 432 MPa, a yield limit of 250 MPa and an elongation of 13%. The results in relation to the form-filling capacity displayed, as previously during the thixocasting tests, the excellent suitability of the Al–Li–X alloys developed. Only at the critical transformation from the 5 mm to the 2 mm step could some small (hot) cracking be identified on the component. For the subsequent real component production, this effect should be avoided by choosing a lower processing temperature and therefore a higher solid:liquid ratio. An adapted tool geometry and locally adapted heating–cooling strategy can further reduce this tendency for hot tearing.

4.7
Production of Al–Li Demonstrators by Rheocasting

In order to check the transferability of the laboratory results under production conditions, a tie-rod was selected as a real component and was produced at the Foundry Institute of RWTH in Aachen using the RCP on the basis of the Al–Li–X raw material billets [35]. This component has previously been proved using the alloy A356 [33]. The results of these studies are given in more detail in Chapter 9. Using the overpressure melt process, the Al–Li–X alloys were reproducibly produced on a 3 kg scale with Li contents of 1.5–4 wt%. In order to prevent contamination by Na and K, it is necessary to employ prealloys and metals of high purity. The maximum Li losses compared with the initial weighing amounted to ~5%. Using the permanent mould system available it was not yet possible, however, to prevent chemical non-homogeneity of the raw material billet. In particular, the elements Cu and Mg are concentrated at the surface of the billet and are below the target values in the centre of the billet. DTA tests confirmed the results of the thermochemical prediction calculations that for the AlLi4Cu4, AlLi4Mg8 and AlLi2.1Mg5.5 alloy systems the largest process window for processing in the semi-solid range can be expected. The evaluations for the microstructure in the semi-solid range for quenched samples indicate that in particular the Ti and Zr grain-refined alloys AlLi4Cu4 and AlLi4Mg8 and also the Sc and Zr microalloyed alloys AlLi2.1Mg5.5 achieve very good results with globular and fine-grained grain structures in the range 38–60 μm. Apart from that, these alloys have a high proportion of premelting eutectic phase, which supports thixotropic flow during forming.

The alloys AlLi4Cu4Ti and AlLi4Mg8Ti with billet formats of 160 × 76 mm were able to be transformed into the semi-solid state by inductive reheating without any problems. Excessive oxidation of the surface of the billet could be minimized by wrapping it with aluminium foil. The primary grain sizes achieved in the thixocast step samples were ~65 μm, where this was very globular and homo-geneously distributed over all step thicknesses. The thixotropic flow characteristics can in this respect be described as 'excellent'. The Sc and Zr microalloyed A1420 (AlLi2.1Mg5.5) alloys were successfully processed using the RCP at the Foundry Institute. Also in this case grain sizes of ~60 μm were achieved in the 'as cast' state. As indicated by the chemical analyses, with the RCP a slight Li burn-off must be taken into account due to the nature of the process. No tendency for mix-separation of the solid–liquid phases was identified. Only in the immediate shell was the liquid phase concentrated. After suitable heat treatment (solution an-nealed at 450 °C for 24 h, quenched in water, precipitation heat treated at 160 °C for 17 h) and the virtually complete dissolution of the residual intermetallic phases, in the principle component (step sample) made from the alloy A1420 (AlLi2.1Mg5.5 + ScZr) mechanical characteristic values were achieved that are within the range of rolled sheets and profiles of the same alloy, and in the case of elongation even exceed them. Due to the high-purity raw material used here, it was possible to safeguard the mechanical properties by keeping the Na content below 6 ppm. In direct comparison with the rolled alloy 1421 (Sc added), it appears that for a semi-solid cast component, already in this early step of development, ~80% of the yield limit, 88% of the tensile strength and equivalent elongations can be achieved. The anisotropy of mechanical properties occurring in rolled sheets does not occur in SSM Al–Li components due to the homogeneous microstructure achieved. By semi-solid forming of the reactive Al–Li alloys, which are difficult to process in conventional casting processes, a completely new production method and an innovative component quality for near-net-shaped production is being achieved.

4.8
Recycling of Aluminium–Lithium Alloys from Thixoforming Processes

4.8.1
State of the Art

Since the commercial introduction of Al–Li moulding alloys in the late 1980s, the question of recycling has presented a challenge in many respects [42]:

- The lithium contained in Al–Li residues represents a value far in excess of the aluminium content.
- The residues can cause problems if they enter the usual recycling route, especially in the foil sector.
- To date, the specifications for secondary casting alloys do not indicate any lithium content and an agglomeration of Al–Li residues is not without consequences, because in moist conditions it can react with the hydrogen being released.

In order to secure a wider use of this group of alloys, a complete recycling concept must be developed for the residues that would facilitate 100% reuse of the recycled Al–Li residues. For recycling, first as far as possible a closed system ('closed-loop recycling') is desirable, in which the residues can be returned directly to the existing process and need not be diverted through the entire secondary cycle for aluminium. In the case of scrap recycling this would probably not be feasible: because of the problems involved in sorting of separated types of scrap, the lithium would be lost.

In principle, there are two possibilities for recycling of the thixo-materials. The first and by far the most sustainable solution would be to retain the valuable alloying elements within the alloy. This is primarily of interest in the case of clean scrap materials that could be melted down in molten salt. The second alternative would be to transfer the alloying elements (Mg, Li) into molten salt in order to reclaim them subsequently from there if required. Such a procedure could be applied for both clean and low-quality scrap materials. From the point of view of economy, the same molten salt should be used that is used in the secondary aluminium industry (70% NaCl + 30% KCl with 2–3% CaF_2 additive), in contrast to the very hygroscopic mixture of LiCl and KCl that has been proposed [43, 44]. Then recycling of the salt slag produced could be performed in a largely conventional manner.

4.8.2
Thermochemical Metal Salt Equilibria

In order to decide which salt mixture compositions come into consideration for the two above-mentioned recycling alternatives, thermochemical calculations were performed in respect of the equilibria between the relevant alloys and various molten salts, using the FACTSAGE program. Four alloy types were combined with salt mixtures, on the one hand based on NaCl and KCl with added CaF_2, $MgCl_2$, LiF and on the other based on KCl and LiCl.

In the equilibria with KCl–NaCl–LiCl molten salts, the calculations indicate for AlLi2Mg5.5 that the Mg and Li contents of the metal melt would remain unchanged, but the Na and K contents, with an LiCl content of up to 50%, would rise substantially but would then fall steeply with a further increase in the LiCl content of the salt mixture.

For a proportion of just 10% LiCl in the salt mixture, the rise in the process temperature would lead to higher K contents in the metal melt, when the Na content remains unchanged. An increase in the LiCl content from 10 to 90% in the salt mixture would reduce the Na and K contents in the metal melt by that factor. All calculations indicate that by melting an Al–Li alloy using NaCl–KCl–LiCl salts, generally no recyclable AlLi alloys can be reclaimed, because the Na and K contents would become unacceptably high. When using binary KCl–LiCl molten salts the calculated K content of the metal melt at 650 °C, with 90% LiCl in the molten salts, would reach a far lower value (Figure 4.27).

The degree of Li and Mg reclamation amounts to ∼98–99.5% (Figure 4.28), whereby the acceptable K content in the alloy is only reached if the LiCl content of the salt mixture is well above 90%.

4.8.3
Experimental Validation of the Thermochemical Calculations

In order to minimize any possible diffusion inhibitions arising in the equilibrium setup, intensive agitation was also provided for. Each salt mixture was melted in a firebrick–graphite crucible (diameter 80 mm, height 100 mm) and brought to the attempt temperature. The agitator was rotated (400 min^{-1}) and swarf of the applicable alloy was added. After melting and in further steps of 1, 3, 6, 10, 20, 30, 40, 50 and 60 min, salt samples were taken for ICP (Inductive Coupled Plasma) analysis and metal samples were taken before and after the attempt.

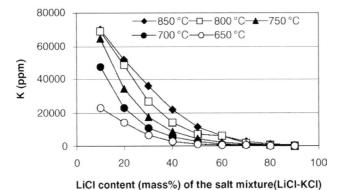

Figure 4.27 Calculated K content equilibria of AlLi2Mg5.5 after melting under molten KCl–LiCl (metal:salt ratio 1:1).

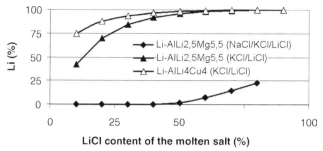

Figure 4.28 Calculated Li yield of Al–Li alloys at 750 °C after contact with NaCl–KCl–LiCl depending on the LiCl content of the salt mixture.

As an example, Figure 4.29 shows that during treatment of an AlLi2.5Mg4 metal melt using 70NaCl–20KCl–10MgCl$_2$ molten salts, Li was completely transformed by MgCl$_2$ into LiCl (see Section 4.7.2). The equilibrium is attained in this case after a short period of 3–4 min. Also for the treatment of AlLi4Mg8 alloy using 70NaCl–27KCl–3LiF molten salts, the Li is completely transformed by the NaCl into LiCl and a small amount of Mg by LiF into MgF$_2$. The experimental testing in its entirety largely confirms the thermochemical calculations. The use of an NaCl–KCl salt mixture with additives of LiF, CaF$_2$ or MgCl$_2$ leads to transfer of the Li from the metal alloy into the molten salts. When using MgCl$_2$ as an additive, the Mg content increases in correspondence with the transfer of Li into the molten salts, which could partially be compensated for by the addition of LiF. By the addition of CaF$_2$ a transformation of Mg into the molten salts is achieved, and also that of the Li. As calculated in Section 4.7.2, the lithium is nearly completely transferred to the molten salts (residual content in the metal melt <0.1%), so that Al–Li alloys cannot be recycled using this salt mixture.

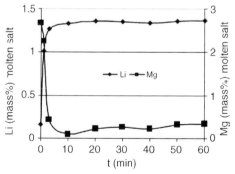

Figure 4.29 Kinetics of equilibrium tests for interaction of AlLi2.5Mg4 with 70NaCl–20KCl–10MgCl$_2$ molten salt at 750 °C.

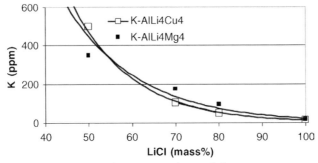

Figure 4.30 Experimental K contents in the metal after treatment of AlLi4Mg4 and AlLi4Cu4 melts using KCl–LiCl salt mixtures at 750 °C.

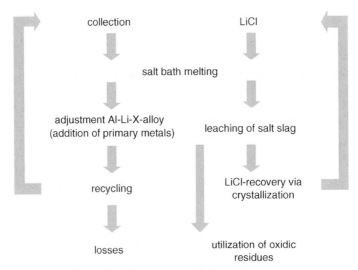

Figure 4.31 Almost closed-loop recycling of AlLi residues.

The tests using a salt mixture of KCl and LiCl for melting residues from AlLiMg and AlLiCu did, however, indicate that it is possible to retain the Li and/or Mg completely in the metal melt, provided there is an LiCl content of more than 90% in a KCl–LiCl salt mixture. In this case, the experimental values for the K content in the metal are even lower than those predicted by the thermochemical calculations (Figure 4.30). In the course of the treatment of the AlLiMg metal melt with pure LiCl, a K content of 14 ppm in the metal melt was achieved, with an Li and Mg yield of over 99.4% in both cases. The reason for the lower experimental K contents in contrast to the theoretical values is that the solubility of K in solid and liquid aluminium was not available.

In order to close the loop in the recycling of Al–Li residues from thixoforming, it would be necessary to process the salt slag produced. When using pure water-soluble LiCl for recycling, it is possible to treat the salt slag by means of the conventional dissolution–crystallization process that is standard within the secondary aluminium industry. This closed loop for the recycling of Al–Li residues (Figure 4.31) permits nearly complete recovery of reusable Al–Li alloys, in addition to returning the used LiCl back into the recycling process.

References

1 Fehlbier, M. (2002) Verarbeitung teilflüssiger metallischer Werkstoffe, Dissertation, RWTH, Aachen, 9.

2 Wagener, W. and Hartmann, D. (2000) Feedstock material for semi-solid casting of magnesium. Proceedings of the 6th International Conference on Semi-solid Processing of Alloys and Composites, Turin, 301–306.

3 Flemings, M.C. (1991) Behavior of metal alloys in the semisolid state, The 1990 Edward Campell Memorial Lecture. *Metallurgical Transactions A*, **22A**, 957–980.

4 Müller-Späth, H. (1999) Legierungs-entwicklung unter Einsatz des SSP-Verfahrens und Umsetzung intelligenter Materialkonzepte beim Thixogiessen, Dissertation, RWTH, Aachen,

5 Wan, G. and Witulski, R. (1993) Thixoforming of aluminium alloys using modified chemical grain refinement for billet production. International Conference on Aluminium Alloys: New Process Technologies, Marina di Ravenna; Brusethaug, S. and Voje, J. (2000) Manufacturing of feedstock for semi-solid processing by chemical grain refinement. Proceedings of the 6th International Conference of Semi-solid Processing of Alloys and Composites, Turin, 451–456.

6 Noll, T. (2003) Die anwendungsgerechte Weiterentwicklung des Al-Werkstoffes EN AC-AlSi7Mg0,3 (A356) mit chemischer Kornfeinung, Dissertation, RWTH, Aachen.

7 Hirt, G. and Sahm, P.R. (1994) Möglichkeiten zur Produktivitätssteige-rung, Qualitätsverbesserung und Materialeinsparung durch endabmessungsnahe Fertigung mit kombinierten Ur/Umformverfahren, EFU-Mitteilungen 1-1994.

8 Loue, W.R. (1996) Metallurgische Aspekte des Thixogiessens der Aluminium-legierungen AlSi7Mg0,3 und AlSi7Mg0,6. *Giesserei-Praxis*, (13/14), 251–260.

9 Decker, R. (1990) Casting semi-solids by thixomolding, Foundry Focus, Autumn/ Winter, 22–23.

10 Decker, R. and Carnahan, R.D. (1991) Thixomolding. *Advanced Materials Technology*, **8**, 174–179.

11 Dworog, A. *et al.* (2000) Formfüllvorgänge beim Magnesiumspritzgiessen, 8. Magnesium Abnehmer Seminar, Aalen, 14–15 June 2000.

12 Decker, R.F. and Carnahan, R.D. (1991) Thixomolding, Thixomat Inc., Ann Arbor, MI, 47th Mg Conference, 106–116.

13 Gullo, G.C. (2001) Thixotrope Formgebung von Leichtmetallen – neue Legierungen und Konzepte, Dissertation, ETH Zürich, Dissertation ETH Nr 14154.

14 Prikhodovsky, S. (2000) Modellierung von Reifungsprozessen und Anwendung auf das Thixoforming, Deutsche Dissertation

15 Steinhoff, K., Gullo, G.C., Kopp, R. and Uggowitzer, P.J. (2000) A new integrated production concept for semi-solid processing of high quality Al products, Proceedings of the 6th International

Conference on Semi-solid Processing of Alloys and Composites, 27–29 September 2000, Turin, Edimet, Brescia, 121–127.

16 Harris, S.J., Noble, B. and Dinsdale, K. (1983) Proceedings of the 2nd International Conference on Aluminium–Lithium Alloys II (eds T.H. Sanders and E.A. Starke), TMS, Warrendale, PA, 219.

17 Kammer, C. (1995) *Aluminium-Taschenbuch, Band 1, Grundlagen und Werkstoffe*, 15. Auflage, Aluminium-Verlag, Düsseldorf, 303.

18 Davis, J.R. (1998) In Davis *et al.* (eds), *ASM Specialty Handbook, Aluminum and Lithium Alloys*, ASM International, Materials Park, OH,

19 Starke, E.A. and Sanders, T.H., Jr (1986) Proceedings of the 1st International Conference on Aluminium Alloys: Their Physical and Mechanical Properties.

20 Bick, M. and Markwoth, M. (1980) Thoughts on the removal of lithium from primary aluminium and experiments to carry out the process on a technical scale. *Metall*, **34** (12), 1095–1098.

21 Bofarini, C. (1991) Prozessentwicklung des Feingiessens von Al–Li Legierungen sowie Untersuchungen ihrer giesstechnischen und mechanischen Eigenschaften *Giessereiforschung*, **43**, (2).

22 Polmear, I.J. (1989) *Light Alloys – Metallurgy of the Light Metals*, Edward Arnold, London, pp. 130–132.

23 Vekateswara, R., Yu, W. and Ritchie, R.B. (1988) *Metallurgical Transactions A*, **19**, 563.

24 Miller, W.S., White, J. and Eloyd, D.J. (1987) Proceedings of the 4th International Conference on Aluminium–Lithium Alloys IV (eds G. Champier, B. Dubost, D. Miannay and E. Sabetay), Paris, *J. de Phys Colloque*, **48**, C3, 139.

25 Webster, D. (1994) Aluminium–lithium alloys, the next generation. *Advances in Materials and Processes*, **145**, (5), 18–24.

26 Sweet, E.D. *et al.* (1994) Proceedings of the 4th International Conference on Al Alloys, Georgia Institute of Technology, Atlanta, GA, 231.

27 Kopp, R. *et al.* Sonderforschungsbereich SFB 289: Formgebung metallischer Werkstoffe im teilerstarrten Zustand und deren Eigenschaften, Arbeits- und Ergebnisbericht 2002/2003/2004 für die Deutsche Forschungsgemeinschaft DFG, 54–56.

28 Balitchev, E. (2004) Thermochemische und kinetische Modellierung zur Legierungsauswahl mehrphasiger Systeme für das Thixoforming und zur Optimierung ihrer Formgebungsprozesse, Dissertation, RWTH, Aachen.

29 Balitchev, E., Hallstedt, B. and Neuschütz, D. (2005) Thermodynamic criteria for the selection of alloys suitable for semi-solid processing. *Steel Research International*, **76** (2/3), 92–98.

30 Fridlyander, I.N., Rokhlin, L.L., Dobatkina, T.V. and Nikitina, N.I. (1993) Phase equilibria in aluminium alloys containing lithium. *Metallovedenie i Termicheskaya Obrabotka Metallov*, (10), 16–19 (in Russian).

31 Hemminger, W. and Cammenga, H. (1989) *Methoden der Thermischen Analyse*, Springer, Berlin.

32 Höhne, G., Hemminger, W. and Flammersheim, H.-J. (1996) *Differential Scanning Calorimetry – an Introduction for Practioners*, Springer, Berlin.

33 Kopp, R. *et al.* (2008) Sonderforschungsbereich SFB 289: Formgebung metallischer Werkstoffe im teilerstarrten Zustand und deren Eigenschaften, Arbeits- und Ergebnisbericht 2002/2003/2004 für die Deutsche Forschungsgemeinschaft DFG, 165.

34 Sauermann, R., Friedrich, B., Grimmig, T., Bünck, M. and Bührig-Polazcek, A. Development of Al–Li alloys processed by the Rheo Container Process, Proceedings of the 9th International Conference on Semi-solid Processing of Alloys and Composites, Busan, Korea.

35 Sauermann, R., Friedrich, B., Püttgen, W., Bleck, W., Balitchev, E., Hallstedt, B., Schneider, J.M., Bramann, H., Bührig-Polaczek, A. and Uggowitzer, P.J. (2004) Aluminium–lithium alloy development for thixoforming. *Zeitschrift für Metallkunde*, **95** (12), 1097–1107.

36 Antrekowitsch, H. (1998) Grundlagenuntersuchungen bei der Erstarrung der Nichteisenmetalle Al und Cu im rotierenden Magnetfeld sowie Anwendungsmöglichkeiten in der Industrie, Dissertation am Institut für Technologie und Hüttenkunde der Nichteisenmetalle an der Montanuniversität Leoben.

37 Sommerhofer, H. *et al.* (2001) Druckabhängigkeit des Gleichgewichts im System Al–Si. *Giessereiforschung*, **53** (1), 25–29.

38 Ohm, L. and Engler, S. (1989) Treibende Kraft der Oberflächenseigerungen beim NE-Strangguss. *Metall*, **43**, 520–524.

39 Buxmann, K. (1977) Mechanismen der Oberflächenseigerung von Strangguss. *Metall*, **31**, 163–170.

40 Cernov, D.B. and Schinyaev, A. (1974) Influence of high pressure on the composition diagram of the Al–Si system, *Struktura FAZ*, Nauka, Moscow, 80–84.

41 Batischev, A.I. (1977) *Crystallization of Metals and Alloys Under Pressure*, Verlag Metallurgie, Moscow (in Russian).

42 Krone, K. (2000) *Aluminium Recycling – Vom Vorstoff bis zur fertigen Legierung*, Aluminium-Verlag Marketing und Kommunikation, Düsseldorf.

43 Mulate, K. (2000) Recycling von Mg–Li-Legierungen: Gleichgewichte zwischen Metall- und Salzschmelzen, Dissertation, Technische Universität Clausthal.

44 Schwerdfeger, K., Mulate, K. and Ditze, A. (2002) Recycling of magnesium alloys: chemical equlibria between magnesium–lithium-based melts and salt melts, *Metallurgical and Materials Transactions*, **33**, 335–364.

5
Thermochemical Simulation of Phase Formation[*]

Bengt Hallstedt and Jochen M. Schneider

5.1
Methods and Objectives

5.1.1
General

In this chapter, we will explore the use of thermochemical simulation methods to support alloy selection and processing in the semi-solid state. Alloy selection is discussed more in detail in Chapters 2–4. Processing-related issues are discussed in detail in Chapters 9 and 11. The use of thermochemical simulation for alloy selection has also been treated in, for example, [1–4]. The focus at RWTH over the last few years has been on semi-solid processing of steels and, therefore, only steels will be discussed here. The steels of main interest here are the relatively high-carbon steels X210CrW12 and 100Cr6. Their compositions are given in Table 5.1. A key property for semi-solid processing is the fraction of the liquid phase as a function of temperature (and possibly time), and this will be dealt with in some detail. It is necessary to know the fraction of liquid phase in order to be able to control the process and in order to simulate the viscous flow during various forming operations (Chapters 6 and 7). It is possible to determine the fraction of liquid phase experimentally using, for example, differential thermal analysis (DTA) or metallography (more details are given in Chapter 3), but this is far from trivial [5]. The approach used here is to calculate the fraction of liquid phase from thermodynamic (and sometimes diffusion) data.

If the thermodynamic properties of a system are known, then any phase diagram can be calculated. This is illustrated in Figure 5.1, which also shows that the thermodynamic description actually contains much more information than the phase diagram alone, since the thermodynamic properties of the included phases are defined for the complete composition range, and not only for the range where they are stable. Not only the stable phase diagram, as shown in Figure 5.1, and any thermodynamic property of the stable state, but also any metastable state can be

[*] A List of Abbreviations and Symbols can be found at the end of this chapter.

Thixoforming: Semi-solid Metal Processing. Edited by G. Hirt and R. Kopp
Copyright © 2009 WILEY-VCH Verlag GmbH & Co. KGaA, Weinheim
ISBN: 978-3-527-32204-6

Table 5.1 Standard chemical composition (mass%) of the steels, used for the calculations in the present chapter.

Alloy	C	Mn	Si	Cr	Mo	V	W
100Cr6	1.00	0.30	0.30	1.50	—	—	—
X210CrW12	2.15	0.30	0.30	12.0	—	—	0.7
HS6-5-2	0.84	0.30	0.30	4.15	4.95	1.9	6.30

calculated. This can be used, as just one example, for calculating the driving force for the precipitation of the stable phase from a metastable state. The coupling of thermodynamics and phase diagrams has been realized within the Calphad method [6–8], which was essentially developed in the 1970s and is now used to deal with a wide variety of alloy systems, oxide systems, salt systems and so on. In the Calphad method, the Gibbs energy of each phase is parameterized in terms of temperature and composition and, in principle pressure, but this is rarely used, based on crystallographic and compositional data. The Gibbs energy functions contain variable coefficients, which are optimized to experimental phase diagrams and thermodynamic data. Crystallographic and thermodynamic data from *ab initio* calculations are increasingly being used to support and extend the experimental data. In this chapter, we use the Thermo-Calc software [9] for thermodynamic calculations with the TC-Fe 2000 [10] and TCFE4 [11] steel databases. The more recent TCFE4 database essentially gives the same results as TC-Fe 2000 for the steels treated here, but additionally contains molar volume data. In the general case, the coverage and quality of the database(s) used are a concern, but for the steels dealt with

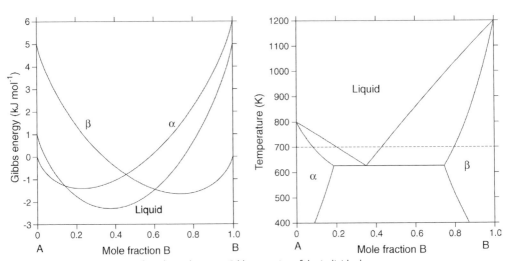

Figure 5.1 The relation between Gibbs energies of the individual phases and the resulting phase diagram. The dashed line shows the temperature at which the Gibbs energy curves are drawn.

here it can be assumed that the databases used are reasonably accurate, particularly at the relatively high temperatures of interest here.

5.1.2
Equilibrium Calculations

Using standard equilibrium calculations is a quick and easy way to obtain an overview of the phases expected to appear in a particular material, their amounts and temperature range in which they are stable. It is also useful for calculating various thermodynamic properties of the material provided that it does not deviate too much from the equilibrium state. In connection with semi-solid processing, the partially liquid state and the solidification behaviour are obviously of major interest. The actual solidification path, in principle, never follows the equilibrium path, due to limited diffusion in the solid phase(s). This results in microsegregation and possibly non-equilibrium solid phases forming. The actual fraction of solid phases at a particular temperature is (almost) always smaller than the equilibrium fraction. For steels, the equilibrium solidification path is often a reasonable first approximation since C is a fast-diffusing element. In thixoforming, where the material is heated from the solid state into the semi-solid state, the melting behaviour of the material is also of importance. However, the melting behaviour of alloys has been far less studied than the solidification behaviour. From the point of view of equilibrium, there is no difference between solidification and melting. In order to see a difference, it is necessary to take diffusion into account and in the case of melting the microstructure of the solid starting material will have a decisive influence on the melting behaviour.

5.1.3
Scheil–Gulliver Calculations

The diffusion in the liquid phase is much faster than in the solid phase(s) and it is often a useful approximation to assume that there is no diffusion at all in the solid phase(s) and infinitely fast diffusion in the liquid phase during solidification. This is called Scheil–Gulliver solidification [12, 13]. In the rest of this chapter we will use the shorthand Scheil to mean Scheil–Gulliver. Numerically, Scheil solidification can be realized as a series of equilibrium calculations, starting with the liquid phase and decreasing the temperature until the liquid has (almost) disappeared. During each temperature step the solid phase(s) formed in that temperature step is simply removed from the calculation. Due to its simplicity, it can be easily realized with standard Calphad software. However, Scheil solidification in this simple form is not a good approximation for steels, because C diffuses fairly fast also in the solid phase(s), ferrite and austenite in this case. For steels it is a better approximation to assume that C diffuses fast enough also in the solid phase(s) to reach global equilibrium. This modified Scheil solidification can also be realized numerically as a series of equilibrium calculations, although not as easily. It has, however, been implemented in the more recent versions of Thermo-Calc. No information on diffusion rates or characteristic length scales is needed for these calculations and consequently the cooling rate is irrelevant. Melting cannot be studied as a separate process in this way, but has to be assumed to be the reverse of solidification.

5.1.4
Diffusion Simulations with DICTRA

In order to verify the equilibrium and Scheil calculations, to obtain a more detailed picture and to investigate, for example, the influence of cooling rate, it is necessary to use some form of diffusion simulation. There are very few reasonably general software packages available for diffusion simulation. Diffusion simulation is numerically a very demanding task, in particular when a moving phase interface is present, as in solidification. Here we use the software package DICTRA [9], which uses Thermo-Calc as a kind of subroutine to provide thermodynamic data. DICTRA treats multi-component diffusion in one dimension with the possibility to have one or more sharp moving interfaces. At the interface(s) local equilibrium is assumed. In addition to the thermodynamic database TC-Fe 2000 [8], the mobility database MOB 2 [14] is used here. This database is suitable for steels and the diffusion data for Fe-rich ferrite and austenite can be considered reliable, but the diffusion data for the liquid are only approximate. A diffusion problem in DICTRA is set up by defining a cell with the phase(s), compositions and temperature at the start of the simulation. The cell size corresponds to the characteristic diffusion length, which in the case of semi-solid processing is half the average distance between the centre of two neighbouring globules. In a normal solidification problem it would typically be half the secondary dendrite arm spacing. If the simulation is started in the liquid, the position where the solid phase will appear has to be specified. It is possible to require that the solid phase forms only when there is a defined driving force (undercooling) available, thus mimicking an energy barrier for nucleation. This possibility was not used in the present simulations. DICTRA does not take interface energy into account. In order to do this and in order to simulate or predict morphology in two or three dimensions, it is necessary to use other methods, in particular phase field methods [15, 16].

5.2
Calculations for the Tool Steel X210CrW12

5.2.1
Phase Diagram

A calculated temperature versus mass% C phase diagram for the tool steel X210CrW12 is shown in Figure 5.2. It is clear that austenite forms as the primary phase during solidification and there is a narrow field where austenite + M_7C_3 (where M is mostly Cr) form eutectically. Below about 800 °C, ferrite and $M_{23}C_6$ (where M is again mostly Cr) become stable, but usually austenite is retained metastably below that temperature and transforms martensitically at a much lower

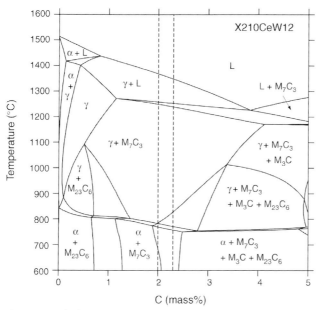

Figure 5.2 Calculated temperature versus mass% C phase diagram for X210CrW12. The dashed lines show the allowed interval for C.

temperature. More details about microstructure evolution of semi-solid processed X210CrW12 are given in Chapter 3 and by Uhlenhaut *et al.* [17].

5.2.2
Solidification

X210CrW12 forms austenite as the primary phase during solidification and, when about 40% liquid remains, an austenite $+ M_7C_3$ eutectic forms. The cooperation between austenite and M_7C_3 is not very good so the eutectic structure becomes rather irregular and, depending on detailed conditions, more or less coarse M_7C_3 particles can be found in the microstructure. The liquid fraction as a function of temperature calculated with different methods is shown in Figure 5.3. The eutectic formation starts at the knee. For the DICTRA simulation a cell size of 20 μm and a cooling rate of 20 K min^{-1} were used. The liquid fraction is given as mole fraction. It would be better to use volume fractions, but in order to do this molar volume data are needed, which are not generally available in thermodynamic databases. In the case of steels, volume data have only very recently become available (in TCFE4 [11]). Mole fractions can in most cases be assumed to be reasonably close to volume fractions, more so than weight fractions, which can also be easily calculated. In Figure 5.3, The DICTRA simulation can be assumed to be the most realistic. The equilibrium curve is very close to the DICTRA simulation above 20% liquid and the modified Scheil calculation, assuming that C is a fast-diffusing element, is practically coincident with

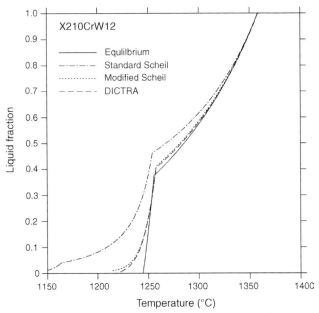

Figure 5.3 Molar liquid fraction as a function of temperature for X210CrW12 using equilibrium (solid), standard Scheil (dash-dotted), modified Scheil (dotted) and DICTRA (dashed) calculation.

the DICTRA simulation. The correctness of these curves is probably determined more by the accuracy of the thermodynamic database used, which for many steels, including tool steels, can be assumed to be fairly high, than by anything else. It is also evident that a standard Scheil calculation is not a good approximation. It is far away from the other curves even at relatively high liquid fractions. At low liquid fractions, there is a pronounced tail in the DICTRA simulation (and in the modified Scheil calculation). This is mostly caused by W segregation.

5.2.3
Determination of Liquid Fraction

To determine the liquid fraction in a semi-solid sample experimentally is far from straightforward. Two methods are frequently used: metallography and DTA. By quenching the sample quickly, one can hope to be able to retain an image of the original state and, thus, to distinguish the originally liquid and solid phases in a metallographic section. This has worked well for a number of Al alloys, but for steels it is only possible in particular cases. This is discussed in some detail in Chapter 3 and by Püttgen *et al.* [5]. The major reason can be seen in Figure 5.4, which shows a DICTRA simulation of the quenching of a X210CrW12 sample from the semi-solid state (50% liquid in this case). At the former solid–liquid interface there is only a gradual change in composition. There is no step that may have been expected. For

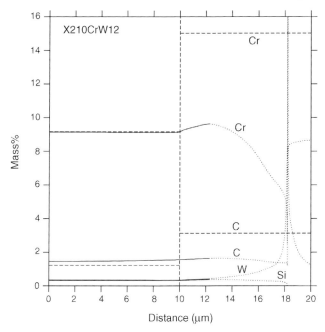

Figure 5.4 Composition profiles of X210CrW12 after quenching at 300 K s^{-1} from 1285 °C (50% liquid) simulated using DICTRA. The dashed vertical line shows the original solid–liquid interface before quenching. The dashed horizontal lines show the C and Cr contents before quenching. The composition profiles for the eutectic region are not realistic and are therefore shown with dotted lines. The M$_7$C$_3$ carbide is to the right, starting at about 18.3 μm.

that, much higher cooling rates are needed, which cannot be realized in a normal quenching experiment. Here, the already present austenite simply continues to grow and it will be practically impossible to distinguish the original austenite from the austenite formed during quenching. There is a change in growth morphology [5], but this does not produce a sharp boundary. The eutectic formed during quenching is much finer than the eutectic formed during normal cooling, so that liquid fractions below about 40% can be evaluated fairly accurately. DTA is also frequently used to evaluate liquid fraction as function of temperature (see Chapter 3). It is far from trivial, however, to evaluate the fraction of liquid from the DTA signal even if the instrument is well calibrated and the measurement is carefully done, since the DTA signal depends on many factors which are difficult to determine and control (see, e.g., [18]).

5.2.4
Origins and Consequences of Compositional Variations

If temperature is used to control the heating (or cooling) of the material to the appropriate semi-solid state, it is necessary to know accurately the composition of the sample, since the composition will determine at which temperature a certain fraction of liquid is present. For steels, C will usually have the strongest influence. Therefore, it makes sense to draw the phase diagram in Figure 5.2 as a function of C content and not some other element. There are many possible origins of compositional variations. Each charge will in principle have a different average composition so that billets coming from different charges have different compositions, but even billets coming from the same charge can have different compositions due to macrosegregation. The centre of the billet can have a different composition from the surface region, also due to macrosegregation or, for example, decarburizing during hot working or heat treatment. There will also be compositional variations on a local scale due to microsegregation. These will depend on the solidification and thermal history of the material. In any case, the average composition of a billet should always be within the limits defined by the standard for a particular steel grade. Large compositional variations are also possible in parts produced by semi-solid processing due to phase separation during processing. This is strongly dependent on the processing conditions and the geometry of the part produced. This is discussed in detail in Chapters 6 and 9. This can lead to strongly varying properties within a produced part.

As far as phase equilibria are concerned, the influence of compositional variations can be quantified in a straightforward manner. In the following, the influence of the composition on a number of key points for X210CrW12 is considered. These are linearized around the standard composition (given in Table 5.1) and given in mass%. They are roughly valid within the composition limits given by, for example, the DIN standard. Additionally, some indication of the influence of Mo, V, Ni and Cu is given. The liquidus is given by:

$$T_L = 1358°C - 67\Delta\%C - 30\Delta\%Si - 4.8\Delta\%Mn - 1.3\Delta\%Cr - 3.8\Delta\%W$$
$$(-5.5\Delta\%Mo - 1.9\Delta\%V - 4.2\Delta\%Ni - 7.8\Delta\%Cu) \tag{5.1}$$

The temperature at which there is 50% liquid is given by:

$$T_{50} = 1285°C - 105\Delta\%C - 31\Delta\%Si - 7.7\Delta\%Mn - 1.4\Delta\%Cr - 7.2\Delta\%W$$
$$(-11\Delta\%Mo - 6.4\Delta\%V - 6.7\Delta\%Ni - 12\Delta\%Cu) \tag{5.2}$$

The temperature where eutectic (M_7C_3) formation starts is given by:

$$T_{M7C3} = 1256°C - 18\Delta\%C - 1.7\Delta\%Si - 5.3\Delta\%Mn + 6.0\Delta\%Cr - 2.8\Delta\%W$$
$$(-9\Delta\%Mo - 11\Delta\%V - 1.8\Delta\%Ni - 2.7\Delta\%Cu) \tag{5.3}$$

The corresponding liquid fraction at that temperature is given by:

$$f_L = 38\% + 37\Delta\%C + 9\Delta\%Si + 1.4\Delta\%Mn + 3.0\Delta\%Cr + 2.2\Delta\%W$$
$$(+1.9\Delta\%Mo \pm 0\Delta\%V + 2.2\Delta\%Ni + 3.9\Delta\%Cu) \tag{5.4}$$

The (equilibrium) solidus temperature is given by:

$$T_S = 1245°C - 34\Delta\%C - 6.1\Delta\%Si - 6.8\Delta\%Mn + 6.9\Delta\%Cr - 4.7\Delta\%W$$
$$(-19\Delta\%Mo - 17\Delta\%V - 2.8\Delta\%Ni - 3.7\Delta\%Cu) \tag{5.5}$$

It is clear that C has by far the strongest influence, but that also Si has an appreciable influence.

5.2.5
Enthalpy and Heat Capacity

The enthalpy of X210CrW12 around the semi-solid range is shown in Figure 5.5 and the heat capacity is shown in Figure 5.6. The enthalpy curve reaches zero close to, but not exactly at, room temperature for the stable state. There is some ambiguity in the definition of heat capacity when several phases with changing compositions and amounts are present. It makes most sense to take the heat capacity to mean the derivative of the actual enthalpy curve [19]. Then the heat capacity is only undefined at an invariant reaction, where it goes to infinity. In Figure 5.6, the heat capacity has not been included in the semi-solid range. There it takes very large values, as is evident from the slope of the enthalpy curve in Figure 5.5. It is strongly sensitive to changes in temperature and is difficult to describe with a single equation. For practical purposes, for example for process control (Chapter 10) or as input into a simulation program (Chapter 6), it is then better not to use the heat capacity directly, but to use an approximation of the following type:

$$\Delta H \approx 0.84\Delta T + 207\Delta f_{\mathrm{L}} \tag{5.6}$$

where f_{L} is the liquid fraction. If needed, a more accurate approximation can be chosen. At low temperature the heat capacity is different depending on whether the

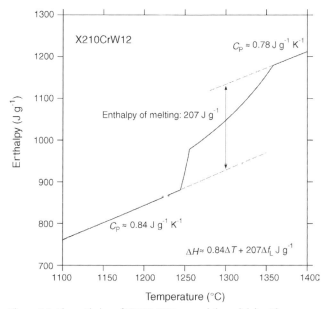

Figure 5.5 The enthalpy of X210CrW12 around the solid–liquid range.

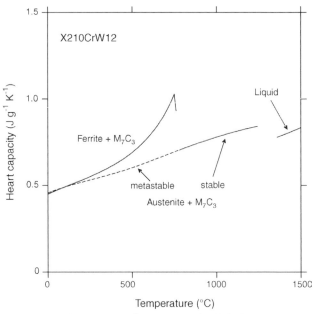

Figure 5.6 The heat capacity of X210CrW12. The dashed curve shows the heat capacity in the case when austenite is retained metastably below its stable range.

material is in a ferritic/martensitic or austenitic state. The heat capacity maximum in the stable (ferrite) curve is due to the ferromagnetic transition, where the maximum corresponds to the Curie temperature.

5.2.6
Density

If the thermodynamic database contains molar volume data, the molar volume or density can be calculated as a function of temperature (Figure 5.7). It is also possible to calculate the densities of the individual phases.

5.2.7
Melting

During thixoforming, the billet is heated to the semi-solid state, usually by induction heating (Chapters 9 and 10). Although melting is often assumed to be the reverse of solidification, this is not necessarily the case. One difference to consider is that whereas during solidification the coarseness of the microstructure formed is essentially determined by the local cooling rate, the coarseness during melting is determined by the starting microstructure and not by the heating rate. In the solid

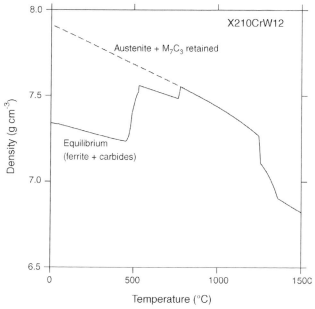

Figure 5.7 The density of X210CrW12. The dashed curve shows the density in the case when austenite is retained metastably below its stable range.

state there is usually a rich spectrum of particles, grain boundaries and so on where the liquid phase can nucleate on heating. In X210CrW12, the most favourable locations are at the phase interfaces between austenite and M_7C_3. In addition, these phases react to form the liquid phase. The initial globule size can therefore be expected to correspond the distance between M_7C_3 particles, which usually means that the globules will be fairly small. In order to investigate the melting process further, the melting process was simulated using DICTRA (see also [20]). The result is shown in Figure 5.8. The heating was started at 1100 °C, either with a completely equilibrated sample or a sample solidified and cooled to 1100 °C, that is, still showing considerable microsegregation. The heating rate was 240 K min^{-1}. There are some points worth noting. The melting curve is not the same as the solidification curve, but is fairly close if the just solidified sample is reheated. The temperature where the first liquid appears (incipient melting) depends strongly on the initial state of the sample and, for a well-homogenized sample can even be considerably above the equilibrium solidus. This melting curve is different from the solidification curve, also concerning the temperature and liquid fraction where M_7C_3 is completely dissolved. The melting is complete very close to the equilibrium liquidus temperature. This is independent of the initial state and heating rate within some limits. Heating rates of 5–500 K min^{-1} were tested. However, if the heating rate is increased further or if the microstructure is considerably coarser, the slope of the melting curves will decrease due to limited

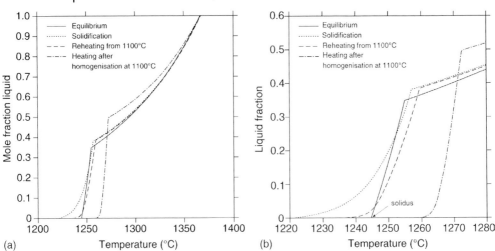

(a)   Temperature (°C)   (b)   Temperature (°C)

Figure 5.8 (a) Amount of liquid for X210CrW12 as a function of temperature. The solid line shows the equilibrium fraction. The dotted line shows a DICTRA simulation of the solidification. The dashed line shows a DICTRA simulation of the melting, directly heating after solidification and cooling to 1100 °C. The dash-dotted line shows a DICTRA simulation of the melting after complete homogenization at 1100 °C. The composition used for the simulation was different from that in Table 5.1, so that the equilibrium and DICTRA solidification curves are slightly different from those in Figure 5.3. (b) Detail of (a).

diffusion in the liquid phase. Complete melting will then occur above the liquidus temperature.

5.2.8
Suitability for Semi-solid Processing

A major criterion to determine if a particular material is suitable for semi-solid processing is usually the temperature sensitivity of the liquid fraction (Chapter 2) or, as suggested by Balitchev *et al.* [1], the temperature interval in which a certain range of liquid fraction can be achieved. Applying this to X210CrW12, the conclusion is that it is well suited provided that the desired liquid fraction is above about 40%. If it is below 40% it is nearly impossible to control if temperature alone is used to control the process (see Figures 5.3 and 5.8). The solidification behaviour of X210CrW12 is actually similar to that of the aluminium alloy A356, except that the temperatures are much higher. They both form a primary metallic phase and around 40% eutectic structure. The eutectic structure is formed in a narrow temperature interval. A356 is currently used extensively for semi-solid processing, suggesting that X210CrW12 is also fairly suitable. At the knee in the liquid fraction curve, the apparent heat capacity changes drastically. This can (and is being) be used to control the process (Chapter 10). Using a combination of temperature and heat flow control, it may even be possible to reach reproducibly liquid fractions below 40%, thus contradicting the standard suitability criterion.

5.3
Calculations for the Bearing Steel 100Cr6

5.3.1
Phase Diagram

A calculated temperature versus mass% C phase diagram for the bearing steel 100Cr6 is shown in Figure 5.9. This steel solidifies completely austenitically (also when microsegregation is taken into account). Below about 900 °C cementite formation is expected and below about 750 °C pearlite or bainite formation is expected, but the extent of these reactions depends strongly on cooling rate, and if the cooling is fast enough they can be completely absent and the austenite transforms martensitically. Due to microsegregation, some austenite may remain in the microstructure. For more details, see Chapter 3 and Püttgen *et al.* [21].

5.3.2
Solidification

The liquid fraction as a function of temperature calculated with different methods is shown in Figure 5.10. For the DICTRA simulation, a cell size of 20 µm and a cooling rate of 20 K min^{-1} were used. The liquid fraction is given as mole fraction. Both the equilibrium calculation and the modified Scheil calculation are close to the DICTRA

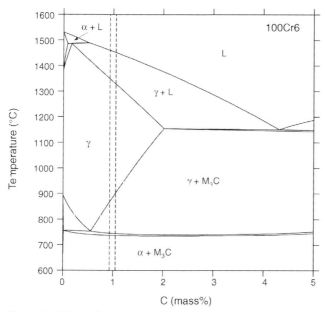

Figure 5.9 Calculated temperature versus mass% C phase diagram for 100Cr6. The dashed lines show the allowed interval for C.

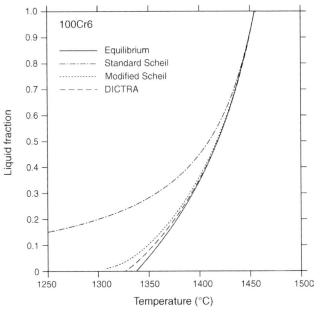

Figure 5.10 Molar liquid fraction as a function of temperature for 100Cr6 using equilibrium (solid), standard Scheil (dash-dotted), modified Scheil (dotted) and DICTRA (dashed) calculation.

simulation. The standard Scheil calculation is completely different. The fact that the equilibrium and DICTRA curve are very close suggests that there is very little segregation, but both Mn and Cr do segregate appreciably. It is not possible to determine the liquid fraction from quenched metallographic samples in this steel [5]. Figure 5.11 shows a DICTRA simulation of quenching from 50% liquid. There is only a gradual change in the composition profiles and there is no possibility of distinguishing the austenite present in the initial state from the austenite formed during quenching. Further details are given in Chapter 3.

5.3.3
Compositional Variations

As for X210CrW12, the influence of compositional variations can be quantified in a straightforward manner. In the following, the influence of the composition on a number of key points for 100Cr6 is considered. These are linearized around the standard composition (given in Table 5.1) and given in mass%. They are roughly valid within the composition limits given by, for example, the DIN standard. The liquidus is given by:

$$T_{\mathrm{L}} = 1455°\mathrm{C} - 72\Delta\%\mathrm{C} - 16\Delta\%\mathrm{Si} - 4.0\Delta\%\mathrm{Mn} - 1.2\Delta\%\mathrm{Cr} \qquad (5.7)$$

The temperature at which there is 50% liquid is given by:

$$T_{50} = 1418°\mathrm{C} - 109\Delta\%\mathrm{C} - 26\Delta\%\mathrm{Si} - 5.3\Delta\%\mathrm{Mn} \pm 0\Delta\%\mathrm{Cr} \qquad (5.8)$$

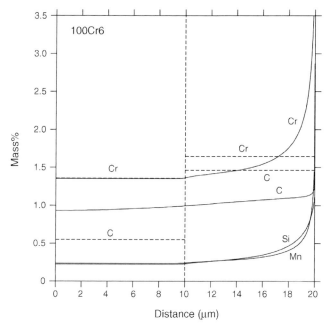

Figure 5.11 Composition profiles of 100Cr6 after quenching at 300 K s⁻¹ from 1418 °C (50% liquid) simulated using DICTRA. The dashed vertical line shows the original solid–liquid interface before quenching. The dashed horizontal lines show the C and Cr contents before quenching.

The (equilibrium) solidus temperature is given by:

$$T_S = 1337°C - 173\Delta\%C - 38\Delta\%Si - 9\Delta\%Mn + 2.9\Delta\%Cr \tag{5.9}$$

As for X210CrW12, C has the strongest influence, followed by Si.

5.3.4
Enthalpy, Heat Capacity and Density

The enthalpy of 100Cr6 around the semi-solid range is shown in Figure 5.12, the heat capacity in Figure 5.13 and the density in Figure 5.14. The strong increase in the heat capacity with temperature for the ferrite + cementite curve is also in this case caused by the ferromagnetic transition, although the Curie temperature is not quite reached in this case.

5.3.5
Suitability for Semi-solid Processing

The liquid fraction curve in Figure 5.10 is fairly steep and without any particular features, suggesting that it could be rather difficult to control the liquid fraction reproducibly. 100Cr6 may be less prone to compositional variations than

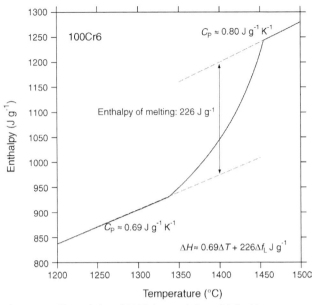

Figure 5.12 The enthalpy of 100Cr6 around the solid–liquid range.

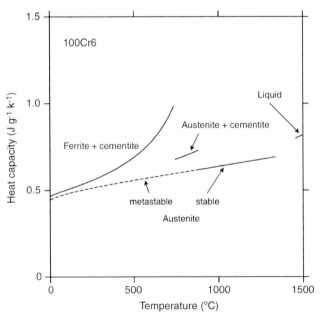

Figure 5.13 The heat capacity of 100Cr6. The dashed curve shows the heat capacity in the case when austenite is retained metastably below its stable range.

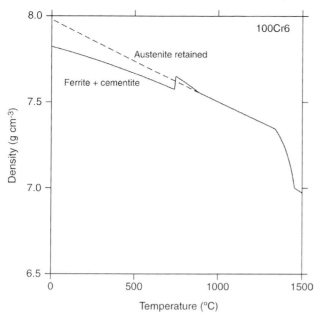

Figure 5.14 The density of 100Cr6. The dashed curve shows the density in the case when austenite is retained metastably below its stable range.

X210CrW12, thus relieving the situation somewhat. The melting behaviour can be expected to be different from that of X210CrW12. It is unclear which sites are more preferred for nucleation of the liquid. The more probable are grain boundaries and non-metallic inclusions. The coarseness of the microstructure is probably controlled by the (austenite) grain size, which can be rather coarse, leading to large globules. Further nucleation of liquid droplets within the globules, in particular when they are large, may be possible by self-blocking remelting as described in Chapter 2 and by Kleiner *et al.* [22]. Reducing the austenite grain size appears to be an effective way to reach an acceptably small globule size with no or limited internal droplets [23].

5.4
Calculations for the High-speed Steel HS6-5-2

So far, only a relatively small amount of work has concerned the high-speed steel HS6-5-2 (or any other high-speed steel). The physical metallurgy of this steel, as for other high-speed steels, is very complex, involving several different carbides. It also shows very strong microsegregation, making it difficult to produce using conventional casting. Here, semi-solid processing could open up new possibilities. A calculated temperature versus mass% C phase diagram is shown in Figure 5.15. The interesting phase field here is the one marked $\gamma + MC + M_6C$. This shows the

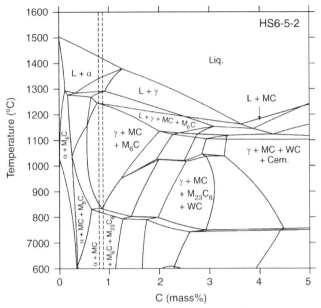

Figure 5.15 Calculated temperature versus mass% C phase diagram for HS6-5-2. The dashed lines show the allowed interval for C.

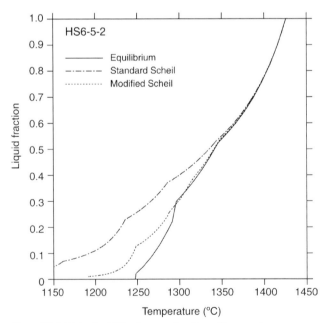

Figure 5.16 Molar liquid fraction as a function of temperature for X210CrW12 using equilibrium (solid), standard Scheil (dash-dotted) and modified Scheil (dotted) calculation.

area which is available for heat treatment and hot working operations. The C content has been chosen to yield the maximum temperature range in this phase field. The solidification sequence is complex, starting with primary ferrite, switching to peritectic formation of austenite followed by the formation of (at least) M_6C and MC carbides. Figure 5.16 shows the liquid fraction as a function of temperature calculated with different methods. As for the other steels, the standard Scheil calculation does not produce useful results. The modified Scheil and equilibrium calculation are very similar above 30% liquid. A DICTRA simulation (not shown) at a slightly different composition gave results similar to the modified Scheil calculation. However, the DICTRA simulation was only possible using a number of tricks.

Acknowledgement

The authors gratefully acknowledge financial support from the Deutsche Forschungsgemeinschaft (DFG) within the Collaborative Research Centre (SFB) 289 'Forming of metals in the semi-solid state and their properties'.

List of Abbreviations and Symbols

DFG	Deutsche Forschungsgemeinschaft
DTA	differential thermal analysis
f_L	liquid fraction
T_{50}	temperature at which there is 50% liquid
T_L	liquidus temperature
T_{M7C3}	temperature where eutectic (M_7C_3) formation starts
T_S	(equilibrium) solidus temperature
ΔH	enthalpy change

References

1 Balitchev, E., Hallstedt, B. and Neuschütz, D. (2005) Thermodynamic criteria for the selection of alloys suitable for semi-solid processing. *Steel Research International*, **76**, 92–98.

2 Hallstedt, B., Balitchev, E., Shimahara, H. and Neuschütz D. (2006) Semi-solid processing of alloys: principles, thermodynamic selection criteria. applicability. *ISIJ International*, **46**, 1852–1857.

3 Liu, D., Atkinson, H.V. and Jones H. (2005) Thermodynamic prediction of thixoformability in alloys based on the Al–Si–Cu and Al–Si–Cu–Mg systems. *Acta Materlulia*, **53**, 3807–3819.

4 Djurdjevic, M.B. and Schmid-Fetzer, R. (2006) Thermodynamic calculation as a tool for thixoforming alloy and process development. *Materials Science and Engineering A*, **417**, 24–33.

5 Püttgen, W., Hallstedt, B., Bleck, W. and Uggowitzer, P.J. (2007) On the microstructure formation in chromium steels rapidly cooled from the semi-solid state. *Acta Materialia*, **55**, 1033–1042.

6 Kaufman, L. and Bernstein, H. (1970) *Computer Calculation of Phase Diagrams*, Academic Press, New York.

7 Saunders, N. and Miodownik, A.P. (1998) CALPHAD (Calculation of Phase Diagrams): a Comprehensive Guide. Pergamon.

8 Lukas, H.L., Fries, S.G. and Sundman B. (2007) *Computational Thermodynamics: the Calphad Method*, Cambridge University Press, Cambridge.

9 Andersson, J.-O., Helander, T., Höglund, L., Shi, P. and Sundman, B. (2002) THERMO-CALC and DICTRA, computational tools for materials science. *Calphad*, **26**, 273–312.

10 TC-Fe 2000–TC Steels/Alloys Database, Thermo-Calc, Stockholm (1999).

11 TCFE4–TC Steels/Alloys Database, Thermo-Calc, Stockholm (2006).

12 Gulliver, G.H. (1913) The quantitative effect of rapid cooling upon the constitution of binary alloys. *Journal of the Institute of Metals*, **9**, 120–157.

13 Scheil, E. (1942) Bemerkungen zur Schichtkristallbildung. *Zeitschrift Fur Metallkunde*, **34**, 70–72.

14 MOB (Mobility) Solution Database v. 2.0, Thermo-Calc, Stockholm (1999).

15 Boettinger, W.J., Warren, J.A., Beckermann, C. and Karma, A. (2002) Phase-field simulation of solidification. *Annual Review of Materials Research*, **32**, 163–194.

16 Steinbach, I., Böttger, B., Eiken, J., Warnken, N. and Fries, S.G. (2007) CALPHAD and phase-field modeling: a successful liaison. *Journal of Phase Equilibria and Diffusion*, **28**, 101–106.

17 Uhlenhaut, D.I., Kradolfer, J., Püttgen, W., Löffler, J.F. and Uggowitzer, P.J. (2006) Structure and properties of a hypoeutectic chromium steel processed in the semi-solid state. *Acta Materialia*, **54**, 2727–2734.

18 Boettinger, W.J., Kattner, U.R., Moon, K.-W. and Perepezko, J.H. (2006) *DTA and DSC Heat Flux Measurements of Alloy Melting and Freezing.* NIST Special Publication 960-15, National Institute for Science and Technology, Washington, DC.

19 Jacobs, M.H.G., Hack, K. and Hallstedt, B. (2007) Heat balances and heat capacity calculations. *Journal of Solid State Electrochemistry*, **11**, 1399–1404.

20 Hallstedt, B. (2008) Melting of a tool steel, in *The SGTE Case Book* (ed. K. Hack), Woodhead, Cambridge, pp. 392–397.

21 Püttgen, W., Hallstedt, B., Bleck, W., Löffler, J.F. and Uggowitzer, P.J. (2007) On the microstructure and properties of 100Cr6 steel processed in the semi-solid state. *Acta Metallurgica*, **55**, 6553–6560.

22 Kleiner, S., Beffort, O. and Uggowitzer, P.J. (2004) Microstructure evolution during reheating of an extruded Mg–Al–Zn alloy into the semisolid state. *Scripta Materialia*, **51**, 405–410.

23 Püttgen, W., Bleck, W., Hallstedt, B. and Uggowitzer, P.J. (2006) Microstructure control and structure analysis in the semi-solid state of different feedstock materials for the bearing steel 100Cr6. *Solid State Phenomena*, **116–117**, 177–180.

Part Two

Modelling the Flow Behaviour of Semi-solid Metal Alloys

Thixoforming: Semi-solid Metal Processing. Edited by G. Hirt and R. Kopp
Copyright © 2009 WILEY-VCH Verlag GmbH & Co. KGaA, Weinheim
ISBN: 978-3-527-32204-6

6
Modelling the Flow Behaviour of Semi-solid Metal Alloys[*]

*Michael Modigell, Lars Pape, Ksenija Vasilić, Markus Hufschmidt, Gerhard Hirt,
Hideki Shimahara, René Baadjou, Andreas Bührig-Polaczek, Carsten Afrath, Reiner Kopp,
Mahmoud Ahmadein, and Matthias Bünck*

6.1
Empirical Analysis of the Flow Behaviour

In recent years, the demand for an industry-related and target-oriented enhancement of the semi-solid processing of metallic alloys and composites has grown. Until then, die design and process parameters such as holding time, pressure and piston velocity had been a matter of trial and error, because design rules from classical casting or forging cannot be transferred to thixoforming [1, 2].

As mentioned in previous chapters, metallic alloys in the mushy state consist of solid particles suspended in a liquid matrix. Hence the flow behaviour is comparable to the one of classical suspensions and in principle the methods to evaluate and model the flow properties can be taken from classical suspension rheology. The main differences for metallic suspensions are related to changes in particle diameter due to Ostwald ripening (see Section 6.1.1) and non-isothermal effects such as solid fraction distribution during die filling (see Section 6.1.5).

The flow properties of suspensions are shear-thinning and thixotropic. They depend on solid content, particle shape and particle diameter. Figure 6.1 shows the viscosity as a function of solid content for ideal particles, titanium oxide and soot. The shear-thinning behaviour is displayed in Figure 6.2 for latex particles suspended in water.

The shear-thinning and time dependent flow properties of semi-solid metal alloys were first discovered by Flemings and co-workers [3] in the early 1970s for a low-melting tin lead alloy. Since then, numerous groups (e.g. [2, 4–9]) have determined the rheological characteristics experimentally and set up empirical models which are used for numerical simulation.

The latest investigations concentrated on the flow properties of industrially relevant aluminium and steel alloys whose metallurgical characteristics are described in Chapter 2. In the following, we focus on the flow behaviour of the steel alloy

[*] A List of Symbols can be found at the end of
this chapter.

Thixoforming: Semi-solid Metal Processing. Edited by G. Hirt and R. Kopp
Copyright © 2009 WILEY-VCH Verlag GmbH & Co. KGaA, Weinheim
ISBN: 978-3-527-32204-6

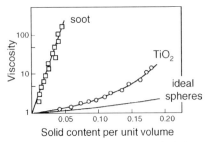

Figure 6.1 Viscosity as a function of solid content and particle shape.

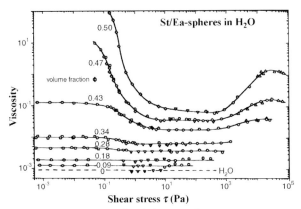

Figure 6.2 Shear-thinning flow behaviour of a suspension.

X210CrW12 complemented by investigations on the aluminium alloy A356. The low-melting tin–lead alloy is just used for supplementary analysis in terms of yield stress of metallic suspensions and structural evolution during rest.

6.1.1
Structural Phenomena Influencing the Flow Behaviour

The complex flow behaviour of suspensions is directly related to their internal structure. Figure 6.3 exemplarily shows for different states of the solid phase how structural changes depend on shear rate and result in changes in viscosity. With increasing shear rate, the particle size may change or particle agglomerates break up. Emulsions (left field in the boxes in Figure 6.3) even allow for a change in particle shape with increasing shear rate. Depending on the shear rate, the structural changes are caused either by structure forces or by hydrodynamic/viscous forces.

In metallic suspensions, the same phenomena can be observed. By decreasing shear rate agglomeration occurs because particles collide and bonds between them are formed. The strength of the bonds increases with time. A further lowering of the shear rate intensifies this process whereas an increase breaks the bonding and the average agglomeration size diminishes. The inner network of a semi-solid alloy billet

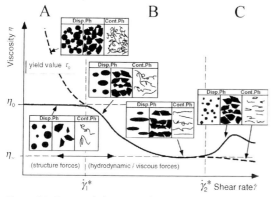

Figure 6.3 Structural changes and corresponding viscosity changes as a function of shear rate.

can reach such stability that it is able to withstand its own weight and can be handled like a solid. As soon as shear is applied, the network in the billet breaks up and the material starts to flow. These structural changes of the solid phase are reflected by the time-dependant flow properties. Quaak *et al.* [10] proposed a two-step model to describe the structural evolution during a shear rate jump and a shear rate drop.

The influence of particle diameter on the viscosity is clearly visible on taking a closer look at the procedure to generate the metallic suspension in a rotational rheometer, subsequently called material preparation (Figure 6.4). First, the alloy is completely liquefied. The subsequent cooling down to the semi-solid state (section I) is followed by an isothermal shearing period of 3 h (sections II and III). During this time, Ostwald ripening (larger particles grow at the expense of smaller ones) takes place, leading to a continuous decay in viscosity. This phenomenon just occurs in a rotational rheometer where the experimental time is orders of magnitude higher than in capillary rheometers or compression tests. Ostwald ripening overlaps each

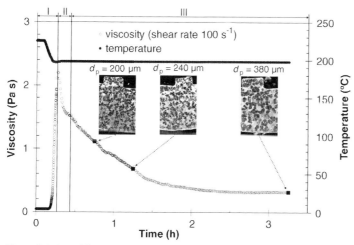

Figure 6.4 Ostwald ripening during material preparation, exemplarily shown for tin–lead.

rheological measurement. Koke and Modigell [11, 12] introduced an approach to predict the influence of convective Ostwald ripening on the viscosity evolution during material preparation in a rotational rheometer. This approach is used here to eliminate particle growth from experimental results just to reflect rheological phenomena.

6.1.2
Mathematical Modelling of the Flow Behaviour

The mathematical approach used to describe experimentally discovered flow phenomena depends on the solid fraction of the sample under investigation. Semi-solid material used for thixocasting with solid fraction between 40 and 60% is described using fluid mechanics whereas material used for thixoforging with higher solid fraction up to 90% requires approaches from solid-state physics [13].

6.1.2.1 Approach from Fluid Mechanics
It was mentioned earlier that in the semi-solid state the globular solid particles are dispersed in a liquid Newtonian matrix and consequently modelling approaches known from classical suspension rheology can be applied. In the equilibrium state, semi-solid alloys are shear-thinning and the most common descriptions are as follows:

Ostwald de Waele:	$\tau = k\dot{\gamma}^n \Rightarrow \eta = k\dot{\gamma}^{n-1}$
Herschel–Bulkley:	$\tau = \tau_0 + k\dot{\gamma}^n \Rightarrow \eta = \dfrac{\tau_0}{\dot{\gamma}} + k\dot{\gamma}^{n-1}$
Cross model:	$\tau = \tau_\infty + \dfrac{\tau_0 - \tau_\infty}{1 + k\dot{\gamma}^n} \Rightarrow \eta = \eta_\infty + \dfrac{\eta_0 - \eta_\infty}{1 + k\dot{\gamma}^n}$

The first two models just differ in the yield stress τ_0, which is a property thoroughly discussed in the literature. Barnes and Walters [14] doubt the existence of a yield stress and claim it to be the stress which cannot be measured due to device-related limits. Contrary to this, Cheng [15] defines a static and a dynamic yield stress, both of which are influenced by the internal structure of the material. In Section 6.2, recent investigations on the low-melting tin–lead alloy using creep tests at very low stresses show the necessity to distinguish between isostructural, dynamic and static yield stress. In addition to the yield stress, the models contain the consistency factor k and the flow exponent n. The stress approaches infinity for very low shear rates and zero for very high shear rates.

In the Cross model, it is assumed that thixotropic fluids effectively behave like a Newtonian fluid in the case of very low or very high shear rates. The viscosity tends towards η_0 for shear rates $\rightarrow 0$ and towards η_∞ for shear rates $\rightarrow \infty$. Atkinson [2] fitted several experimental data for tin–lead to the Cross model but stated that experimental values at the transition to the plateau regions are sparse or even missing.

To account for thixotropic effects in metallic suspensions, the common theories can be divided into three groups [2, 16]:

- Introduction of a scalar parameter (e.g. κ) to describe the degree of agglomeration of solid particles. The limits are $\kappa = 0$ for a completely deagglomerated structure and $\kappa = 1$ for a completely agglomerated structure.
- Direct description of the temporal change of the microstructure by taking into account the number of bonds between solid particles [17].
- Use viscosity–time data to base a theory on Ref. [18].

In the present work, the chosen constitutive equation is a combination of the Herschel–Bulkley law and a thixotropic model, and belongs to the first group in the above list. It is based on an approach initially suggested by Moore [19]. The model, which includes a finite yield stress, is composed for semi-solid alloys at different solid fractions and is fitted to experimental data (see below and e.g. Ref. [11]). The shear stress is defined as a function of shear rate and a time-dependent structural parameter, describing the structural influence on the flow behaviour, such as the current state of agglomeration. Furthermore, the shear stress is assumed to grow exponentially with increasing solid fraction. The equation of state is assumed to be

$$\eta = \frac{\tau}{\dot{\gamma}} = \left[\frac{\tau_0(f_S)}{\dot{\gamma}} + k^*(f_S)\dot{\gamma}^{m(f_S)-1} \right]\kappa + \eta_L \tag{6.1}$$

In the strict sense, η_L does not represent the liquid viscosity. If κ equals zero, the suspension consists of non-interacting particles suspended in the liquid phase. As a consequence, η_L rather describes the viscosity of this suspension, although its contribution to the total stress is negligible.

The yield stress τ_0, the consistency factor k^* and the exponent m are dependent on the solid fraction. The generalized shear rate $\dot{\gamma}$ is defined on the basis of the second invariant of the deformation rate tensor:

$$\dot{\gamma} = \sqrt{2II_D} = \sqrt{2\text{tr}\left(\underline{\underline{D}}^2 \right)} \text{ with } \underline{\underline{D}} = \frac{1}{2}\left(\nabla\underline{u} + \nabla^T\underline{u} \right) \tag{6.2}$$

The structural changes in time follow the deformation history of the material. In the special case of the step change of shear rate experiment (see Section 6.1.5.1), the structural parameter will approach an equilibrium value κ_e corresponding to the current shear rate. It is assumed that the equilibrium flow curve is also a Herschel–Bulkley curve but with another flow exponent n:

$$\eta_e = \left[\frac{\tau_0(f_S)}{\dot{\gamma}} + k^*(f_S)\dot{\gamma}^{m(f_S)-1} \right] \quad \kappa_e = \left[\frac{\tau_0(f_S)}{\dot{\gamma}} + k(f_S)\dot{\gamma}^{n-1} \right] \tag{6.3}$$

By rearranging Equation 6.3, we obtain the equilibrium structural parameter in the form

$$\kappa_e(\dot{\gamma}) = \frac{\tau_0(f_S) + k(f_S)\dot{\gamma}^n}{\tau_0(f_S) + k^*(f_S)\dot{\gamma}^{m(f_S)}}; \quad 0 < \kappa_e < 1 \tag{6.4}$$

The temporal evolution of the structural parameter κ is described by a kinetic differential equation, which is assumed to follow first-order reaction kinetics in the following form (on the left-hand side the substantial derivative of the structural parameter is given):

$$\frac{D\kappa}{Dt} = c(\dot{\gamma})[\kappa_e(\dot{\gamma}) - \kappa(t)] \tag{6.5}$$

where c is the rate constant of approaching the equilibrium value. The rate constant appears from the experiments to be lower for high shear rates and is modelled as

$$c = ae^{-b\dot{\gamma}} \tag{6.6}$$

6.1.3
Approach from Solid-state Mechanics

The essential properties of a material that behaves as a plastic are the existence of a yield point and the occurrence of permanent deformation when the stresses have reached the material's yield point. The simplest description of the plastic behaviour is expressed by the elastic–ideal plastic model (Prandtl and Reuss) and rigid–ideal plastic (Lévy and von Mises) material model. Based on these constitutive equations in the metal forming the elastic-plastic also the viscoplastic constitutive equations are used, which are primarily based on the plasticity theory of von Mises. By modifying these models, the processes with work hardening or softening effects can also be calculated.

The ideal plastic constitutive equation of Lévy and von Mises describes the correlation between stress and deformation by a linear relationship between the strain rate $\dot{\varepsilon}_{ij}$ and the deviatoric stress s_{ij}:

$$\dot{\varepsilon}_{ij} = \dot{\lambda}\, s_{ij} \quad \text{with} \quad \dot{\lambda} = \frac{1}{\tau_0}\sqrt{\frac{3}{2}\dot{\varepsilon}_{ij}\dot{\varepsilon}_{ij}} \tag{6.7}$$

The location-dependent proportionality factor $\dot{\lambda}$ is a function of all components of the strain rate tensor $\underline{\dot{\varepsilon}}$ and the yield stress τ_0. The yield stress is described by the so-called flow curve and depends on the current state of deformation, strain rate and temperature.

Since the yield stress is defined for the uniaxial stress condition, but in most real processes a triaxial stress condition is present, an equivalent stress σ_V is calculated from the stress tensor. This equivalent stress can be compared with the yield stress. If the equivalent stress reaches the value of the yield stress ($\sigma_V = \tau_0$) at a given equivalent strain ε_V and equivalent strain rate $\dot{\varepsilon}_V$, plastic deformation occurs.

$$\sigma_V = \sqrt{\frac{1}{2}\left[(\sigma_{11}-\sigma_{22})^2 + (\sigma_{22}-\sigma_{33})^2 + (\sigma_{33}-\sigma_{11})^2\right] + 3\left(\tau_{12}^2 + \tau_{23}^2 + \tau_{31}^2\right)} \tag{6.8}$$

$$\dot{\varepsilon}_V = \sqrt{\frac{2}{3}\left[\dot{\varepsilon}_{11}^2 + \dot{\varepsilon}_{22}^2 + \dot{\varepsilon}_{33}^2 + \frac{1}{2}\left(\dot{\gamma}_{12}^2 + \dot{\gamma}_{23}^2 + \dot{\gamma}_{31}^2\right)\right]} \tag{6.9}$$

$$\varepsilon_V = \sqrt{\frac{2}{3}\left[\varepsilon_{11}^2 + \varepsilon_{22}^2 + \varepsilon_{33}^2 + \frac{1}{2}\left(\gamma_{12}^2 + \gamma_{23}^2 + \gamma_{31}^2\right)\right]} \tag{6.10}$$

The description of the yield stress is therefore of central relevance since it defines when plastic deformation happens. Furthermore, it represents an important parameter for the calculation of metal forming properties.

At high temperatures, the yield stress is especially dependent on the strain rate. The viscoplastic constitutive equation describes the velocity dependence of plasticity. The theory of the viscoplasticity dates back as early as 1922 and was developed by Bingham [20]. This theory could only be applied to metal forming analyses when the finite element-method (FEM) was introduced as a modification of the rigid-ideal plastic model of Lévy and von Mises [21].

6.1.4
Experimental Investigations to Determine Model Parameters

Experimental investigations of the rheological behaviour of metallic suspensions have been conducted in different types of rheometers, which are mainly capillary or rotational types, and using compression tests. Their range of application for semi-solid alloys together with advantages and disadvantages are given in Table 6.1.

6.1.5
Experimental Setups

6.1.5.1 Rotational Rheometer
The experiments were performed with two self-developed high-temperature Couette rheometers – one for aluminium and the other for steel alloys. The steel rheometer setup is shown schematically in Figure 6.5. The principle setup of the aluminium rheometer is comparable and given elsewhere [22].

When developing the steel rheometer, the main challenge was device-related, because a combination of steel and ceramic elements had to be employed in the rheometer assembly. The measuring device consists of a rotating grooved rod (to

Table 6.1 Advantages and disadvantages of common rheometer for analysing the rheological behaviour of semi-solid alloys.

Capillary rheometer	High shear rates up to $10\,000\,\mathrm{s}^{-1}$
	Conditions as during thixocasting ($f_S < 60\%$)
	Δp measurements difficult
	Material used is reheated into semi-solid state
Rotational rheometer	Large variation of experimental procedure possible
	(e.g. shear rate jumps, shear stress ramps, creep, oscillation)
	Medium shear rates $<1000\,\mathrm{s}^{-1}$
	Solid fraction $f_S < 60\%$
	Material used is cooled to semi-solid state
Compression test	Conditions as during thixoforging ($f_S > 60\%$)
	No exact viscosimetric flow can be operated with wall slip as elongation flow or without wall slip as shear flow
	Simulation necessary to evaluate experimental data

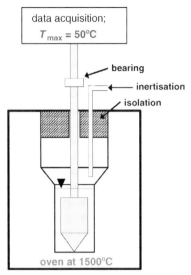

Figure 6.5 Schematic experimental assembly of the steel rheometer.

prevent wall slip) with a diameter of approximately 30 mm and a fixed cylinder with a diameter of approximately 38 mm, both made of aluminium oxide (Figure 6.6). The cup is fabricated using slip casting whereas the rod is made by isostatic pressing.

The measuring device is placed in an oven allowing a maximum temperature of 1700 °C. The rotating rod sticks to a ceramic spindle which in turn is connected to a metallic shaft and joined to the measuring head (data acquisition). Thus the thermal

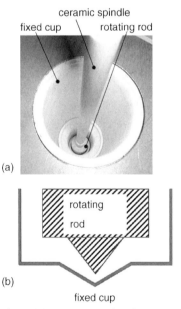

Figure 6.6 (a) View into the Al_2O_3 measuring system and (b) bearing at bottom of cup.

load on the temperature-sensitive measurement module is minimized. The whole spindle of length ~550 mm needs to be pivoted to prevent an eccentric or vibrating rotation. Additionally, the rod is centred in the fixed cup as the tip of the rod (Figure 6.6) rotates in a small depression in the bottom of the cup. The momentum of the bearing is predetermined and subtracted from each measurement. Oxidation of the sample surface is avoided by flushing the void above with preheated argon. The measuring system was calibrated using silicon oil of known viscosity.

During the so-called material preparation (see also Section 6.1.1), the semi-solid material in the rheometer was generated by first completely liquefying solid material followed by cooling to the desired temperature in the mushy state at a fixed cooling rate and a shear rate of $100\,\mathrm{s}^{-1}$. After isothermal shearing for 1 h, the experiments described in Section 6.1.5.1 were carried out.

6.1.5.2 Capillary Rheometer

In capillary rheometry, the fluid is extruded through a slot capillary of a circular or rectangular cross-section at a constant flow rate. When the flow reaches steady-state conditions, the pressure drop Δp along the capillary length L is measured experimentally. Hence the true shear stress at the capillary wall Δ_w can be determined by the following equation:

$$\tau_w = \frac{H \Delta p}{2L} \tag{6.11}$$

In terms of a slit capillary of rectangular cross-section with a flow rate $Q = uHW$, where u is the average velocity of the slurry, H is the height and W is the width of the capillary, the apparent shear rate $\dot{\gamma}_{ap}$ can be determined by

$$\dot{\gamma}_{ap} = \frac{6Q}{WH^2} \tag{6.12}$$

The apparent viscosity η_{ap} of the slurry is defined as

$$\eta_{ap} = \frac{\tau_w}{\dot{\gamma}_{ap}} \tag{6.13}$$

According to Equation 6.11, the viscosity is proportional to the measured pressure drop Δp and inversely proportional to the steady-state flow rate Q. Capillary rheometry is a single-point method, that is, it only provides one viscosity value and its corresponding shear rate for each experiment. The results are extracted under highly transient conditions since the measuring time for these experiments varies in the range of seconds. Hence no steady-state values can be extracted, although the information gathered from these experiments is probably close to isostructural data. Compared with rheological studies which have been carried out using concentric cylinder rheometers (see Section 6.1.4), the length of the experimental time in capillary rheometers is orders of magnitudes smaller and the applied shear rates are orders of magnitudes higher (Table 6.1). Furthermore, experiments in rotational rheometers can be conducted under steady-state conditions and time-dependent effects can be measured using discontinuous or linear variances of the shear rate [23].

The determination of time dependency in capillary flows requires the use of several pressure transducers along the capillary length.

Nevertheless, the experimental concept is advantageous as it is closely related to the thixocasting process. If similar flow conditions occur in real semi-solid processing due to comparable geometric boundary conditions, the results extracted from these experiments are relevant in terms of viscosity, morphology of the flow front and phase separation under laminar flow conditions [24].

Two different types of capillary viscometers have been designed in order to study the rheological behaviour of both Sn–15%Pb alloy and aluminium alloys in the semi-solid state under processing conditions similar to those in real semi-solid metal forming. In capillary rheometry, the partially liquefied alloy is extruded through a slot capillary and the pressure difference and the corresponding flow rate are measured in order to evaluate the viscosity and the shear rate. The capillary must be kept under isothermal conditions to realize appropriate experimental conditions.

6.1.5.3 Vertical Capillary Rheometer

An experimental setup has been designed using a Buhler shot-controlled high-pressure die-casting machine (H-630 SC) with a vertical slot-capillary built into a modular tool, Figure 6.7. Two piezoelectric pressure sensors have been installed along the capillary length. These type f sensors and also an additional inductive flow front sensor are in common use in high-pressure die-casting technology. Regarding the sensitivity of the pressure sensor, the measured flow pressures vary in the region of 1×10^5 Pa while the cavity pressure sensors applied exhibit a maximum of about 2×10^8 Pa with an error in linearity of $\leq$2%. The vertical capillary viscometer can be heated with conventional heat transfer oil to a maximum temperature of 250 °C, which is necessary to ensure isothermal conditions for the experiments with the Sn–15%Pb alloy. Analysing the signal of the two piezoelectric sensors or the signal of

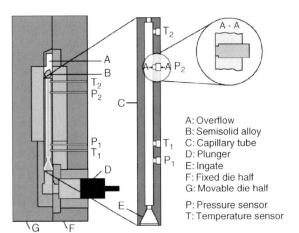

A: Overflow
B: Semisolid alloy
C: Capillary tube
D: Plunger
E: Ingate
F: Fixed die half
G: Movable die half

P: Pressure sensor
T: Temperature sensor

Figure 6.7 Illustration of the experimental setup. The viscometer is integrated in a modular high-pressure die-casting tool using a vertical design of the slot capillary.

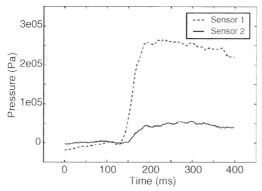

Figure 6.8 Experimental results exhibit a quasi-stationary flow in pressure–time curve although a downward drift is observable for the first sensor located near the ingate. (mean velocity of the slurry $u = 6\,\mathrm{m\,s}^{-1}$).

one pressure sensor and the effective plunger pressure, a quasi-stationary flow is assumed, although a downward drift is observable in the pressure–time curve for the first sensor located close to the ingate system (Figure 6.8).

6.1.5.4 Horizontal Capillary Rheometer

Since the operating temperature of the vertical capillary viscometer is limited to a maximum temperature of 250 °C, the experimental setup is not appropriate for rheological measurements of semi-solid alloys with an excessive solidus–liquidus interval. A horizontal capillary viscometer has been designed particularly for rheological measurements of higher melting alloys such as aluminium and magnesium. The application of electric heating cartridges ensures significantly higher operating temperatures of up to 650 °C while special attention is paid to the heat balance of the tool assembly. The modular design of the viscometer in combination with the design, position and power dimensioning of the electric heating system ensures a continuous flow of the material under investigation through the capillary without solidification occurring during the experiment, and also short setup times through a simple assembly. Compared with the vertical viscometer tendencies of time-dependent viscosity behaviour can be detected with the installation of several sensors over the flow length (Figure 6.9).

Based on the design experience and the thermal requirements, especially the design of the measurement system had to be changed. Quartz sensors, directly mounted on the capillary surface, are in common use in pressure die casting. For sensitivity reasons, these sensors are replaced with highly sensitive force sensors. Because the operating temperature is limited to a maximum of 250 °C, the sensors are mounted in a water-cooled rack on top of the tool assembly. They are connected to the capillary surface with pins of a specific diameter for load transmission (Figure 6.10). In order to detect time dependency in the rheological behaviour, the capillary is equipped with four pressure sensors in an equidistant arrangement over the flow length. If time dependency occurs, the sensors will detect a non-linear

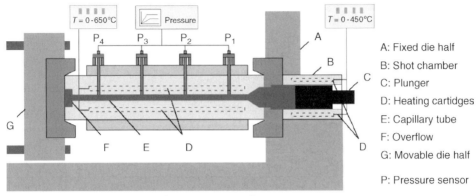

A: Fixed die half

B: Shot chamber

C: Plunger

D: Heating cartidges

E: Capillary tube

F: Overflow

G: Movable die half

P: Pressure sensor

Figure 6.9 Illustration of the horizontal capillary viscometer designed to study the rheological behaviour of higher melting alloys (aluminium, magnesium) and to detect tendencies in time-dependent rheological behaviour.

pressure drop with the measuring length. The entire experimental setup of the horizontal capillary viscometer is shown in Figure 6.11 and results are presented in Figure 6.12 for a flow velocity of $3.55 \, \text{m s}^{-1}$.

The indirect force measurement system has turned out to be a weak point, likely to reduce the performance of the experiment significantly. A sufficient impermeability is required with excellent tribological performance at the same time. Most of the evaluated material combinations could not provide sufficient impermeability due to drilling tolerances or non uniform heat expansion. When performing an experiment, either the system failed in terms of friction or the semi-solid metal infiltrated the still existing gap and blocked the load transmission system as it solidified. Several

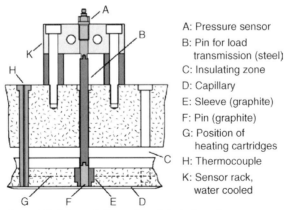

A: Pressure sensor

B: Pin for load transmission (steel)

C: Insulating zone

D: Capillary

E: Sleeve (graphite)

F: Pin (graphite)

G: Position of heating cartridges

H: Thermocouple

K: Sensor rack, water cooled

Figure 6.10 Schematic illustration of the indirect force measurement system. Highly sensitive force sensors are mounted outside the capillary tool. Flow pressure is transmitted by either monolithic or assembled pins.

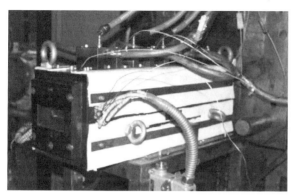

Figure 6.11 Experimental setup. High-temperature capillary viscometer integrated in a Buhler H-630 SC high-pressure die-casting machine with a horizontally designed slot capillary of rectangular cross-section.

systems have been evaluated, such as monolithic steel and carbon pins, multicomponent pins combining a steel body mounted on a carbon or boron nitride tip running in a steel, carbon or boron nitride sleeve. In addition to the pin–sleeve systems, a membrane system was evaluated which provides a complete separation of the load transmission from the semi-solid material. Due to the low flow pressures and the heat impact, the thin membranes of only 50 µm wall thickness failed due to

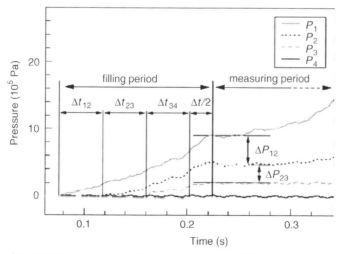

Figure 6.12 Experimental results using four sensors exhibit a quasi-stationary flow in pressure–time curves only with cumulative distance from the ingate system whereas sensors nearby exhibit a continuously increasing flow pressure (velocity $u = 3.55 \, \mathrm{m \, s^{-1}}$).

plastic deformation and facture caused by the heat extension. Each of the characterized systems had specific advantages and disadvantages. As a consequence, the transmission system had to be changed or mechanically reworked after each experiment leading to an extensive experimental effort.

6.1.5.5 Compression Test

The compression test is a conventional testing method to acquire flow stress (yield stress) data for cold and warm forming operations [25]. Generally, cylindrical specimens are compressed and the resulting values for the required force subjected to the reduced height characterize the mechanical properties of the utilized material [26].

The compression test is carried out with a servo-hydraulic press. Figure 6.13 shows the experimental setup of the compression test with the three steps of the complete procedure. In the first step (1), the testing specimen is heated to the required temperature interval in the induction furnace. Thus it is inserted into the preheated testing chamber (2), where it is placed on the lower compression plate. After compression, the specimen can be quenched in a water basin (3). To achieve adequate data for the thixoforming process of X210CrW12, the tests are carried out at different temperatures (900–1320 °C) and strain rates (0.1, 1.0 and 10 s^{-1}).

Cooling of the test specimen during its transport to the upsetting chamber cannot be avoided and is a drawback of this experimental setup. This drawback is being dealt with by slightly overheating the test specimen during the initial heating up, so that the temperature of the test specimen is slightly above the experimental temperature after the transport. The upsetting is conducted in the moment when the temperature of the test specimen has dropped to the desired experimental temperature. The overheating required for this procedure is about 30–40 °C.

The consistency of the semi-solid material requires special preparation of the testing samples above 1220 °C. To avoid crumbling of the specimen that occurs in the molten phase of carbides at the grain boundaries, the specimen is encapsulated in thin tube shells made of the higher melting low-carbon steel S235JR (1.0038) (cylinder, 20 mm diameter × 30 mm height; tube shell, thickness 1.5 mm) [26].

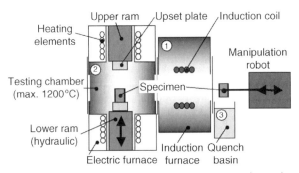

Figure 6.13 Experimental setup of compression test (numbers indicate the steps of the procedure).

6.1.6
Experimental Results and Modelling

6.1.6.1 Rotational Rheometer

The experimental procedures conducted to analyse the flow behaviour of the metallic suspensions include all the varieties listed in Table 6.1.

Shear Rate Jump Experiments   The shear rate jump experiments served to identify steady-state flow curves and also thixotropic effects after a sudden change in shear rate. The influence of solid fraction on the flow behaviour of X210CrW12 is shown in Figure 6.14. After material preparation, the shear rate was decreased to $50\,\mathrm{s}^{-1}$ followed by consecutive shear jumps up to $225\,\mathrm{s}^{-1}$. The shear rate jump experiments show the shear-thinning and time-dependent flow properties of semi-solid X210CrW12. Furthermore, it can be seen that the agglomeration process of solid particles occurring during a decrease in shear rate is slower than the deagglomeration process when shear rate increases.

Taking a closer look at the flow behaviour immediately after a change in shear rate (Figure 6.15), isostructural attributes, a phenomenon discussed in the literature,

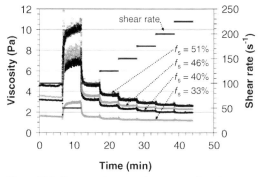

Figure 6.14 Shear rate jumps and corresponding viscosity curves for different solid fractions. Material: X210CrW12.

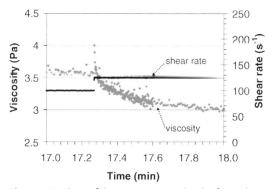

Figure 6.15 Slope of the viscosity immediately after a change in shear rate.

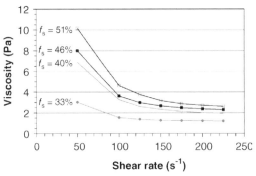

Figure 6.16 Equilibrium viscosity curves for different solid fractions. Material: X210CrW12.

become apparent. The agglomeration state is assumed to remain constant momentarily if the shear rate is suddenly changed from a basis shear rate to a higher or lower one. As a result, Newtonian flow behaviour is anticipated, but, as already observed for tin–lead [12, 27] and aluminium alloys [13, 22]. So far, this phenomenon cannot be explained. In the model approach (Equation 6.3) the flow exponent m describes the isostructural behaviour when κ remains constant.

Figure 6.16 shows the steady-state flow curves generated from Figure 6.14. Figure 6.17 more clearly displays the flow curve for the liquid material ($f_s = 0\%$), which shows as expected Newtonian properties with a viscosity of $\sim$50 mPa s. The high fluctuations in the liquid measurements result from the bearing of the rod as the momentum of the bearing is in the range of that from the liquid metal.

The shear rate jump experiments are used to calculate initial model parameters from Equation 6.3, which in turn can be utilized for the more exact inverse parameter determination described in Section 6.2.4.2. Figure 6.18 exemplarily shows a comparison between experimental data and the modelling approach for a solid fraction of 33%. The model is able to predict the flow behaviour in a reasonable range. The main disagreement is observed for the jump downwards immediately after material preparation. This phenomenon can be discerned also for aluminium or tin–lead alloys but cannot be explained at the moment.

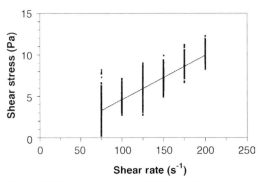

Figure 6.17 Flow curve for liquid X210CrW12.

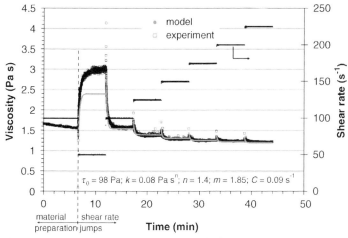

Figure 6.18 Comparison between experimentally determined flow behaviour and modelled flow behaviour.

The parameters k, m and n depend on the solid fraction. Its relationship is plotted in Figure 6.19. The flow exponents n and m show a linear decrease with increase in solid fraction, whereas the consistency factor k increases potentially with f_s.

A comparison of the experimentally obtained flow behaviour of steel alloy X210CrW12 and aluminium alloy A356 for a solid fraction of 46% is given in Figure 6.20. As expected, the principal flow behaviour is the same, just the absolute values of the viscosity are higher for the steel alloy. From precise studies on tin–lead [12], it is known that the viscosity increases with decrease in particle size.

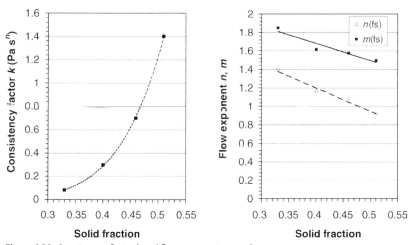

Figure 6.19 Consistency factor k and flow exponents m and n as a function of solid fraction for the steel alloy X210CrW12.

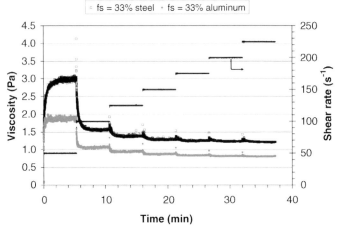

Figure 6.20 Comparison between the flow behaviour of steel alloy X210CrW12 and aluminium alloy A356 for the same solid fraction.

Consequently, it is assumed that the solid particles in the steel alloy are smaller than those in aluminium, which in the present case – according to quenching tests – are around 260 μm. Due to device-related limits, quenching experiments for the steel alloy to prove this assumption are not possible so far.

Yield Stress of Metallic Suspensions As mentioned above, the yield stress is a property extensively discussed in the literature. Creep experiments using a tin–lead alloy performed with the UDS200 Couette rheometer from Anton Paar Germany allowed a closer look at cases in which yield stress can be expected. Due to the presence of the additional bearing of the rod in the aluminium and steel rheometer, these investigations with very small stresses and deformations could not be conducted for the high-melting alloys so far. Anyhow, the results are expected to be qualitatively the same as those obtained for tin–lead, because in principle the rheological behaviour of metallic suspensions is independent of the alloy under investigation.

After material preparation, the yield stress is usually measured using shear stress ramps. During the ramp time, the thixotropic properties of metallic suspensions lead to a change in microstructure. Hence the measured yield stress depends on the slope of the ramp and also on the resting time immediately after material preparation, as illustrated schematically in Figure 6.21a. The change of the internal structure can be measured with oscillation experiments. It can be observed that the evolution of the storage modulus (indicating the increase in internal strength) versus resting time is similar to the slope of the yield stress measured with stress ramps (Figure 6.21b). The build-up of this so-called dynamic yield stress is based on the formation of a particle skeleton in the slurry. This hardening process approaches the static yield stress when the skeleton forms a stable network. Further details on oscillation experiments are given in the next section.

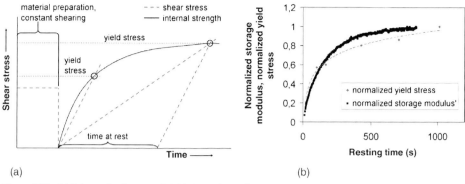

(a)   (b)

Figure 6.21 (a) Schematic diagram of yield stress depending on
the internal structure of the material and (b) comparison
between measured yield stress and measured storage modulus
(internal strength) for Sn–15%Pb alloy.

In addition to Cheng's conclusion [15] one has to differentiate between isostructural, dynamic and static yield stress. The isostructural yield stress occurs immediately
after shearing, since the slurry's structure remains unchanged and it is equivalent to
the structure during shearing. Creep tests with loads up to 10 Pa were performed in
order to determine the yield stress immediately after shearing. As illustrated in
Figure 6.22a, the deformation increases continuously up to a maximum value. This
indicates that the stress at the beginning of the creep test is higher than the stability of
the internal structure. As hardening proceeds, the deformation rate decreases until the
internal strength dominates (equals constant deformation). The time span *t* required
to reach the plateau value increases with increase in shear stress. Plotting the shear
stress against the corresponding time span (Figure 6.22) shows the time-dependent

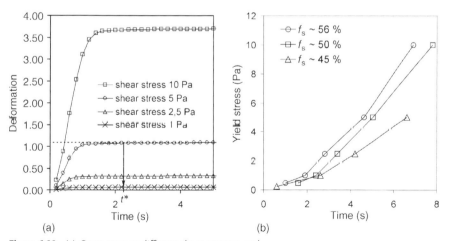

(a)   (b)

Figure 6.22 (a) Creep tests at different shear stresses and
(b) yield stress versus t^* extracted from the creep tests. Material:
Sn–15%Pb.

magnitude of the dynamic yield stress. This enables us to predict the isostructural yield stress which appears at $t = 0$. If the curves are extrapolated to zero, the isostructural yield stress also tends to zero, independent of the solid fraction. As a consequence, no stable particle network exists directly after stirring and therefore semi-solid alloys do not exhibit an isostructural yield stress.

Oscillation Experiments   Oscillation experiments are applied in suspension rheology to gain information about structural changes over a wide range of shear rates. It is known that these changes depend on particle size. In the case of a metallic suspension, the occurrence of Ostwald ripening (see above) is used to adjust different particle diameters at a constant solid fraction. The material preparation shown in Figure 6.4 is stopped after a certain time and oscillation is conducted for about a 30 min. After this, shearing is continued until another oscillation period at a different particle diameter took place. Oscillation was performed within the linear viscoelastic regime with an angular velocity of $25\,s^{-1}$ and an amplitude of $8\,\mu rad$ for the sinusoidal shear rate. Quenching experiments at the beginning and end of the oscillation period showed that particle growth during this time is negligible. The loss modulus G'' and the storage modulus G' were monitored and the loss angle was calculated from the phase shift between signal and stimulation.

The development of the storage modulus together with the loss angle is presented in Figure 6.23. The particle diameter of the metallic suspension was adjusted up to $550\,\mu m$. The storage modulus increases with time during the oscillation period. The loss modulus behaves in the same manner. At the beginning, the slope of the curve is very high. This is an indicator of rapid structural changes of the material. With time the slope decreases and approaches a steady-state value.

The loss angle is the ratio between storage and loss modulus. For pure viscous materials it is $90°$ and for pure elastic materials it is $0°$. Figure 6.24 shows that the material properties change from a 'more viscous-like' to a 'more elastic-like' flow

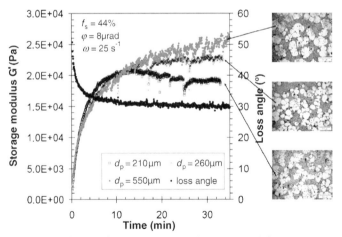

Figure 6.23 Influence of particle diameter on the storage modulus G' and the loss angle δ. Material: Sn–15%Pb.

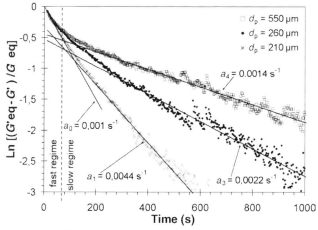

Figure 6.24 Influence of particle diameter on the storage modulus G'. Material: Sn–15%Pb.

behaviour with increasing resting time. Furthermore, it can be seen that the steady-state value of the storage modulus increases with increasing particle diameter, whereas the evolution of the loss angle is independent of the particle diameter.

If the storage modulus is normalized regarding its equilibrium value and plotted in a half-logarithmic diagram versus time, two agglomeration processes can be observed (Figure 6.24). In the first few seconds, a fast process independent of particle diameter with an absolute slope of $a_0 = 0.001\,s^{-1}$ occurs. This is followed by a diameter-dependent slower process with higher absolute slopes between 0.0014 and $0.0044\,s^{-1}$ ($a_1 - a_3$). During the fast process, bonds between particles are formed and the slower process is characterized by hardening of the initially generated bonds.

6.1.6.2 Capillary Rheometer

Experimental Results and Discussion (Sn–15%Pb alloy) The experiments were carried out under isothermal conditions and were designed to verify rheological modelling in terms of closely process-related conditions. At the beginning of each experiment, the alloy, supplied in cylindrical billets of 75 mm diameter and 160 mm length, was heated to the desired experimental temperature using a convection furnace. While heating the billet, the core temperature was recorded using a type K thermocouple. The heating rate was derived from the time–temperature curve as shown in Figure 6.25. The first sets of experiments, for which the Sn–15%Pb alloy was used, were carried out at a solid fraction $f_S = 0.6$. The flow velocity of the slurry u was varied from 2 to $7\,m\,s^{-1}$, which corresponds to a variance in the true shear rate between 1800 and $9200\,s^{-1}$ (Figure 6.26).

Experimental Results and Discussion (Aluminium–Silicon alloy A356) Rheological experiments on semi-solid aluminium alloys are characterized by extremely high tool temperatures. At the beginning of each experiment, the alloy, supplied again in

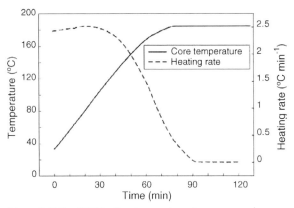

Figure 6.25 Sn–15%Pb alloy: temperature–time curve and heating rate during convection heating of a 75 mm billet of 160 mm length.

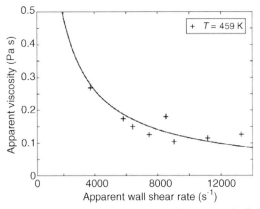

Figure 6.26 Experimental results using the Sn–15%Pb alloy. Apparent viscosity as a function of the apparent shear rate. The regression follows the Herschel–Bulkley model.

cylindrical billets 75 mm in diameter and 160 mm in length, was heated to the desired experimental temperature using a horizontal inductive heating system. While heating the billet, the core temperature was recorded using a type K thermocouple and the heating rate was derived from the time–temperature curve as shown in Figure 6.27. First rheological experiments were carried out using an MHD aluminium–silicon alloy of A356 standard composition (AlSi7Mg0.3). The experiments were performed for two different solid fractions, 0.4 and 0.35. The flow velocity of the slurry u was changed from 1.4 to 3.55 m s^{-1}, which corresponds to a variance in the true shear rate between 1300 and 3200 s^{-1} (Figure 6.28).

In addition to the determined flow curves, the samples which had been removed from the capillary tool after solidification completely demonstrate the flow characteristics of the semi-solid material under the specific experimental conditions

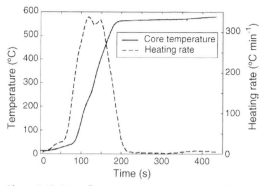

Figure 6.27 A356 alloy: temperature–time curve and heating rate during inductive heating of a 75 mm billet of 160 mm length.

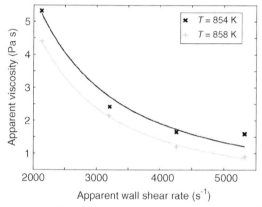

Figure 6.28 Flow curves. Apparent and representative viscosity of an AlSi7Mg0.3 alloy is presented as a function of the apparent shear rate for two different temperatures.

(Figure 6.29). With low flow velocities, the material running in the overflow maintains the shape of the capillary cross-section and crumples, being pressed against the front wall of the die tool (Figure 6.29a). With increasing flow velocity, the continuous flow breaks and loses its specific shape (Figure 6.29b and c).

Two-phase Phenomena The experimental investigations of semi-solid materials or suspensions in general are affected by segregation phenomena. In a capillary rheometer, segregation occurs due to an inhomogeneous heat distribution during induction heating. Additionally, liquid-enriched material is squeezed out from the billet during its initial compression within the shot chamber and concentrates at the flow tip. As a consequence of a strong shear strain of the material when passing the ingate system and entering the capillary, a continuous phase separation occurs during the experimental period. This type of segregation leads to continuous enrichment of solid phase, consequently attended by increasing viscosity. Indicated by a continuously increasing flow pressure, the drifting flow properties interfere with

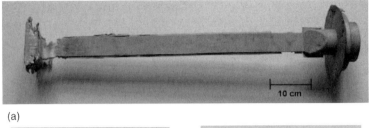

(a)

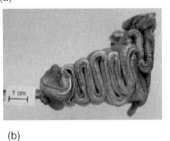

(b)

(c)

Figure 6.29 (a) Samples of the aluminium alloy after solidification. Due to geometric boundary conditions the ingate system consumes most of the billet material, which limits the experimental time for quasi-steady-state flow significantly. (b, c) Material running in the overflow keeps the shape of the capillary cross-section and exhibits a specific flow characteristic at certain shear rates. Examples are given for flow velocities of (a) 2.1 and (b) 3.6 m s^{-1}.

experimentally necessary steady-state conditions. The measurement time is shortened to an analysable constant pressure level lasting for 0.04 s. Second, wall slip is caused by hydrodynamically induced segregation due to pressure and shear velocity gradients perpendicular to the flow direction (Figure 6.30). The above-mentioned

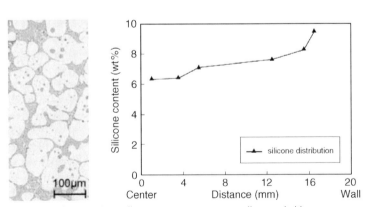

Figure 6.30 Capillary experiments are generally attended by segregation phenomena. Phase distribution can be correlated with the chemical distribution. For an AlSi7Mg0.3 alloy and an initial $f_S = 0.45$ the silicone content increases perpendicular to the flow direction.

Ostwald ripening can be neglected in capillary experiments, because the experimental time is orders of magnitude shorter that the time necessary for particle growth.

Wall slip causes the core flow to slip along the wall on a thin pure liquid layer and falsify the measured relation of rotation speed and turning moment and the flow rate and pressure drop considering rotational and capillary rheometers. Conventional evaluation yields an apparent shear stress gradient which is less steep than it would be in the case of true wall adhesion. In order to reduce or even avoid wall slip, the walls of the rheometer can be grooved evenly (as mentioned in Section 6.1.4). Thereby micro-eddies and thus near-wall mixing are induced and the slip layer is destroyed. However, it is important to note that by that at least near the wall viscosimetric flow can no longer be assumed, which also may influence the measurement results for small capillary heights. It has been observed that wall adhesion increases with an applied treatment to the capillary surface. Roughening of the surface can be achieved either by a mechanical treatment such as grinding or milling of defined structures or by coating the surface. However, the surface treatments reduce but do not completely destroy the liquid layer. Second, wall slip can also be accounted for by subsequent correction of smooth wall measurement results. Beforehand, a correlation of shear stress and slip velocity has to be determined. applying well-established methods. These are based on comparing measurement results from different geometric setups for both capillary and rotational rheometers [23]. However, this procedure is applicable only to steady-state measurements and not to transient measurements.

6.1.6.3 Compression Tests

The material is first pressed via a conventional experimental technique with a standard cylinder specimen starting with hot forming temperatures up to temperatures above the solidus. As shown in Figure 6.31, the flow stresses of the compressed samples decrease drastically when the solidus is passed. Beginning with expected flow behaviour, the material shows less formability when the solidus ($T_s \approx 1220\,^{\circ}\text{C}$) is passed. Metallography confirms this behaviour with the resulting microstructure. Below the solidus, the microstructure shows an austenitic matrix with small areas of dropped out chromium carbides. Depending on the position, a forming structure appears. Above the solidus, eutectics occur at the grain boundaries, which grow with increasing temperature and cause debilitation of the microstructure against tensile stresses. The test specimen disintegrates in this condition, even when only slightly loaded mechanically, so that a precise flow curve cannot be calculated because of the missing cross-sectional area. To avoid the disintegration of the test specimen, it is covered with a thin layer of a higher melting material.

This method to determine flow curves of such materials is the so-called 'coat test' [29]. The early development of cracks is inhibited by the hydrostatic pressure caused by the shell. During evaluation of flow curves, the effect of the shell on the measured force is determined by calculation (inverse modelling). The prerequisite for this is the knowledge of the flow curve of the shell material.

Figure 6.32 shows the samples and the shell in its initial geometry and after forming. Even a compression with a high deformation can be carried out without disintegration of the material.

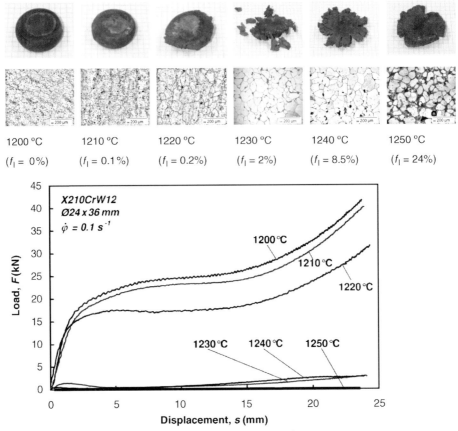

1200 °C	1210 °C	1220 °C	1230 °C	1240 °C	1250 °C
(f_l = 0%)	(f_l = 0.1%)	(f_l = 0.2%)	(f_l = 2%)	(f_l = 8.5%)	(f_l = 24%)

Figure 6.31 Results of the compression test for steel grade X210CrW12 around the solidus temperature: shape of the compressed specimens with their microstructures and the resulting load curves [28].

Figure 6.32 Specimen and coat for avoiding disintegration, compressed sample and cross-section of a compressed sample (X210CrW12/S235JR).

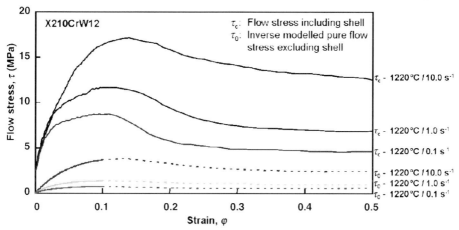

Figure 6.33 Experimentally determined flow stress τ_c of shell specimen and inverse modelled flow stress τ_0 of X210CrW12 for three strain rates.

The resulting flow curves of this modified method still include the influence of the bulging and the shell material, therefore they are indicated as τ_c (c = composite) (Figure 6.33). To obtain a precise flow curve without these influences, a further data processing step is necessary. For this purpose, the inverse modelling method using a FEM simulation with LARSTRAN/SHAPE (see Section 6.2.3) is applied. In the simulation, the unknown flow stress of the semi-solid material is given as mathematical equations with a set of parameters ($m1$–$m8$) and the flow stress of the shell material is given as a table form from the separately executed measurements. Two equations are used for different strain ranges (φ) (low strain range 0–0.18 and high strain range 0.19–0.7) to describe the characteristic form of the flow curves:

$$\tau_{1(\varphi:0-0.18)} = m1\dot{\varphi}^{\,m2}\varphi^{m3}e^{\varphi m4}, \quad \text{for } 0 < \varphi < 0.18 \tag{6.14}$$

$$\tau_{2(\varphi:0.19-0.7)} = m5\dot{\varphi}^{\,m6}\varphi^{m7}e^{\varphi m8}, \quad \text{for } 0.19 < \varphi < 0.7 \tag{6.15}$$

The values of thermophysical material properties are based on the literature [30]. The initial parameter values of the equations for the inverse modelling are determined by regression of τ_c for each test condition. For the inverse modelling, the upsetting simulations are repeated until the simulated load curve of the shell specimen matches that of the experiment while the material parameters of the equations are varied step by step. The determined pure flow curves τ_0 of X210CrW12 are shown in Figure 6.33 together with the resulting τ_c [29].

The flow stress shows a stationary value after the completion of the structural breakdown. Moreover, the series of upsetting tests reveals the main influence factors of the flow stress. The flow stress depends predominantly on the state of the microstructure. However, the dependence of the temperature over the range of hot and semi-solid forming plays a very important role for technical practice. The direct comparison of the values of the flow stress in the semi-solid state and the hot forming

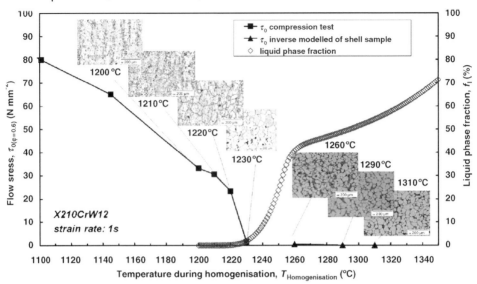

Figure 6.34 Overview about the flow stress (steady values) and liquid phase fraction depending on temperature with related micrographs.

area emphasizes the significant dependence of the globular microstructural formation (Figure 6.34).

6.2
Numerical Modelling of Flow Behaviour

6.2.1
Motivation

When filling the die cavities, the semi-solid alloy shows complex shear-thinning thixotropic flow behaviour, influenced by the occurrence of yield stress and segregation effects. Additionally, the flow is greatly affected by the change in the slurry temperature causing changes in solid fraction distribution, and also solidification effects in the contact of the material with the die wall (see Section 6.1.1). Since the mechanical properties of the thixoformed components are strongly influenced by the spatial distribution of the solid content, the mentioned phenomena directly influence the quality of the final product, The result of the work of Noll *et al.* [31] is that the degree of segregation is influenced by the geometry and the process conditions, and therefore the die geometry and also the process conditions have to be adjusted in such a way that segregation is minimized. So far, the development of new dies and thixoforming processes has been done experimentally, by trial and error. This way of designing new dies leads to high development costs, which set a demand for improvement of the process by use of numerical modelling, in terms of parameter determination and process optimization.

Based on the solid fraction of the metallic suspension, one can distinguish between the two groups of slurries and of corresponding models. When the solid fraction in the semi-solid slurry ranges from 20 to 90%, suspensions above 60% behave as non-linear viscoplastic solids and are used in forging processes, and metal slurries below a solid fraction of 60% are used in casting processes and exhibit a non-Newtonian, history-dependent flow behaviour that is referred to as pseudoplasticity and thixotropy. Consequently, there are basically two groups of models to describe semi-solid slurries: for highly concentrated suspensions behaving as solid bodies, models based on solid body mechanics are used; and for slurries with solid content less than 60% that behave like non-Newtonian suspensions, models based on computational fluid dynamics (CFD) are use. The models based on solid body mechanics are explained in Section 6.1.2.

Models based on CFD can be classified according to whether the modelling is one-phase, where the material is assumed to be continuum, or two-phase, where both liquid phase and globular particles are taken into account. Furthermore, classification can also be made based on discretization methods, that is, methods of approximating the differential equations by a system of algebraic equations such as finite element (FE), finite difference (FD) and finite volume (FV) models.

Another important issue when dealing with such flows is the proper determination of the interface position in time. There are basically two methods for dealing with this issue: moving mesh strategies and fixed mesh approaches. The first group uses some form of the mesh moving strategy, such as Lagrangian or arbitrary Lagrangian–Eulerian (ALE) techniques. Due to the possibility of mesh distortion, such strategies often need to be complemented by a remeshing strategy. The second approach uses a fixed fluid mesh, supplemented by an appropriate strategy for tracking the fluid–solid (fluid–air) interfaces.

The main difficulty when modelling thixoforming processes is to take into account all the complex phenomena occurring within the flow. Some of the models refer to the thixotropic behaviour only, where metals are treated as homogeneous materials [32]. The others take into account the two-phase flow, but neglect the thixotropy. Other authors, while considering thixotropy, neglect the heat transfer during the process. Alexandrou *et al.* [33] implemented the heat transfer and the phase change together with thixotropy in a commercial simulation package, but did not take the phase segregation into account.

The literature on numerical models of thixoforming was recently reviewed by Atkinson [2].

6.2.2
Numerical Models Used

6.2.2.1   Material Models
The basic point of any numerical method is the mathematical model, that is, the set of partial differential or integro-differential equations and boundary conditions (an appropriate model for the target application, incompressible, inviscid, turbulent; two-

or three-dimensional, etc.) [34]. Here, the one- and two-phase models based on the Herschel–Bulkley approach will be presented.

One-phase Model The thixotropic model applied is based on an approach initially suggested by Moore [19] and described in Section 6.1.2.1. The model, which includes a finite yield stress, is composed of semi-solid alloys at different solid fractions and is fitted to experimental data [35]. The shear stress is defined as a function of shear rate and a time-dependent structural parameter, describing the structural influence on the flow behaviour, such as the current state of agglomeration. Furthermore, the shear stress is assumed to grow exponentially with increasing solid fraction.

Two-phase Model This numerical approach is based on a two-phase modified Herschel–Bulkley model, considering the yield stress, shear thinning and thixotropic behaviour (see Section 6.1.2.1). It describes the metal alloy as a mixture of a continuous matrix liquid phase and dispersed solid particles regarded as a pseudo-continuum. The mathematical model uses the two-phase approach, with two sets of conservation equations for the bulk and liquid phases, where the relation between solid and liquid phase is modelled by Darcy's law [9]. The mass and momentum conservation equations are solved for the bulk and liquid phases. The mass conservation equation of the bulk phase is

$$\nabla \cdot \underline{u} = 0 \tag{6.16}$$

The velocity u is defined as the sum of the velocities of individual phases (u_S, solid phase velocity; u_L, liquid phase velocity) weighted by the corresponding fractions of solid, f_S, and liquid, f_L:

$$\underline{u} = f_S \underline{u}_S + f_L \underline{u}_L \tag{6.17}$$

The bulk flow is governed by the general momentum conservation equation in the classical form:

$$\rho \frac{D\underline{u}}{Dt} = -\nabla p + \nabla \cdot \underline{\underline{S}} + \rho \underline{g} \tag{6.18}$$

with the solid-phase density ρ_S, the liquid phase density ρ_L and the isotropic pressure p; g is the acceleration due to gravity. The non-Newtonian flow behaviour is expressed by the extra stress tensor S:

$$\underline{\underline{S}} = 2\eta_B \underline{\underline{D}} \tag{6.19}$$

where D is the deformation rate tensor and η_B describes the viscosity of the bulk phase (Equation 6.1). For the liquid phase, the momentum conservation equation has the following form:

$$\rho_L \frac{D\left(f_L \underline{u}_L\right)}{Dt} = -f_L \nabla p + \nabla \cdot \left(f_L \underline{\underline{S}}_L\right) + \rho_L f_L \underline{g} + \underline{q} \tag{6.20}$$

with the liquid density ρ_L, the isotropic pressure p the extra stress tensor for the liquid phase S_L and the Newtonian viscosity of the liquid phase η_L. The specific interface friction force q is proportional to the 'slip' velocity difference:

$$\underline{q} = C_{LS}(f_S)\left(\underline{u}_S - \underline{u}_L\right) = \frac{C_{LS}(f_S)}{f_S}\left(\underline{u} - \underline{u}_L\right) \tag{6.21}$$

For the high solid fractions occurring during thixoforming, the momentum conservation equation of the liquid phase reduces to Darcy's law:

$$\left(\underline{u}_S - \underline{u}_L\right) = \frac{f_L}{C_{LS}(f_S)}\nabla p \tag{6.22}$$

The non-isothermal phenomena are described by the energy equation, coupled with a temperature–enthalpy and solid fraction relation, which is used in the following form:

$$\frac{Dh}{Dt} = \nabla(\lambda \nabla T) + \eta_B \dot{\gamma}^2, h(T) = f_S \int_{T_{ref}}^{T} c_S \rho_S dT + (1 - f_S)\int_{T_{ref}}^{T} c_L \rho_L dT + (1 - f_L) \tag{6.23}$$

with enthalpy h, thermal conductivity λ and temperature T; c_S and c_L are the specific heat capacities of the solid and the liquid phase, respectively. The time evolution of the solid fraction, f_S, is described by the mass conservation equation for the solid phase:

$$\frac{\partial f_S}{\partial t} + \nabla\left(f_S \underline{u}_S\right) = \frac{d\tilde{f}_S}{dT}\frac{DT}{Dt} \tag{6.24}$$

where the right-hand side describes the rate of phase change based on the Scheil equation. Influencing the solid fraction, the thermal changes are then reflected in the rheological Equation 6.1, which affects the flow of both phases.

6.2.2.2 Discretization Methods (Numerical Solution Techniques)

Discretization methods are numerical techniques, approximations for solving partial differential equations. that is, methods of approximating the differential equations by a system of algebraic equations for the variables at some set of discrete locations in space and time. There are three distinct streams of numerical solution techniques: finite difference (FD), finite volume (FV) and finite element (FE) methods.

Finite difference methods describe unknowns ϕ of the flow problem by means of point samples at the node points of a grid coordinate lines. The truncated Taylor series expansions are often used to generate finite difference approximations of derivatives of ϕ in terms of point samples of ϕ at each grid point and its immediate neighbours. Those derivatives appearing in the governing equation are replaced by finite differences. yielding an algebraic equation for the values of ϕ at each grid point.

The finite volume method is based on the control volume formulation of analytical fluid dynamics, where the domain is divided into a number of control volumes, while the variable of interest is located at the centroid of the control volume. The governing equations in differential form are then integrated over each control volume. Interpolation profiles are then assumed in order to describe the variation of the concerned variable between cell centroids. The resulting equation is called the discretized or discretization equation. In this manner, the discretization equation expresses the conservation principle for the variable inside the control volume.

The method is used in many computational fluid dynamics packages. The most compelling feature of the FVM is that the resulting solution satisfies the conservation of quantities such as mass, momentum, energy and species. This is exactly satisfied for any control volume and also for the whole computational domain and for any number of control volumes. Even a coarse grid solution exhibits exact integral balances. Therefore, FVM is the ideal method for computing discontinuous solutions arising in compressible flows. Since finite volume methods are conservative, they automatically satisfy the jump conditions and hence give physically correct weak solutions. FVM is also preferred while solving partial differential equations containing discontinuous coefficients.

In the *finite element method* approach, the actual structure is assumed to be divided into a set of unstructured discrete subregions or elements which are interconnected only at a finite number of nodal points. The distinguishing feature of the finite element method is that the equations are multiplied by a weight function before they are integrated over the whole domain. The properties of the complete structure are found by evaluating the properties of the individual finite elements and superposing them appropriately. Due to the ability to model and solve large complicated structures and to deal with arbitrary geometries, the method is widely used in computational science.

6.2.3
Software Packages Used

Within this work, the well-known commercial software packages FLUENT, Flow-3D and MAGMASOFT were used, in addition to the 'home-made' numerical solver PETERA and LARSTRAN/SHAPE.

- *FLUENT* is a CFD software package to simulate fluid flow problems. It uses the finite-volume method to solve the governing equations for a fluid. It provides the capability to use different physical models such as incompressible or compressible, inviscid or viscous, laminar or turbulent and so on. To determine the phase boundary in multiphase flows, and in particular free surface (or wave) flows, the so-called volume of fluid (VOF) model is used. Motion of fluid interfaces based on the solution of a conservative transport equation for the fractional volume of fluid in a grid cell is the method employed in FLUENT.

- *Flow-3D* is a general-purpose CFD program with many capabilities, claiming to be particularly good for free surface flows. Finite-difference or finite-volume approximations to the equations of motion are used to compute the spatial and temporal evolution of the flow variables. In FLOW-3D, free surfaces are modelled with the VOF technique, incorporating some improvements beyond the original VOF method to increase the accuracy of boundary conditions and the tracking of interfaces. One refers to this implementation as TruVOF.

- *MAGMASOFT*, based on the finite difference method, is a numerical simulation tool for castings. It features the full process from mould filling to solidification. In addition to the filling behaviour, volume shrinkage, thermal stresses and also distortions can be calculated. It allows an iterative design of different casting processes such as high-pressure die casting, thixocasting, sand casting and chill casting, whereas both geometry (cast, feeder, cores, etc.) and casting parameters (temperature, velocity, etc.) can be easily varied. MAGMASOFT is the most established casting-simulation software worldwide.

- *PETERA* numerical code is based on the Lagrange–Galerkin approach that uses a special discretization of the Lagrangian material derivative along particle trajectories with a Galerkin finite element method [36]. The numerical model uses an analytical solution of rate equations along segments of particle trajectories in addition to backward particle trajectories tracking in time. The thermal energy conservation equation is coupled with the mass conservation equation for the solid phase and solved simultaneously in the time stepping process of the numerical algorithm. The equations are solved for the prime unknowns in the model equations: temperature, enthalpy and solid fraction. The use of f_S as an independent variable allows additional complicated equations to describe temperature–enthalpy relations for metals in the mushy state to be avoided. The calculated f_S is used in the flow equations of the bulk and liquid phases. The numerical algorithm is based on the fixed-grid methods, in which the free surface is at an unknown position between the nodes. The determination of the free surface position in time is based on a modified pseudo-density approach. The procedure of activating/deactivating mesh elements in the transient process of filling/emptying the elements makes the algorithm less time consuming and more resistant to numerical diffusion.

- *LARSTRAN/SHAPE* is a specialized FEM code for the simulation of metal forming. It is based on a Lagrange formulation and implicit time integration. Both elastic–plastic and rigid–plastic material behaviour can be modelled, including full thermomechanical coupling (dissipation and phase changes) and heat transfer.

6.2.4
Numerical Examples

6.2.4.1 One-phase, Finite Difference, Based on Flow-3d
The selected example is a die filling of a steering axle using aluminium alloy A356. A steering axle is tested in order to study the influence of different parameters on the die filling. It was found that above a critical inlet velocity, the filling is no longer laminar. In Figure 6.35, a comparison of the numerical simulation and the experiments performed in terms of flow front development for two different inlet velocities ($v_1 = 250\,\mathrm{mm\,s^{-1}}$ and $v_2 = 500\,\mathrm{mm\,s^{-1}}$) is shown. The results show good agreement

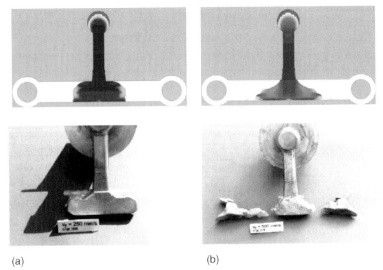

(a)   (b)

Figure 6.35 Experiment versus one-phase simulation: comparison of the experimentally obtained and simulated flow front for two piston velocities, (a) $v_1 = 250\,\mathrm{mm\,s^{-1}}$ and (b) $v_2 = 500\,\mathrm{mm\,s^{-1}}$.

in the terms of contour of the front, and the transition between laminar and turbulent filling can be represented reasonably well.

Although the developed model should be suitable for any globular semi-solid material, the disadvantage of this approach is that the parameters have to be adjusted for each different filling case. In the present simulation, the parameters for the model were adjusted empirically during the simulation study. The simulation gives reasonable results for flow front, but reasonable results for the pressure cannot be obtained. Another disadvantage when using a single fluid simulation is the inability to simulate structural changes within material caused by segregation and solidification. Therefore, two-phase simulation is necessary.

6.2.4.2 Two-phase, Finite Element Based on PETERA

As a two-phase numerical example, simulation of the simple T-shaped die filling experiment is presented. The specially designed die filling experiment (Figure 6.36) permits the continuous observation of the flow pattern. The pressure is measured with a pressure sensor, which is placed near the inlet of the vertical bar. The die is placed inside an oven, which guarantees isothermal conditions in the die and allows the adjustment of different die temperatures. Figure 6.36 also shows the container, carrying the metallic alloy, which is connected to the lower side of the die.

The experiments were performed at different filling velocities and die temperatures. After each experiment, a chemical analysis of the spatial distribution of the lead content in the produced part was made. With this lead content and the two-phase diagram for tin–lead alloys, the local solid fraction is determined.

Before being able to simulate the filling process, parameter determination has to be carried out.

Figure 6.36 Experimental setup, T-shaped die.

Model Parameter Determination and Adjustment   The model parameters m, n, τ_0, k and C (see Equation 6.1) were determined on the basis of the step change shear rate experiments and also shear rate ramp experiments at a reference solid content $f_{S,ref.}$ of 52% The rheological experiments led to reliable results for the yield stress τ_0, the flow exponent n and the consistency coefficient k. These model parameters are related to steady-state conditions. The determination of the isostructure flow exponent m and the rate constant c (Equation 6.5), which describe the transient behaviour of the material, suffers from a slight uncertainty.

Ordinary die filling processes do not last longer than approximately 0.5 s. During this time, the semi-solid alloy undergoes a lot of shear rate changes and jumps. Hence the time span for a shear rate jump is in the range of milliseconds. So far there is no possibility of acquiring viscosity values at these time scales with conventional rheometers. A further problem arises from the permeability coefficient k_P used in the two-phase model.

In order to overcome these problems, a special approach of parameter adjustment has been developed (Figure 6.37). The aim is to achieve congruence between numerical and experimental results. The filling experiments are carried out under isothermal conditions. Varying the filling velocity and the initial solid fraction of the metallic suspension leads to different filling pressures, flow patterns and spatial distributions of solid content. The simulation starts with a set of parameters determined on the basis of the rheological experiments performed and a first guess of the permeability coefficient. As long as no congruence of the flow front contour, the solid fraction distribution and the filling pressure is achieved, a new simulation with modified flow exponent m, rate constant c and permeability coefficient k_P is carried out.

Two-phase Simulation   Using the obtained set of material parameters, the T-shaped die-filling experiment described above was simulated using the two-phase model implemented into the solving algorithm PETERA. In order to show that the flow is mainly influenced by segregation effects, one-phase simulations considering yield

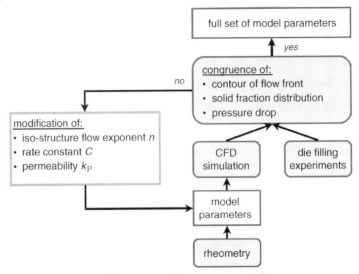

Figure 6.37 Procedure for parameter determination and adjustment.

stress, shear thinning and thixotropic behaviour are also performed. Figure 6.38 shows a comparison of the experimentally obtained flow patterns, and the flow patterns obtained using one- and two-phase simulations. The isothermal experiments were carried out at piston velocities of 10, 50 and 100 mm s^{-1}, an initial solid

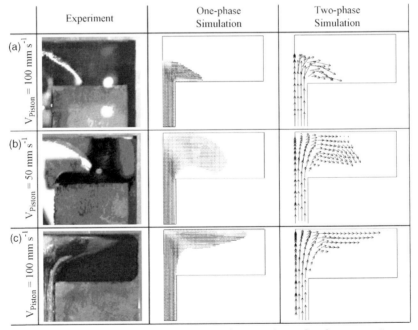

Figure 6.38 Isothermal experiment, one- and two-phase simulation: flow front comparison.

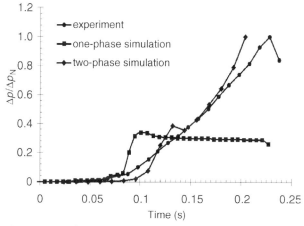

Figure 6.39 Isothermal experiment, one- and two-phase isothermal simulation: pressure comparison.

fraction of 52% and material and die temperature of 192 °C. One can see that the calculated shapes of the flow front at different filling velocities are in good agreement with the observed flow fronts (Figure 6.38). Problems arise, however, in the case of one-phase simulation, where using one set of material parameters it was not possible to achieve satisfactory results for different piston velocities. The initial set of parameters gave reasonable results only for one piston velocity, and for other velocities there was no agreement between the experimental and numerical results and the parameters have to be readjusted.

Figure 6.39 shows a comparison of the calculated and measured pressure. Although it was possible to achieve good agreement between the one-phase simulation and the observed flow front, this simulation gives no reliable results for pressure. In contrast to the experiment, the simulation yields no further increase in pressure during filling of the horizontal part but rather a nearly constant pressure. This is in accordance with the theoretical assumption for free jet flows but not with the experimental results. The wrong calculation of the pressure during the filling of the horizontal duct is caused by the disregard of the segregation.

The two-phase simulation of this filling process shows good agreement of the measured and calculated pressure course (Figure 6.39). The pressure course increases until the end of the filling process because segregation effects were taken into account. Segregation first occurs when the material enters the vertical duct. Then the solid content of the flow front is rather low and no significant segregation effects in the flow front appear while it moves through the die. Hence the solid fraction of the material increases while flowing through the vertical duct, which results in higher viscosity values and thus in a continuous increase in pressure (Figure 6.39).

The analysis of the solid fraction distribution showed that significant segregation occurs at high initial solid fraction (Figure 6.40). At the beginning of the filling process, the liquid phase is squeezed out of the billet like water out of a sponge. Hence

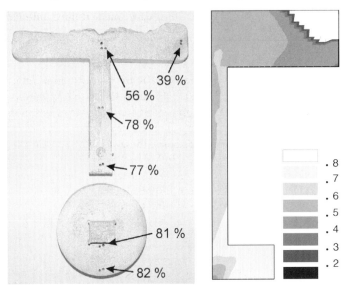

Figure 6.40 Isothermal experiment versus two-phase isothermal simulation: phase segregation.

the flow front is formed by material with a low solid content, which leads to the observable less smooth surface of the contour.

Another important issue is solidification occurring on contact of the material with the die walls. To investigate this phenomenon, non-isothermal experiments and simulations are carried out. Non-isothermal experiments were carried out at a piston velocity of $50\,\mathrm{mm\,s^{-1}}$, an initial solid fraction of 52%, a material temperature of $192\,°C$ and a die temperature of $40\,°C$. Figure 6.41 shows the flow patterns

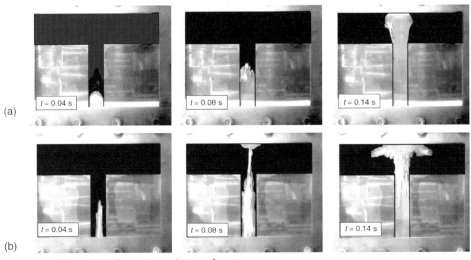

Figure 6.41 Flow patterns during a forming process:
(a) isothermal and (b) non-isothermal experiment.

of (a) an isothermal and (b) a non-isothermal filling experiment at three different times.

Under isothermal conditions, the surface of the flow front is relatively smooth. At the point where the flow direction changes from vertical to horizontal, the material temporarily detaches from the die walls (Figure 6.41a). In the non-isothermal filling process, the material forms a less smooth surface (Figure 6.41b). At the beginning, the semi-solid alloy flows through the vertical part of the die as a thin jet with almost no contact with the die walls. It reaches the upper wall of the die earlier than in the isothermal case. The contour of the flow front leads to the conclusion that the solid fraction in this area is rather low. Starting at the inflow to the vertical duct, the hot material solidifies at the colder walls, which causes a reduction in the free cross-section. Hence the flow velocity rises and a jet is formed. Additionally, the pressure drop increases, which leads to a higher degree of segregation.

The observed results were compared with numerical simulations. Figure 6.42 shows the simulated solid fraction and the flow front in the lower vertical duct for both isothermal and non-isothermal cases. The difference in the shape of the flow front for isothermal and non-isothermal simulation is comparable to the experimental results. It can also be seen that solidification occurs in contact with the cold walls and that the thickness of the solidified layer increases during the die filling. Figure 6.43 shows the influence of the solidification on the velocity field in this region. Due to the small free cross-section, the degree of segregation is considerably large. In the simulation, material with a low solid content flows through the vertical duct and fills the horizontal part. This corresponds to the measured low solid content in the horizontal duct. With the optimized set of model parameters, the simulated flow front and the predicted segregation agree qualitatively with experimental data.

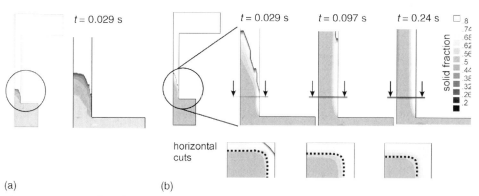

Figure 6.42 Numerical results: (a) isothermal simulation and (b) non-isothermal simulation with solidification effects near the die walls. The dotted lines indicate the growth of the solid layer thickness during the filling.

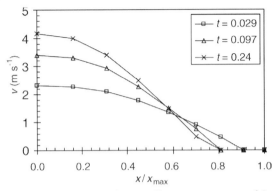

Figure 6.43 Simulated velocity profile in a cross-section of the vertical duct.

6.2.5
Summary

The one-phase computational model based on the Herschel–Bulkley approach takes into account yield stress, thixotropy and shear-thinning behaviour, and can be used in some cases to simulate the flow front of the die filling experiments, but the parameters have to be adjusted for each set of process conditions. Since this model makes the approximations that material is homogeneous, the results cannot always be physically meaningful. The mathematical model based on the two-phase approach, in contrast, takes into account all time-dependent phenomena occurring during casting but, most importantly, it accounts for non-isothermal phase change in addition to phase segregation, which is one of the important issues and often the main reason for an unsuccessful casting.

6.3
Simulation of Cooling Channel

Simulation of feedstock formation for thixoforming has attracted little interest compared with the simulation of thixotropic-behaviour, and thixoprocessing of semi-solid metals [2]. The aim of this work was to study the effect of the different process parameters on the quality of the slugs obtained from the cooling channel [37]. Grain density and size, solid fraction and macro-solute distributions were used as quality measures.

Simulation was carried out with the CFD software FLUENT 6.3 combined with the multiphase flow and solidification simulation code MeSES, which was first developed at the Foundry Institute at Aachen University within the scope of SFB370. Application of the MeSES code requires many thermophysical parameters which were obtained from the literature, while others that were not available were determined experimentally.

6.3.1
Model Description

Three phases are involved in the cooling channel semi-solid process: the liquid melt, the solidifying grains and air. The volume averaging approach is employed to formulate the conservation equations of mass, momentum, species and enthalpy for the three phases. More details can be found elsewhere [38]. The air has no mass transfer with the other phases, therefore the source term for the mass and species exchange with air was set to zero. The description of the transport equations and the definitions of the corresponding source terms are available in detail in the literature [39–41].

An undercooling-dependant grain-density model according to Thevoz and Rappaz [42, 43] was adopted, where the grain density can be calculated using the Gaussian distribution, according to the following equation:

$$\frac{dn}{d(\Delta T)} = \frac{n_{max}}{(2\pi)^{\frac{1}{2}}\Delta T_\sigma} e^{-\frac{1}{2}\left(\frac{\Delta T - \Delta T_N}{\Delta T_\sigma}\right)^2} \tag{6.25}$$

where the characteristic parameters ΔT_N, ΔT_σ and n_{max} are the mean undercooling, standard deviation of undercooling, and maximum grain density, respectively. The parameters were determined experimentally as explained later. The under cooling value, ΔT, is calculated for each control volume at each time step. If ΔT of the current time step is greater than that of the preceding one, then nucleation is allowed and the newly formed nuclei will be added to the source term.

The presence of air in this model increases its flexibility as it can substitute metal shrinkage, heat and drag force exchange, which makes the model closer to the real metal flow and solidification. On the other hand, including air increases the computational time and requires a finer mesh to account for the distinct interface topology. The densities of both solid and liquid phases are temperature dependent to account for the thermal convection that takes place during solidification and particularly in the container.

6.3.2
Determination of Grain Nucleation Parameters

The aim of this part is to determine the characteristic parameters required for the grain nucleation model implemented in MeSES and to investigate the other models provided in the simulation code. An experimental setup was built according to Refs [42, 44]. Pre-experiments coupled with casting simulation using MAGMASOFT were conducted to optimize the construction of the experimental setup for the grain refined aluminium alloy A356 with the average composition given in Table 6.2. The liquidus temperature of A356 was 614.6 °C after Scheil simulation using Thermo-Calc. More details of the experimental description and results can be found elsewhere [45].

From the cooling curves, which were recorded at different cooling rates, the maximum undercooling ΔT_{max} corresponding to the recalescence temperature was

Table 6.2 Results of spectrometric analysis for aluminium alloy A356.

	Element											
Si	Fe	Cu	Mn	Mg	Zn	Ni	Cr	Ti	Ag	Sr	V	Ga
wt% 7.19	0.148	0.014	0.010	0.358	0.012	0.007	0.005	0.082	0.001	0.035	0.009	0.008

determined. With the help of image analysis, the grain density was deduced at each thermocouple. The accumulative density function was fitted to the n–ΔT_{max} plot and correspondingly the Gaussian curve was generated. The characteristic nucleation parameters obtained are summarized in Table 6.3.

The nucleation parameters obtained, together with the experimental boundary conditions, were used to investigate the models provided within MeSES code numerically. An axisymmetric grid cell was constructed and the Eulerian multi-phase approach implemented in FLUENT was applied. The simulation results exhibited good qualitative and quantitative agreement with the experimental results.

6.3.3
Process Simulation

Based on the previous simulation work of Wang *et al.* [46], a new 2D grid (Figure 6.44) was built with a finer mesh. Two solid zones (channel and container walls) and two fluid zones (channel and container domains) were employed. The thermophysical properties provided in Ref. [45] were utilized together with the boundary conditions from the experiments.

The process variables during experiments and simulation were the inlet temperature, T_i, inlet velocity v_i, channel wall temperature $T_{w(ch)}$ and container wall temperature $T_{w(cont)}$. For instance, if $v_i = 0.15 \, \mathrm{m \, s}^{-1}$, $T_i = 630 \,°\mathrm{C}$, $T_{w(ch)} = 120 \,°\mathrm{C}$ and $T_{w(cont)} = 60 \,°\mathrm{C}$, the calculated pouring time to fill the container was found to be 4.3 s and, after that pouring time, pouring at the inlet is turned off. The rest of the metal over the channel is allowed to drain off for an additional 0.4 s to complete the container filling. After these two stages, simulation is continued in the container only up to ~550 s. Figure 6.45 shows the resulting liquid fraction after a 3.7 s flow time.

The profiles of temperature, solid fraction, grain density, grain diameter and solute concentration in the container after a solidification time of 553 s are illustrated in Figure 6.46 . The results reveal the great macro-homogeneity of all quantities inside

Table 6.3 Characteristic grain nucleation parameters for A356 aluminium alloy.

Liquidus temperature (°C)	ΔT_N (°C)	ΔT_σ (°C)	n_{max} (m^{-3})
614.6	7.889	0.346	2.17×10^{11}

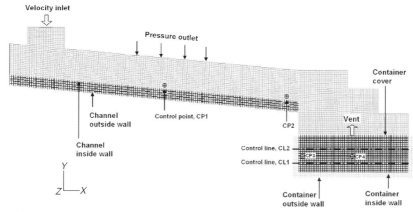

Figure 6.44 Simulation grid of cooling channel process.

the container. This is due to the convection eddies (Figure 6.46f) and the low temperature gradient inside the container. The temperature and correspondingly the liquid fraction are slightly high at the upper middle of container because of the dual insulation at the metal surface due to the air and the isolating cover of the container.

The temperature, solid fraction and grain density were monitored during fluid flow over the channel and recorded as shown in Figure 6.47. At the middle of the channel (CP1 in Figure 6.44), the temperature of the metal is slightly higher, and the corresponding solid fraction is considerably lower, compared with those at the channel outlet (CP2), where the metal cut a longer distance on the cooled channel. The average temperature difference between CP1 and CP2 after 1.5 s was about 16 °C, whereas this difference decreased to about 2 °C as the inlet velocity increased to 0.2 m s^{-1}. The overall temperature rise for both points after 4.5 s was about 3 °C due to heating of the channel inside wall. The impact of this temperature increase on the solid fraction is also obvious in Figure 6.47a. The grain density in Figure 6.47b has a

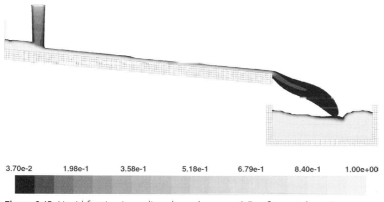

| 3.70e-2 | 1.98e-1 | 3.58e-1 | 5.18e-1 | 6.79e-1 | 8.40e-1 | 1.00e+00 |

Figure 6.45 Liquid fraction in cooling channel process 3.7 s after metal pouring.

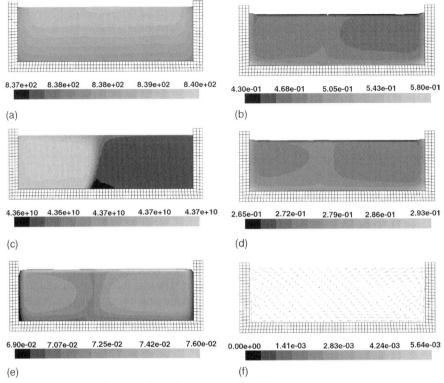

(a)

8.37e+02   8.38e+02   8.38e+02   8.39e+02   8.40e+02

(b)

4.30e-01   4.68e-01   5.05e-01   5.43e-01   5.80e-01

(c)

4.36e+10   4.36e+10   4.37e+10   4.37e+10   4.37e+10

(d)

2.65e-01   2.72e-01   2.79e-01   2.86e-01   2.93e-01

(e)

6.90e-02   7.07e-02   7.25e-02   7.42e-02   7.60e-02

(f)

0.00e+00   1.41e-03   2.83e-03   4.24e-03   5.64e-03

Figure 6.46 Simulation results in the container after 553 s:
(a) temperature, K, (b) solid fraction, (c) grain density, m^{-3},
(d) grain size, mm, (e) solute concentration, wt%, (f) liquid
velocity, m s^{-1}, after 93 s.

sudden rise at about 0.6 s as a result of the high supercooling that occurred when the hot liquid metal reached CP1 and CP2. It fluctuates afterwards as a result of latent heat release, flow drift and grain transport. The slight decrease in cooling rate at CP1 as per Figure 6.47a results in gradual reduction in grain production rate and n (Figure 6.47b).

The values of T, f_S and n were tracked in the container and the results (Figure 6.48) manifested the role of the container in this process. The cooling curves in Figure 6.48a show a relatively higher temperature at the container top compared with the bottom and the difference diminishes as time proceeds. These results agree with the experiments of Bünck [47]. The grain size (Figure 6.48b) exhibits a continuous increase due to temperature losses and grain growth. The grain density reaches a steady state value of 4.39×10^{10} m^{-3} after complete filling of container at about 5 s, where the cooling rate is slower and the expected undercooling is much lower. If the grain density in the container is compared with the outgoing one from the channel (Figure 6.47b), the former was found to have 1.0×10^{10} m^{-3} more grains. This means that further nucleation took place during container filling, where its initial temperature was 60 °C.

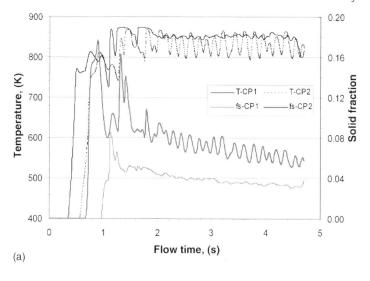

(a)

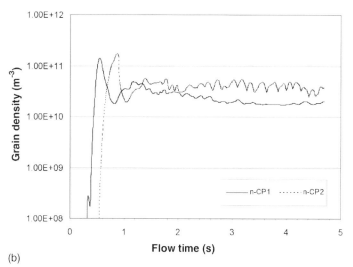

(b)

Figure 6.47 Variation of (a) temperature and solid fraction and
(b) grain density at a middle point and outlet point over the
channel versus flow time.

Validation experiments were conducted under the same conditions as simulation.
Figure 6.49 shows the resulting microstructure after holding for 120 s in the
container. The average measured temperature inside the container after 120 s was
863.60 K while the computed value in Figure 6.48a lies between 864.88 and 867.27 K,
which is an indicator of the good agreement of heat flow models and boundary
condition application. On the other hand, the experimental grain density
$(1.731 \times 10^{12}\,\mathrm{m}^{-3})$ and the average grain size (0.0558 mm) deviated from the

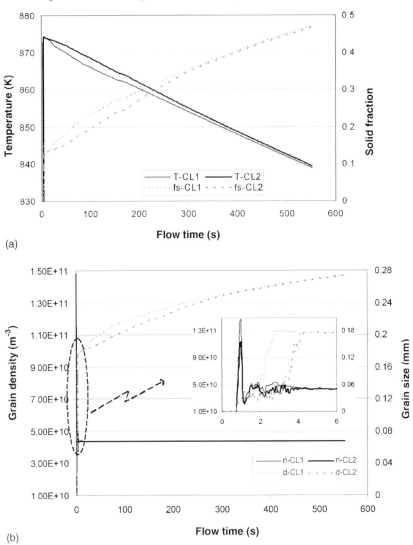

(a)

(b)

Figure 6.48 (a) Computed cooling curves inside the container for in A356 aluminium alloy; (b, c) grain density and grain size versus solidification time.

computed values ($4.37 \times 10^{10} \, \mathrm{m}^{-3}$ and 0.21 mm, respectively). The most probable reason is the grain multiplication due to continuous cooling/shearing of the solidifying particles over the channel. This effect was not considered in MeSES. The calculated local shear rate values ($\mathrm{d}v/\mathrm{d}y$) near CP2 at 2.3 s are shown in Figure 6.50. The shear rate can reach more than $630 \, \mathrm{s}^{-1}$. The grain nucleation model implemented in MeSES is based on cooling rate and demonstrated good results [44] in the absence of shear.

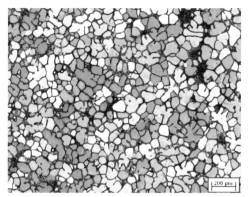

Figure 6.49 Microstructure of A356 feedstock produced by the cooling channel after holding for 120 s in the container.

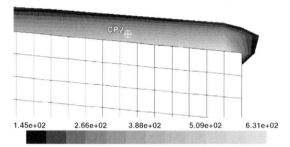

| 1.45e+02 | 2.66e+02 | 3.88e+02 | 5.09e+02 | 6.31e+02 |

Figure 6.50 The computed shear rate (s^{-1}) in the melt over the channel near CP2 at 2.3 s.

6.3.4
Summary

The MeSES code was applied successfully to simulate the formation of semi-solid feedstock using the cooling channel process. The models implemented within the code exhibited a satisfactory response to the variation of different process parameters. The results agree with expectations and have the same trend as the experimental results in the literature [37, 46]. After application of the experimentally determined grain nucleation parameters, the results were improved and became more reliable in the absence of melt convection and shear. In the presence of shear flow, the grain density and grain diameter values deviated from the experiments. Further study of the effect of shear and vigorous melt convection during continuous cooling of molten alloys on grain nucleation and multiplication is highly recommended.

List of Symbols

$\underline{\underline{T}}$ tensor
$\underline{v}$ vector

u, T	scalar
η	viscosity [Pa s]
τ	shear stress [Pa]
$\dot{\gamma}$	shear rate [s^{-1}]
κ	consistency factor [Pa s^n]
n	power law coefficient (flow exponent)
τ_0	yield stress; zero shear stress [Pa]
τ_∞	infinite shear stress [Pa]
η_0	zero shear viscosity [Pas]
η_∞	infinite shear viscosity [Pa s]
f_S	solid fraction (volume fraction solid)
κ	degree of agglomeration
κ_e	degree of agglomeration equilibrium
II	second invariant
$\underline{\underline{S}}$	stress tensor
$\underline{\underline{D}}$	deformation rate tensor
$\underline{u}$	velocity vector
u_S	solid velocity [m s^{-1}]
u_L	liquid velocity [m s^{-1}]
subscript S	solid phase
subscript L	liquid phase
subscript B	bulk phase
c	rate constant for κ
τ_w	wall shear stress [Pa]
Δp	pressure drop [Pa]
L	length [m]
a	volumetric flow rate [m^3 s^{-1}]
H	height [m]
W	width [m]
C	diameter [m]
R	radius [m]
r	radial coordinate [m]
η_{ap}	apparent viscosity [Pa s]
n	rotational speed [s^{-1}]
ω	angular velocity [rad s^{-1}]
G''	loss modulus [Pa]
G'	storage modulus [Pa]
$\underline{\underline{S}}$	stress tensor
S_{ij}	components of the stress tensor [Pa]
$\underline{\underline{\varepsilon}}$	strain tensor
ε_{ij}	components of the strain tensor
$\dot{\underline{\underline{\varepsilon}}}$	strain rate tensor
$\dot{\varepsilon}_{ij}$	components of the strain rate tensor [s^{-1}]

References

1 Atkinson, H.V. (1999) Proceedings of the International Conference on the Synthesis of Lightweight Metals III (eds F.H. Froes, C.M. Ward-Close, P.G. McCormick and D. Eliezer), Minerals, Metals and Materials Society, pp. 205*ff*.

2 Atkinson, H.V. (2005) Modelling the semi-solid processing of metallic alloys. *Progress in Materials Science*, **50** (3), 341.

3 Spencer, D.B., Mehrabian, R. and Flemings, M.C. (1972) Rheological behavior of Sn–15 Pct Pb in the crystallization range. *Metallurgical and Material Transactions B*, **3** (7), 1925.

4 Loué, W.R., Landkroon, S. and Kool, W.H. (1992) Rheology of partially solidified AlSi7Mg0.3 and the influence of SiC additions. *Materials Science and Engineering A*, **151** (2), 255.

5 Kirkwood, D.H. (1994) Semi-solid metal processing. *International Materials Review*, **39** (5), 173.

6 Mada, M. and Ajersch, F. (1996) Rheological model of semi-solid A356–SiC composite alloys. Part I: dissociation of agglomerate structures during shear. *Materials Science and Engineering A*, **212** (1), 157.

7 Suéry, M., Martin, C.L. and Salvo, L. (1996) Overview of the rheological behaviour of globular and dendritic slurries. Proceedings of the 4th International Conference on Semi-solid Processing of Alloys and Composites, Sheffield, 19–21 June 1996, pp. 21–29.

8 Brabazon, D., Browne, D.J. and Carr, A.J. (2003) Experimental investigation of the transient and steady state rheological behaviour of Al–Si alloys in the mushy state. *Materials Science and Engineering A*, **356** (1–2), 69.

9 Hufschmidt, M., Modigell, M. and Petera, J. (2006) Modelling and simulation of forming processes of metallic suspensions under non-isothermal conditions. *Journal of Non-Newtonian Fluid Mechanics*, **134** (1–3), 16–26.

10 Quaak, C.J., Katgerman, L. and Kool, W.H. (1996) Viscosity evolution of partially solidified aluminium slurries after a shear rate jump. Proceedings of the 4th International Conference on Semi-Solid Processing of Alloys and Composites, Sheffield, 19–21 June 1996, p. 35.

11 Koke, J. (2001) Rheologie teilerstarrter Metalllegierungen, in Fortschritt-Berichte, VDI. Reihe, 5(620).

12 Koke, J. and Modigell, M. (2003) Flow behaviour of semi-solid metal alloys. *Journal of Non-Newtonian Fluid Mechanics*, **112** (2–3), 141.

13 Kumar, P., Martin, C.L. and Brown, S. (1993) Shear rate thickening flow behaviour of semi-solid slurries. *Metallurgical and Materials Transactions A*, **24** (5), 1107.

14 Barnes, H.A. and Walters, K. (1985) The yield stress myth? *Rheologica Acta*, **24** (4), 323–326.

15 Cheng, D.V.-H. (1986) Yield Stress: a time-dependent property and how to measure it. *Rheologica Acta*, **25** (5), 542–554.

16 Barnes, H.A. (1997) Thixotropy. A review. *Journal of Non-Newtonian Fluid Mechanics*, **70** (1–2), 1.

17 Denny, D.A. and Brodkey, R.S. (1962) Kinetic interpretation of non-Newtonian flow. *Journal of Applied Physics*, **33** (7), 2269.

18 Mewis, J. and Schryvers, J. (1996) Unpublished International Fine Particle Research Report.

19 Moore, F. (1959) The rheology of ceramic slips and bodies. *Transactions of the British Ceramic Society*, **58**, 470.

20 Bingham, E.C. (1922) *Fluidity and Plasticity*, 1st edn, McGraw-Hill, New York, p. 125.

21 Oh, S.I., Rebelo, N.M. and Kobayashi, S. (1978) Finite-element formulation for the analysis of plastic deformation at rate-sensitive materials in metal forming. IUTAM Symposium, Tutzing, Germany, pp. 273*ff*.

22 Modigell, M., Pape, L. and Hufschmidt, M. (2004) The rheological behaviour of metallic suspensions. *Steel Research International*, **75** (9), 506.

23 Modigell, M. (2004) Modelling/rheology. Keynote lecture. Proceedings of the 5th International Conference on Semi-solid Processing of Alloys and Composites, Limassol, Cyprus, 21–23 September 2004.

24 Lou,é W.R. and Suéry, M. (1998) Key problems in rheology of semi-solid alloys. Proceedings of the 5th International Conference on Semi-solid Processing of Alloys and Composites Godlen, CO, USA, 23–25 June 1998.

25 Kopp, R., Luce, R., Leisten, B., Wolske, M., Tschirnich, M., Rehrmann, T. and Volles, R. (2001) Flow stress measuring by use of cylindrical compression test and special application to metal forming processes. *Steel Research*, **72** (10), 394–405.

26 Shimahara, H. and Kopp, R.(2004) Investigations of basic data for the semi-solid forging of steels. Proceedings of the 8th International Conference on Semi-solid Processing of Alloys and Composites, Limassol, Cyprus, Session 22, 419–429.

27 Hufschmidt, M., Modigell, M. and Petera, J. (2004) Two phase simulations as a development tool for thixoforming processes. *Steel Research International*, **75** (8–9), 513.

28 Meuser, H. (2003) Ermittlung und Bewertung von Werkstoffparametern für das Thixoforming von Aluminium und Stahllegierungen, Dissertation, RWTH Aachen University, Shaker Verlag.

29 Shimahara, H., Baadjou, R., Kopp, R. and Hirt, G. (2006) Investigation of flow behaviour and microstructure on X210CrW12 steel in the semi-solid state. Proceedings of the 9th International Conference on Semi-solid Alloys and Composites (S2P), Solid State Phenomena, Volumes 116–117, pp. 189–192.

30 Shimahara, H. and Hirt, G. (2006) Thermo-physical properties and boundary conditions for the semi-solid forging of steels and their applications by means of FEM Proceedings the 9th ESAFORM, Glasgow, pp. 811–814.

31 Noll, T., Friedrich, B., Hufschmidt, M., Modigell, M., Nohn, B. and Hartmann, D. (2003) Evaluation and modelling of chemical segregation effects for Thixoforming processing. *Advanced Engineering Materials*, **5** (3), 156–160.

32 Petera, J. and Kotynia, M. (2004) The finite element model of non-isothermal semi-solid fluid flow. *International Journal of Heat and Mass Transfer*, **47** (6), 1483–1498.

33 Alexandrou, A.N., Bardinet, F. and Loué, W. (1999) Mathematical and computational modelling of die filling in semi-solid metal processing. *Journal of Materials Processing Technology*, **96** (1), 59–72.

34 Ferziger, J.H. and Peric, M. (1996) *Computational Methods for Fluid Dynamics*, Springer, Berlin.

35 Modigell, M. and Koke, J. (2001) Rheological modelling on semi-solid metal alloys and simulation of thixocasting processes. *Journal of Materials Processing Technology*, **111** (1–3), 53–58.

36 Hufschmidt, M., Modigell, M. and Petera, J. (2004) Comparison of two phase finite element simulation with experiments on isothermal die filling of semi-solid tin–lead. *International Journal of Forming Processes*, **11** (1–2), 123–140.

37 Grimmig, T., Aguilar, J., Fehlbier, M. and Bührig-Polaczek, A. (2004) *Optimization of the Rheocasting Process under Consideration of the Main Influence Parameters on the Microstructure*, TMS, Cyprus.

38 3. Fluent Inc. (2006) *FLUENT 6.3 User's Guide*, Fluent Inc., Lebanon, NH, pp. 23–40.

39 Ludwig, A. and Wu, M. (2002) Modelling of globular equiaxed solidification with a two-phase approach. *Metallurgical and Materials Transactions A*, **33** (12), 3673–3683.

40 Wang, T., Wu, M., Ludwig, A., Abondano, M., Pustal, B. and Bührig-Polaczek, A.

(2005) Modelling the thermosolutal convection, shrinkage flow and grain movement of globular equiaxed solidification using a three phase model. *International Journal of Cast Metals Research*, **18** (4), 221–228.

41 Fisher, K. (1984) *Fundamentals of Solidification*, Trans Tech Publications, Stafa-Zurich.

42 Rappaz, M. and Gandin, Ch.-A. (1993) Probabilistic modelling of microstructure formation in solidification processes. *Acta Metallurgia et Materialia*, **41** (2), 345–360.

43 Thevoz, Ph. and Rappaz, M. (1989) Modelling of equiaxed microstructure formation in casting. *Metallurgical and Materials Transitions A*, **20** (2), 311–322.

44 Rappaz, M. (1989) Modelling microstructure formation in solidification processes. *International Materials Reviews*, **34** (3).

45 Ahmadein, M., Berger, R., Pustal, B., Subasic, E., Bührig-Polaczek, A. (2008) Multiphase solidification modelling of aluminum alloys based on experimental grain nucleation parameters. Proceedings of the 9th International Conference on Mechanical Design and Production, Cairo, 8–10 January 2008, pp. 907–915.

46 Wang, T., Pustal, B., Abondano, M., Grimmig, T., Bührig-Polaczek, A., Wu, M. and Ludwig, A. (2005) Simulation of cooling channel rheo-casting process of A356 aluminum alloy using three-phase volume averaging model. *Transactions of Nonferrous Metals Society of China*, **15** (2), 389–394.

47 Bünck, M. (2007) Proceedings of the 5th Decennial International Conference on Solidification Processing, Sheffield, July 2007.

7
A Physical and Micromechanical Model for Semi-solid Behaviour*

Véronique Favier, Régis Bigot, and Pierre Cézard

7.1
Introduction

Semi-solid metals exhibit time- and strain rate-dependent behaviour denoted thixotropy: they behave like solids in the undisturbed state and like liquids during shearing provided that the shear rate is high enough (see, e.g., [1]). Semi-solid metal forming, called thixoforming, exploits this thixotropic and shear-thinning behaviour since the semi-solid slug may be handleable but also flow easily in the die. This last feature contributes to form near net-shaped products. The use of finite element method (FEM) simulations to obtain the filling of the dies and to optimize the thixoforming process is clearly of great interest. To carry this out properly, the semi-solid flow and the heat transfer into the die have to be correctly described. In practice, various constitutive equations are used since discrepancies appear between experimental rheological data, even under isothermal and steady-state conditions reached after a period as an equilibrium microstructure is established. In addition to the shear-thinning behaviour, constitutive equations have to describe peculiar phenomena such as the presence or not of plastic threshold, and normal and abnormal behaviour, namely hardening and softening stress–strain rate relationship [2–5]; (Koke and Modigell, 2003); [6]. The objective of this chapter is to propose a new constitutive equation accounting for the mechanical role of liquids and solids and of their spatial distribution within the material on the overall behaviour of semi-solids. To do so, we use micromechanics and homogenization techniques that naturally relate the microstructure and the deformation mechanisms to the overall properties. In the first section, basic concepts of the micromechanics of heterogeneous materials are recalled. Then, we apply these concepts to semi-solids and explain the mathematical equations used in the modelling. Finally, some results concerning the isothermal steady-state and also transient and non-isothermal behaviour are given and compared with experimental data.

---

* A List of Symbols and Abbreviations can be found at the end of this chapter.

Thixoforming: Semi-solid Metal Processing. Edited by G. Hirt and R. Kopp
Copyright © 2009 WILEY-VCH Verlag GmbH & Co. KGaA, Weinheim
ISBN: 978-3-527-32204-6

7.2
Basic Concepts of Micromechanics and Homogenization

Most engineering materials are heterogeneous in nature. They generally consist of different constituents or phases, which are distinguishable at specific scales. Each constituent may show different physical properties (e.g. elastic moduli, thermal expansion, yield strength) and/or material orientations. However, in many engineering applications, a structure component may contain numerous such constituents such that it is impractical or even impossible to account for each of them for engineering design and analysis. When studying the properties of a real material, we need to define the microscopic length scale at which the properties of interest are directly relevant to determine the overall or effective property of the material from which the component is made. The term 'overall properties' means the properties averaged over a certain volume of the heterogeneous material. For such overall properties to be meaningful, the average taken over by an arbitrary volume element comparable to the relevant scale must be the same as for the heterogeneous material sample under consideration. Heterogeneous materials that meet this requirement are said to be macroscopically homogeneous. In addition, if in a heterogeneous material the microscopic length parameter d and the macroscopic length parameter D can be identified for a length scale of interest such as $d/D \ll 1$, then the heterogeneous material is microscopically homogeneous at the length scale D. A volume element with characteristic dimension D is called a representative volume element (RVE) because the overall properties on any RVE would be the same. In other words, the overall properties of each RVE represent the overall properties of the heterogeneous material.

Modelling based on micromechanics and homogenization techniques aims at relating microstructure features and local behaviour defined at a relevant length scale to the overall properties. It is based on the solution of Eshelby, who determined the stress and strain tensors within an inclusion embedded in a homogeneous matrix [8]. Details of the scale transition methodology and micromechanics were given by, for example, Kröner [9, 10], Mura [11] and Zaoui [12]. To sum up, this approach has three steps:

1. *Definition of the representative volume element:* The size of the RVE must be such that it includes a very large number of heterogeneities and also be statistically homogeneous and representative of the local continuum properties, so that appropriate averaging schemes over these domains give rise to the same mechanical properties. These properties correspond to the overall or effective mechanical properties. Because it is impossible to have knowledge of every detail of the microstructure, only relevant statistical (we are not interested in periodic microstructure) information on the microstructure is incorporated in the definition of RVE. These 'average' features allow the identification of certain mechanical 'phases' that dictate the overall behaviour.

2. *Definition of concentration equations:* Local strains are different from the overall strain due to the inhomogeneities within the material. Concentration equations aim to relate local and overall variables such as strain, strain rate or stress fields.

3. *Macroscopic averages or homogenization:* This step aims at determining the overall, also called effective, behaviour. The average stresses and strains over the RVE are defined as

$$\overline{\underline{\underline{\sigma}}\,(r)} = \frac{1}{V}\int_V \underline{\underline{\sigma}}\,(r)\mathrm{d}V \qquad \overline{\underline{\underline{\varepsilon}}\,(\,r)} = \frac{1}{V}\int_V \underline{\underline{\varepsilon}}\,(r)\mathrm{d}V \qquad (7.1)$$

They are equal to the overall stresses and strains when homogeneous boundary conditions are applied at the RVE boundaries.

In this chapter, this methodology is applied to semi-solid behaviour.

7.3
Modelling Semi-solid Behaviour

The behaviour and properties of materials at each length scale are controlled by the observable microstructure at the corresponding length scale. Therefore, whether a material is heterogeneous or not depends on the length scale used in the observation. In this section, we first analyse semi-solid microstructure to define the RVE. Then, we determine strain rate concentration tensors. Finally, macroscopic averages are calculated for the determination of the effective properties.

7.3.1
Definition of the Representative Volume Element

7.3.1.1 Morphological Pattern
Figure 7.1 displays a typical microstructure of a semi-solid Sn–15wt%Pb alloy. The material is a two-phase system with a solid volume fraction equals to f^s. The system exhibits a particular morphology consisting of a more or less globular solid phase bonded by surrounding liquid. A more accurate observation of the picture reveals (i) the presence of liquid entrapped within the globular solid particles and (ii) the presence of connections between the solid particles. These two relevant scales are taken into account in the representation of the microstructure: 'agglomerates' constituted of both solid and liquid are embedded in a contiguous zone composed also of both liquid and solid. In a statistical representation of the microstructure, this complex system can be assumed to be equivalent to one inclusion gathering all the agglomerates surrounded by a coating gathering the contiguous phase (Figure 7.2, A). This representation is all the more relevant where the evolution of the system is concerned. Indeed, it is commonly admitted that the deformation takes place in local sites such as the bonds between the solid grains and the liquid that is not entrapped in the agglomerated solid particles [13–15]. Therefore, the parts of solid and liquid contained in the 'coating' contribute to the deformation and are called the active zone, whereas the inclusions associated with solid grains and entrapped liquid hardly do so.

To individualize the mechanical role of the non-entrapped and entrapped liquid or of the solid bonds and the solid particles in the deformation mechanisms and to give

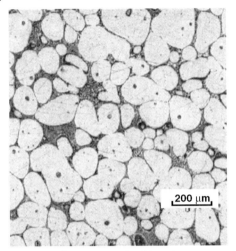

Figure 7.1 Photomicrograph of the microstructure of an
Sn–15wt%Pb alloy after water quenching from the semi-solid
state with 0.6 solid fraction. The microstructure reveals a light
primary solid phase and a dark eutectic phase corresponding to
the solid phase and to the liquid phase present in the semi-solid
state, respectively.

importance to the active zone where the strain is mainly localized, the material is represented by a morphological pattern as follows: thanks to the random and consequently spherical symmetry (isotropic) distribution of the constituents (whatever the real form of the solid and liquid 'particles'), the coated inclusion is considered as spherical (Figure 7.2, B). The inclusion is composed of both solid and liquid with volume fractions f_I^s and f_I^l, respectively, to represent entrapped liquid within solid particles. It is surrounded by a coating (the active zone) composed of the solid bonds and the non-entrapped liquid with volume fractions f_A^s and f_A^l, respectively. From this statistical point of view, when the real material is subjected to either

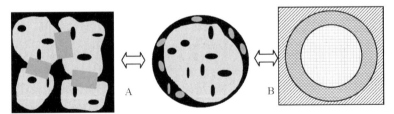

Figure 7.2 (A) Schematic representation of the semi-solid
microstructure exhibiting the presence of one inclusion of solid
and entrapped liquid surrounded by a coating of liquid and solid
bonds. (B) Morphological pattern constituted of a coated
inclusion embedded into a matrix having the effective properties
of the heterogeneous semi-solid material.

displacement boundary conditions or stress boundary conditions, one may envision that the effects of the applied loads (through the boundary conditions) and the interaction with other heterogeneities can be accounted for by assuming that the coating inclusion is placed within a homogeneous 'matrix' having the same properties as the real material, namely having the effective properties [9]. As a consequence, the RVE is defined as a volume containing a coated inclusion embedded in the homogenized medium.

7.3.1.2 Internal Variable

Deformation by shearing breaks the solid bonds because solid bonds mostly carry the deformation. This leads to a deagglomeration process and a release of some liquid [16]. Consequently, the resulting bimodal liquid–solid distribution is different from the initial one (Figure 7.3). To capture the evolution of the microstructure and the bimodal distribution while the strain rate changes, the solid volume fraction of the active zone, f_A^s, is introduced as an internal variable. This is related to the volume fraction of solid bonds and decreases with the strain rate. This internal variable is very similar to the structural parameters introduced first by Kumar *et al.* [14, 17]; (Martin *et al.*, 1994) and more recently by many others [18–22]; (Modigell *et al.*, 2001) [5]. It is equal to one when all particles are connected to each other and zero when all particles are separated. More accurately, percolation theories [23, 24] predict that under a critical volume fraction of solid, the solid phase appears as isolated agglomerates so that f_A^s is equal to zero (no solid bonds). Above a critical volume fraction f^c, a macroscopically connected network of solid is formed and f_A^s is given by a differential equation accounting for competing kinetics for agglomeration and deagglomeration (Equation 7.2). Following the work of Kumar *et al.* [14, 17]; (Martin *et al.*, 1994), the evolution law for f_A^s is given by

$$\dot{f}_A^s = K_{ag} f^s \left(1 - f_A^s\right) - K_{dg}\left(1 - f^s\right) f_A^s (\dot{\gamma})^n \tag{7.2}$$

where K_{ag}, K_{dg} and n are material parameters describing the agglomeration and deagglomeration mechanisms, respectively; $\dot{\gamma}$ is the overall shear rate given by

Figure 7.3 Schematic representation of the evolution of the microstructure with the strain rate.

$\dot{\gamma} = \sqrt{3}\dot{E}_{eq}$, where $\dot{E}_{eq}$ is the macroscopic von Mises equivalent strain rate. As a consequence, the steady-state solid fraction in the active zone is calculated by [25]

$$f_A^s = \frac{f^s}{f^s + \frac{K_{dg}}{K_{ag}}(1-f^s)(\dot{\gamma})^n} \tag{7.3}$$

It is worth noting that Equation 7.3 provides a macroscopic description of the strain-rate dependence of the internal structure.

To improve the description of relationships between local deformation mechanisms and microstructure evolution, we incorporate the fact that the relevant shear rate is not the overall one but the local one within the solid bonds. We postulate that the solid bonds break as soon as the local shear reaches a critical value γ_c. To do so, Equation 7.2 is rewritten as

$$\dot{f}_s^A = K_{ag}f_s(1-f_s^A) - (1-f_s)\frac{\dot{\gamma}_{bonds}}{\gamma_c}f_s^A \tag{7.4}$$

The shear rate within the solid bonds is naturally given by the micro–macro modelling (see Equation 7.11 where the medium B is the active zone A) and is equal to

$$\dot{\gamma}_{bonds} = \sqrt{3}(\dot{\varepsilon}_A^s)_{eq} \tag{7.5}$$

During deformation, it is admitted that the kinetics of the agglomeration process are much lower than the kinetics of the deagglomeration process (Martin et al., 1994). Consequently, for isothermal thixoforming, we can neglect the agglomeration term in comparison with the deagglomeration term, so that

$$\dot{f}_s^A = -(1-f_s)\frac{\dot{\gamma}_{bonds}}{\gamma_c}f_s^A \tag{7.6}$$

For non-isothermal processing when the dies are colder than the slug, an increase in solid fraction related to solidification due to thermal exchanges at the tool–slug interface is observed. As a result, the solid particles are more agglomerated. In order to incorporate this agglomeration–solidification phenomenon, we introduce a phenomenological term in Equation 7.4 accounting for the increase in the solid fraction within the active zone when temperature decreases. It is written as

$$\dot{f}_s^A = -a(1-f_s^A)f_s(T)|\dot{T}|^b - (1-f_s)\frac{\dot{\gamma}_{bonds}}{\gamma_c}f_s^A \tag{7.7}$$

where a and b are two positive material parameters.

7.3.1.3 Local Behaviours
The description of the RVE requires liquid and solid behaviour representation. The liquid phase is regarded as a Newtonian fluid. The solid phase is regarded as a viscoplastic material, the consistency of which may depend on temperature. Solid

and liquid are both assumed to be isotropic and incompressible so that their flow behaviour can be expressed by means of a single modulus, namely the viscosity μ, by the following equations:

$$s_{ij}^{l} = 2\mu^{l}\dot{\varepsilon}_{ij} \quad \text{with} \quad \mu^{l} = K^{l} \quad \text{for the Newtonian liquid} \tag{7.8}$$

$$s_{ij}^{s} = 2\mu^{s}\left(\dot{\varepsilon}_{eq}\right)\dot{\varepsilon}_{ij} \quad \text{with} \quad \mu^{s} = K^{s}\left(\dot{\varepsilon}_{eq}\right)^{m-1} \quad \text{for the viscoplastic solid} \tag{7.9}$$

where the superscripts l and s refer to the liquid and solid phase, respectively, s_{ij} is the deviatoric stress tensor, $\dot{\varepsilon}_{ij}$ is the deviatoric strain rate tensor, K is the consistency and m is the strain rate sensitivity parameter.

Actually, one may distinguish the constitutive equation for the entrapped and non-entrapped liquid and for the solid particles and the solid bonds by using different values for the consistency and the strain rate sensitivity parameter. In the following sections, as a first approximation, we assume that the entrapped and non-entrapped liquids have the same properties. Concerning the solid phase, we assume that the consistencies of the solid bonds and the solid particles are different. This choice implies that the strength of the solid bonds may be adjusted to account for the effect of rest time or temperature on the semi-solid microstructure and mechanical features.

7.3.2
Homogenized Estimate of the Semi-solid Viscosity: Concentration and Homogenization Steps

As illustrated in Figure 7.2 and discussed previously, the semi-solid is represented by a spherical 'coated inclusion'. The inclusion and the coating, also called the active zone, are both composed of liquid and solid. Therefore, the determination of the effective properties of the semi-solid requires first the evaluation of the overall behaviour of the inclusion and of the coating from the behaviour of the solid and liquid phases. The effective viscosity μ_{SS} of the coated inclusion, representing the semi-solid material, is then determined from the previous results. To do so, we use the self-consistent approximation. This approach implies that each particle of liquid (solid or coated inclusion) feels the surroundings (i.e. the neighbouring particles) through a fictitious homogeneous medium having the effective properties of the heterogeneous material (Figure 7.4).

7.3.2.1 Step a
The effective viscosities of the inclusion and the active zone are defined by Equation 7.10 using the same formalism as for local behaviours (Equation 7.9):

$$\Sigma_{B} = 2\mu_{B}\dot{E}_{B} \tag{7.10}$$

The suffix B is equal to I when it refers to the inclusion and to A for the active zone. Σ_{B} and $\dot{E}_{B}$ are the overall stress and strain rate tensors of the medium B. We assume that the medium B is submitted to boundary conditions $\dot{E}_{B}$. It is now required to

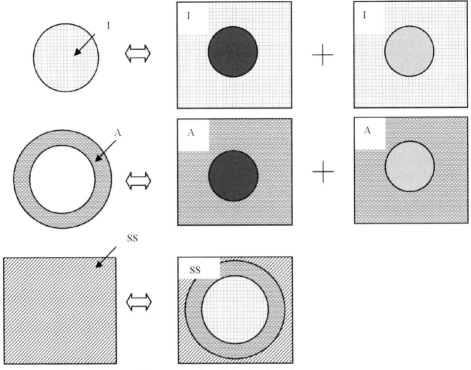

Figure 7.4 Schematic representation of the double scale transition performed to determine the homogenized properties of (1) the inclusion (I) from the solid particles and the entrapped liquid properties, (2) the active zone (A) from the solid bonds and the non entrapped liquid properties, (3) the semi-solid (SS) from the active zone and the inclusion properties through the morphological pattern of the coated inclusion.

relate boundary conditions to local strain rates. Applying the self-consistent scheme [26] to our problem illustrated on Figure 7.4, we find

$$\dot{\varepsilon}_B^s = T_B^s \dot{E}_B \tag{7.11}$$

$$\dot{\varepsilon}_B^l = T_B^l \dot{E}_B \tag{7.12}$$

where

$$T_B^s = \frac{5\mu_B}{3\mu_B + 2\mu_B^s} \tag{7.13}$$

$$T_B^l = \frac{5\mu_B}{3\mu_B + 2\mu_B^l} \tag{7.14}$$

are the strain rate concentration tensors associated with the liquid and solid particles, respectively. The overall response, namely the overall stresses, is calculated by averaging conditions on stresses:

$$\Sigma_B = f_B^s \sigma_B^s + \left(1 - f_B^s\right)\sigma_B^l \tag{7.15}$$

Introducing local constitutive equations and concentrations relations leads to

$$\Sigma_B = f_B^s \sigma_B^s + \left(1 - f_B^s\right)\sigma_B^l \tag{7.16}$$

$$\Sigma_B = \left[f_B^s \mu_B^s T_B^s + \left(1 - f_B^s\right)\mu_B^l T_B^l\right]\dot{E}_B \tag{7.17}$$

The effective viscosity is thus defined by the following implicit equation:

$$\mu_B = f_B^s \mu_B^s \frac{5\mu_B}{3\mu_B + 2\mu_B^s} + \left(1 - f_B^s\right)\mu_B^l \frac{5\mu_B}{3\mu_B + 2\mu_B^l} \tag{7.18}$$

7.3.2.2 Step b

As done previously, the mechanical interactions between the coated inclusions are solved by a self-consistent scheme. However, the mathematical form of the strain rate concentration tensors is different from that given in Equations 7.13 and 7.14 because now we have to account for the specific spatial distribution of the inclusion and the active zone. In other words, we have to account for the fact that the inclusion is surrounded by the active zone. Following the work of Christensen and Lo [27] and Cherkaoui et al. [28], the effective viscosity μ_{SS} of the semi-solid is thus determined by the following equation:

$$\mu_{SS} = f_A \mu_A T_A + f_I \mu_I T_I = \mu_A + f_I(\mu_I - \mu_A)T_I \tag{7.19}$$

where

$$T_I = \frac{5\mu_{SS}}{3\mu_{SS} + 2\mu_I + \left(\frac{1-f_I}{f_I}\right)\frac{6}{5}(\mu_I - \mu_A)\left(\frac{\mu_{SS}}{\mu_A} - 1\right)} \tag{7.20}$$

where $f_I = R^3/(R + \Delta R)^3$, R being the radius of the spherical inclusion and ΔR the thickness of the coating.

7.4
Results and Discussion

7.4.1
Isothermal Steady-state Behaviour

In this section, the micro–macro model is applied to the case of Sn–15wt%Pb alloys having a globular-like morphology for the solid phase and solid fractions ranging from 0.4 to 0.7. Calculations are compared with experimental results found in the literature. These were obtained using isothermal conditions and were carried out on parallel plate compression [29] at shear rates ranging from 10^{-5} to $10\,\mathrm{s}^{-1}$ or on a concentric cylinder viscometer [16, 30] at shear rates ranging from 2 to $1000\,\mathrm{s}^{-1}$. For the following calculations, the viscosity of the liquid phase is taken as equal to $1.81 \times 10^{-3}\,\mathrm{Pa\,s}$ [31]. The critical volume fraction of solid f^c is equal to 0.4. The parameters to be identified are the consistency K^s and the strain rate sensitivity parameter m of the solid phase, the ratio K_{dg}/K_{ag}, the coefficient n in Equation 7.3 and the fraction of the active zone f_A, assumed to be constant during changes of strain rates. The identification of all these parameters results from successive comparisons between the experimental Sn–15wt%Pb alloy data corresponding to a solid fraction of 0.5 found in the literature [16, 29] and the numerical results. Good agreement between the experimental and calculated data is obtained as for both low strain rates and high shear rates (Figures 7.5 and 7.6) with the set of values given in Table 7.1. At low shear rates (Figure 7.5), experimental viscosities in the region of $10^7\,\mathrm{Pa\,s}$ are obtained. These viscosities are many orders of magnitude greater than those obtained for the same initial spheroidal structure after vigorous agitation at high shear rates (Figure 7.6). The present model is able to describe well this experimental shear-thinning behaviour at low and high shear rates with the same set of parameters. In such a modelling, the overall properties are mainly controlled by the active zone behaviour. At low shear rates (from 10^{-5} to $\sim 10\,\mathrm{s}^{-1}$), solid particles can aggregate

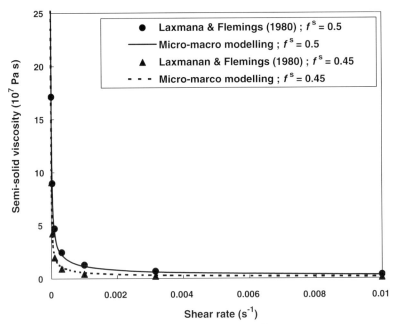

Figure 7.5 Evolution of the semi-solid viscosity as a function of the shear rate at 'low' shear rates for the Sn–15wt%Pb alloy. Experimental data were obtained using parallel plate compression tests.

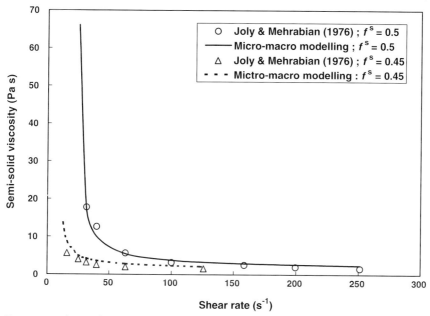

Figure 7.6 Evolution of the semi-solid viscosity as a function of the shear rate at 'high' shear rates for the Sn–15wt%Pb alloy. Experimental data were obtained using concentric cylinder viscometer tests.

into 'clusters' able to form a stiff network. Mechanical loading is transmitted through interglobular bonds and the system behaves like a solid, leading to a high viscosity. In the calculations, the solid fraction in the active zone f_A^s is found to be higher than 0.4. As the shear rate increases, more and more bonds between solid particles are broken, leading to a progressive decrease in f_A^s down to 0.4. A more dispersed suspension having a fluid-like behaviour is thus obtained and, in the calculation, a strong decrease in the active zone viscosity is observed. These results prove that the shear-thinning behaviour of semi-solid alloys is due to an evolution of the liquid–solid spatial distribution within the material. This evolution is attributed to local deformation mechanisms, namely the fact that deformation is mainly carried by the solid bonds and the non-entrapped liquid.

The prediction capability of the model is tested by calculating the semi-solid steady-state viscosity for a solid fraction of 0.45, keeping all the parameters constant. Very good agreement between the calculated and experimental results (Figures 7.5 and 7.6)

Table 7.1 Set of identified parameters for the Sn–15wt%Pb semi-solid alloy simulations of steady-state viscosities for $f^s = 0.5$.

K^s	m	K_{dg}/K_{ag}	n	f_A
1 MPa	0.43	0.8	0.2	0.007

is found. These results demonstrate the interest in such a modelling for predicting the behaviour for different solid fractions, at least in an intermediate solid fraction range.

7.4.2
Isothermal Non-steady-state Behaviour

Industrial thixoforming is a rapid process and the steady state is actually not achieved for such mechanical conditions. In this section, we simulate isothermal compression tests. Now, the evolution of the active solid fraction is accounted for by means of Equation 7.6 in order to capture the transient rheological response. The material parameters are given in Table 7.2. Note that in this study, we just compare qualitatively our calculated results with experimental values found in the literature. Consequently, there is no need to identify the material parameters precisely. Their values are just taken to be consistent with experiments. For the simulated compression tests, the ram speed is constant and equal to $500 \, \text{mm s}^{-1}$. The radius and the height of the initial slug are 15 and 45 mm, respectively. The load–displacement curves for different solid fractions are shown in Figure 7.7. Typically, the load–displacement curve displays a peak followed by a strong increase in the load, as is often observed experimentally (Loué *et al.*, 1992; [4, 32, 33]). In addition, in good agreement with Liu *et al.*'s experiments [34], the model predicts that the height of the peak falls with decreasing solid fraction (or increasing temperature) and the minimum load before and beyond the peaks also decreases with decreasing solid fraction. The presence of the peak is attributed to the breakdown of the solid skeleton. To understand better the relation between local deformation mechanisms and the overall response, we analyse the strain rate distribution within the semi-solid. Figure 7.8 shows the evolution of the strain rate within the overall material, the solid particles and the solid bonds. The evolution of the solid fraction in the active zone is also given in order to highlight relations between the microstructure and the deformation mechanisms. At the beginning of the processing, the solid network is highly connected ($f^s_{A,\text{initial}} = 0.7$ in Table 7.2). The magnitudes of the strain rate within the solid bonds and the solid particles are not far from each other although it is, as expected, slightly higher in the active zone. This means that, in this step, the whole solid phase carries the deformation. Since the solid bonds are less resistant and more deformed than the solid particles, the strain within the bonds reaches the critical

Table 7.2 Set of parameters used for simulations of isothermal and non-isothermal compression tests.

Solid			Liquid	Morphological pattern		Internal variable	
$K^s_{\text{particles}}$	K^s_{bonds}	$m^s_{\text{particles}} = m^s_{\text{bonds}}$	K^d	f_A	f_c	$f^s_{A,\text{initial}}$	γ_c
30 000 Pa	20 000 Pa	0.2	0.8	0.01	0.4	0.7	1

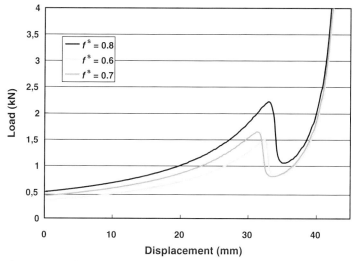

Figure 7.7 Simulated load–displacement curves of isothermal compression tests for different solid fractions.

strain to rupture, leading to a progressive (from a statistical point of view) deagglomeration of the solid network. As the active solid fraction decreases down to 0.4, we observe a strong decrease in the strain rate within the two solid phases, indicating that, at this moment, it is the liquid's turn to carry most of the strain. Since the liquid viscosity is much lower than the solid viscosity, this leads to a load drop. At this stage, since the solid bonds are no longer (or very slightly) deforming, the spatial distribution of the liquid and solid does not change any further. The semi-solid

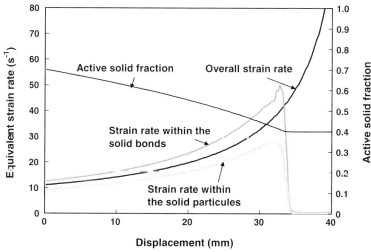

Figure 7.8 Simulated strain rate evolutions within the overall material, the solid bonds and the solid particles and active solid fraction as a function of displacement during a compression test.

structure thus remains the same and is behaving in a Newtonian manner. This behaviour provides the strong increase in the load observed after the peak.

As expected, decreasing the solid fraction decreases the yield stress and more generally the load level, except at the end of the experiments. In this last step, as mentioned previously, the semi-solid behaves as a suspension and its response is slightly less dependent on the solid fraction (in the range between 0.6 and 0.8) than before. In reality, an increase in temperature not only decreases the solid fraction but also changes the degree of agglomeration, the size and the morphology of the particles, maybe the mechanical resistance of the solid bonds, and so on. The modelling allows one to separate all these effects on the overall response and consequently can help to achieve a better understanding.

7.4.3
Non-isothermal and Non-steady-state Behaviour

As discussed previously, semi-solid behaviour depends strongly on the initial volume solid fraction within the material. In addition, this volume solid fraction may change during thixoforming because of viscoplastic dissipation or/and thermal exchanges between the dies and the slug. For example, in the case of steels, Cézard *et al.* [35, 36] highlighted the key role of thermal exchanges on the flow behaviour and the load level during thixoforming. Since it is very difficult to obtain isothermal conditions for industrial thixoforming, accounting for thermal exchanges is of great importance for simulations. This section analyses the effect of increasing overall solid fraction due to temperature loss simulating thermal exchanges between the die and semi-solid. The relationship between temperature and solid fraction is given by the Sheil equation (here we deal with a semi-solid aluminium alloy to obtain calculated data consistent with real experiments, but only qualitative results, that may extended to other alloys, are discussed). The kinetics of the active solid fraction used for these simulations is given by Equation 7.7, where a and b are taken, arbitrarily, as equal to 1. The other material parameters are the same as used in Section 7.2 and are given in Table 7.2. The initial overall solid fraction is 0,8. Compression tests having temperature losses equal to 0, 50 and 80 °C are simulated using a ram speed and slug size identical with those used in Section 7.2. Figure 7.9 shows the load–displacement curves. Clearly, increasing temperature loss leads to an increase in the load peak. This peak is delayed to higher displacement. Here again, these results are related to the microstructure evolution. Material variable analysis reveals that the increase in the overall solid fraction (see Figure 7.10) leads to a reduction in the decrease of the active solid fraction. The presence of a higher quantity of solid able to contribute to the deformation leads to a decrease in the strain rate within the solid phase and in particular within the solid bonds. As a result, the critical strain to bond rupture is reached for higher displacement. It is worth noting that the kinetics of the active solid fraction result from both agglomeration/solidification and deagglomeration. It obviously depends on parameters used for Equation 7.7. In particular, a and b of the agglomeration/solidification are here chosen arbitrarily and further accurate investigations are required to improve the modelling of such a phenomenon.

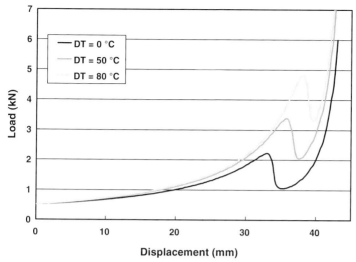

Figure 7.9 Simulated load–displacement curves for non-isothermal compression tests for the same initial solid fraction but for different temperature losses labelled DT.

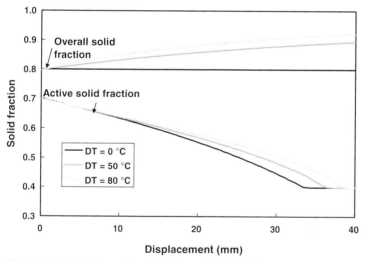

Figure 7.10 Simulated overall solid fraction and active solid fraction during non-isothermal compressions test having temperature losses of 0, 50 and 80 °C.

7.5
Conclusion

A new modelling based on a micro–macro approach is proposed to describe the steady-state but also transient, isothermal and non-isothermal behaviour of semi-solid materials. It is based on the application of the micromechanics concept of 'coated inclusion' to the semi-solid behaviour and the introduction of a bimodal distribution of the liquid and solid phases. The representative morphological pattern allows the deformation mechanisms involving in the non-entrapped liquid and solid bonds gathered in the coating of the 'coated inclusion' to be elucidated. The evolution of the microstructure with the shear rate is captured via an internal variable that modifies the liquid–solid bimodal distribution in the coating and in the inclusion. This variable represents the degree of agglomeration of the solid skeleton. The proposed rate equation of structure breakdown accounts for the time and temperature dependency. The model succeeds in describing the isothermal steady-state viscosities of semi-solid alloys at low and high shear rates. In addition, it provides results in good qualitative agreement for isothermal and non-isothermal compression tests. It demonstrates the key role of both the liquid and the solid phases and their distribution in the overall response and emphasizes the relationship between microstructure and deformation mechanisms. This 3D model provides a useful tool for a better understanding of thermomechanical mechanisms occurring within the material during thixoforming. It is also able to predict the semi-solid response under different conditions of loading paths or thermal history (by accounting for material variables such as the solid fraction, the degree of agglomeration and the resistance of the solid bonds) and can be exploited for semi-solid material design. It has been implemented in the finite element code FORGE2005 [33] to simulate die filling for industrial thixoforming.

List of Symbols and Abbreviations

ΔR	thickness of the coating
$\dot{E}_B$	strain rate tensors of the medium B
$\dot{E}_{eq}$	macroscopic von Mises equivalent strain rate
σ	stress tensor
$\dot{\varepsilon}_{ij}$	deviatoric strain rate tensor
$\dot{\gamma}$	overall shear rate
Σ_B	overall stress of the medium B
μ	viscosity
a, b	positive material parameters
D	macroscopic length parameter
d	microscopic length parameter
f^c	critical volume fraction
FEM	finite element method
f^l	volume fraction (liquid)
f^s	volume fraction (solid)
K	consistency
K_{ag}	material parameter, describing the agglomeration mechanism
K_{dg}	material parameter, describing the deagglomeration mechanism
m	strain rate sensitivity parameter

r radius
R radius of the spherical inclusion
RVE   representative volume element
S_{ij} deviatoric stress tensor
T^s, T^l strain rate concentration tensors (s, solid; l, liquid)
V volume

References

1 Flemings, M.C. (1991) Behaviour of metal alloys in the semi-solid state. *Metallurgical Transactions. A*, **22**, 957–981.

2 Loué, W.R., Landkroon, S. and Kool, W.H. (1992) Rheology of partially solidified AlSi7Mg0.3 and the influence of SiC additions. *Materials Science and Engineering A – Structural Materials Properties Microstructure and Processing*, **151**, 255–262.

3 Quaak, C.J. (1994) Properties of semi-solid aluminium matrix composites. *Materials Science and Engineering A – Structural Materials Properties Microstructure and Processing*, **188**, 277–282.

4 McLelland, A.R.A., Henderson, N.G., Atkinson, H.V. and Kirkwood, D.H. (1997) Anomalous rheological behaviour of semi-solid alloy slurries at low shear rates. *Materials Science and Engineering A – Structural Materials Properties Microstructure and Processing*, **232**, 110–118.

5 Modigell, M. and Koke, J. (2001) Rheological modelling on semi-solid metal alloys and simulation of thixocasting processes. *Journal of Material Processing Technology*, **111**, 53–58.

6 Koke, J. and Modigell, M. (2003) Flow behaviour of semi-solid metal alloys. *Journal of Non-Newtonian Fluid Mechanics*, **112**, 141–160.

7 Atkinson, H.V. (2005) Modelling the semi-solid processing of metallic alloys. *Progress in Materials Science*, **50**, 341–412.

8 Eshelby, J.D. (1957) The determination of the elastic field of an ellipsoidal inclusion and related problems. *Proceedings of the Royal Society London*, **A241**, 376–396.

9 Kröner, E. (1958) Berechnung der elastischen konstanten des Vielkristalls aus den Konstanten des Einskristalls. *Zeitschrift für Physik*, **151**, 504–518.

10 Kröner, E. (1961) Zür plastischen Verformung des Vielkristalls. *Acta Metallurgica*, **9**, 155–161.

11 Mura, T. (1987) *Micromechanics of Defects in Solids*, Martinus, NiJhoff Publishers.

12 Zaoui, A. (2001) Plasticité: approches en champs moyens, in *Homogénéisation en Mécanique des Matériaux (Vol. 2): Comportements Non Linéaires et Problèmes Ouverts* (eds M. Bornert, T. Bretheau and P. Gilormini), Hermès, Paris, Chapter I, pp. 17–44.

13 Doherty, R., Lee, H. and Feest, E. (1984) Microstructure of stir-cast metal. *Materials Science and Engineering*, **65**, 181–189.

14 Kumar, P., Martin, C. and Brown, S.B. (1993) Shear rate thickening behaviour of semi-solid slurries. *Metallurgical Transactions A*, **24**, 1107–1116.

15 Seconde, I.F. and Suery, M. (1984) Effect of solidification conditions on deformation behaviour of semi-solid Sn–Pb alloys. *Journal of Materials Science*, **19**, 3995–4006.

16 Joly, P.A. and Mehrabian, R. (1976) The rheology of a partially solid alloy. *Journal of Materials Science*, **11**, 1393–1418.

17 Kumar, P., Martin, C. and Brown, S.B. (1994) Constitutive modelling and characterization of the flow behaviour of semi-solid metal alloy

slurries – I. The flow response. *Acta Metallurgica and Materialia*, **42**, 3595–3602.

18 Martin, C., Kumar, P. and Brown, S.B. (1994) Constitutive modelling and characterization of the flow behaviour of semi-solid matal alloy slurries– II. Structural evolution under shear deformation. *Acta Metallurgica and Materialia*, **42**, 3595–3602.

19 Zavaliangos, A. and Lawley, A. (1995) Numerical simulation of thixoforming. *Journal of Materials Engineering and Performance*, **4**, 40–47.

20 Wahlen, A. (2000) Modelling the thixotropic flow behaviour of semi-solid aluminium alloys, in Proceedings of the 6th International Conference on Semi-solid Processing of Alloys and Composites (eds G.L. Chiarmetta and M. Rosso), Turin, 27–29 September 2000, pp. 565–570.

21 Kirkwood, D.H., Ward, P.J., Barkhudarov, M., Chin, S.B., Atkinson, H.V. and Liu, T.Y. (2000) An initial assessment of the flow-3D thixotropic model, in Proceedings of the 6th International Conference on Semi-solid Processing of Alloys and Composites (eds G.L. Chiarmetta and M. Rosso), Turin, 27–29 September 2000, pp. 545–551.

22 Burgos, G.R., Alexandrou, A.N. and Entov, V. (2001) Thixotropic rheology of semi-solid metal suspensions. *Journal of Materials Processing Technology*, **111**, 164–176.

23 De Gennes, P.G. (1979) *Scaling Concepts in Polymer Physics*, Cornell University Press, Ithaca, NY.

24 Stauffer, D. (1985) *Introduction to Percolation Theory*, Taylor and Francis, London.

25 Favier, V., Rouff, C., Bigot, R., Berveiller, M. and Robelet, M. (2004) Micro–macro modelling of the isothermal steady-state behaviour of semi-solids. *International Journal of Forming Processes*, **7**, 177–194.

26 Berveiller, M. and Zaoui, A. (1981) A simplified self-consistent scheme for the plasticity of two-phase metals. *Research Mechanica Letters*, **1**, 119–124.

27 Christensen, R.M. and Lo, K.H. (1979) Solutions for effective shear properties in three-phase sphere and cylinder models. *Journal of Mechanics and Physics of Solids*, **27**, 315–330.

28 Cherkaoui, M., Sabar, H. and Berveiller, M. (1994) Micromechanical approach of the coated inclusion problem and applications to composite materials. *Journal of Engineering Materials and Technology*, **116**, 274–278.

29 Laxmanan, V. and Flemings, M. (1980) Deformation of semi-solid Sn–15 pct Pb alloy. *Metallurgical Transactions A*, **11**, 1927–1937.

30 Ghosh, D., Fan, R. and Vanschilt, C. (1994) Thixotropic properties of semi-solid magnesium alloys AZ91D and AM50, in Proceedings of the 3rd International Conference on Semi-solid Processing of Alloys and Composites (ed. M. Kiuchi), Tokyo, 13–15 June 1994, pp. 85–94.

31 Lucas, L.D. (1984) Viscosité des principaux métaux et métalloïdes, Techniques de l'Ingénieur, MB1 M66.

32 Loué, W.R. Landkroon, S., Kool, W.H. (1992) Rheology of Partially solidified AlSi7Mg0.3 and the influence of SiC additions. *Materials Science and Engineering*, **A151**, 255–262.

33 Kirkwood, D.H. (1994) Semi-solid metal processing. *International Material Reviews*, **39**, 173–189.

34 Liu, T.Y., Atkinson, H.V., Kapranos, P., Kirkwood, D.H. and Hogg, S.C. (2003) Rapid compression of aluminum alloys and its relationship to thixoformability. *Metallurgical Materials Transactions A*, **34**, 1545–1554.

35 Cézard, P., Bigot, R., Favier, V. and Robelet, M. (2006) Thixoforming of steel – influence of thermal parameters. *Solid State Phenomena*, **116–117**, 721–724.

36 Cézard, P., Favier, V., Balan, T., Bigot, R. and Berveiller, M. (2005) Simulation of semi-solid thixoforging using a micro–macro constitutive equation. *Computational Materials Science*, **32**, 323–328.

Part Three
Tool Technologies for Forming of Semi-solid Metals

Thixoforming: Semi-solid Metal Processing. Edited by G. Hirt and R. Kopp
Copyright © 2009 WILEY-VCH Verlag GmbH & Co. KGaA, Weinheim
ISBN: 978-3-527-32204-6

8
Tool Technologies for Forming of Semi-solid Metals[*]

Kirsten Bobzin, Erich Lugscheider, Jochen M. Schneider, Rainer Telle, Philipp Immich, David Hajas, and Simon Münstermann

8.1
Introduction – Suitable Tool Concepts for the Thixoforming Process

Semi-solid processing (thixoforming) is an innovative metal forming technology bearing a high potential for cost reduction by reducing forming steps, forming forces and improving work piece quality. These beneficial effects are currently exploited only in light metals to be shaped in a semi-solid state, being in direct competition with highly sophisticated conventional forming technologies such as high-pressure die casting (HPDC). Although the potential of thixoforming technologies is notably more pronounced for ferrous metals, economically in terms of market numbers and technically in view of load-bearing structures, a transfer to high-melting alloys has not yet been accomplished.

One major drawback for industrial implementation is the lack of suitable tools and dies that meet the process demands and exhibit an economically satisfactory service life within the range of tolerated degradation. Tool systems and materials that are applied in established metal forming processes, that is, surface-treated and/or conventionally coated tool steels, hot-working steels and hard metals, suffer from severe deformation, hot tearing and build-up of metal and scale layers on the surface when applied in steel thixoforming. This consequently leads to a rapid decrease in shape accuracy of the manufactured parts and, thus, to short tool changing cycles. The reason for this severe attack on forming dies is to be found in the complex load profile acting on these dies during semi-solid processing. Four main load categories may be distinguished: (1) mechanical, (ii) thermal, (iii) chemical and (iv) tribological impacts (Figure 8.1).

Mechanical loads comprise the forming pressure required for form filling and also the final densification pressure to eliminate porosity in the work pieces. Whereas the forming pressure is comparatively low at approximately 25 MPa according to our own experiments and literature data, the latter is discussed equivocally with values given

---

[*] A List of Symbols and Abbreviations can be found at the end of this chapter.

Thixoforming: Semi-solid Metal Processing. Edited by G. Hirt and R. Kopp
Copyright © 2009 WILEY-VCH Verlag GmbH & Co. KGaA, Weinheim
ISBN: 978-3-527-32204-6

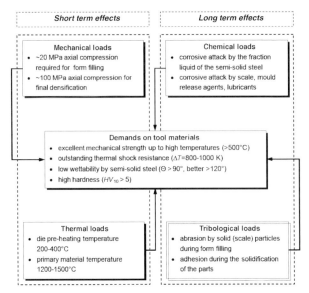

Figure 8.1 Load profile acting on steel thixoforming dies and resulting demands on tool materials.

in the literature ranging from less than 50 MPa to more than 1500 MPa [1, 2]. This diversity is due to the fact that no explicit study has yet been conducted to determine experimentally the required minimum forming pressure to obtain fully dense steel parts, which may be due in part to the lack of suitable tools. However, a minimum densification pressure of 100 MPa seems to be a reasonable approach [1, 2]. Thermal loads are exerted by the preheated primary material in contact with the mould and the amount of heat that is transferred from the semi-solid slurry to the die during form filling and solidification. The thermal process window ranges from the temperature of the primary material as an upper boundary to the die preheating temperature defining the lower boundary. The forming pressure is directly dependent on the work alloy, typically being in the range 1200–1500 °C. The latter may be varied in a certain range within the operating temperatures of the die material applied. Maximum tool temperatures in conventional metal forming processes are restricted to the annealing temperature of the metallic tool frames, being typically 500–550 °C. Hence severe thermal shocks act on the dies in each forming cycle.

Chemical attack on the die surface is exerted mainly by the liquid fraction of the semi-solid slurry during form filling. Scale residues on the preheated primary material also have a strong influence on the chemical interaction between the die and the metal during forming. In order to avoid oxide inclusions in the formed parts, the entire process chain preferably is covered with a protective atmosphere. Although in first experimental setups an encapsulated forming process was envisaged, this concept was abandoned in view of the process requirements in industrial production. Consequently, provisions were made with the target of reducing the contact of the steel with ambient air to a minimum by continuous flushing with argon during preheating, transport and insertion of the steel billets. However, this is ineffective during die closing and opening, leading to strongly varying oxygen partial pressures on the die surface that are virtually impossible to determine experimentally. Tribological loads consist of abrasive attack by the solid particles of the semi-solid slurry during material flow and by the solidified part during ejection. The different loads may be related to failure effects on a time scale: mechanical and thermal loads result in stress states that evolve in the parts during the different process steps. If peak stresses exceed the material strength, cracks are initiated, causing rupture of the parts. Since this will happen after only a few forming cycles, these are classified as short-term effects. In contrast, chemical and tribological attacks result in surface degradation during steady-state operation, hence they are classified as long-term effects.

The combination of these impacts yields a multifaceted load profile, the severity of which is significantly increased compared with the already challenging load profiles acting on typical forging and casting dies. Contrary to first expectations derived from light metal thixoforming, the decrease in process temperatures compared with sand casting and investment casting does not lead to reduced demands on tool materials [2, 3]. The targeted combination of the beneficial aspects of forging and casting consequently causes a superposition of the respective loads on forming dies: mechanical impacts are superior to pressureless casting, while chemical attack is drastically increased by the presence of a ferrous melt in comparison with hot forging. Therefore, instead of a plain transfer of state-of-the-art tool solutions applied in conventional metal forming processes, the development of adapted tool systems is mandatory for the semi-solid processing of high-melting alloys. In comparison with light metal thixoforming, the significantly increased forming temperature for steels is the most critical aspect, affecting the entire process chain from preheating and material manipulation in the semi-solid state to the impact on forming dies. In Figure 8.2, the evolution of the die temperature during cyclic thixoforming without any temperature control is depicted schematically.

Starting at a typical die preheating temperature of $350\,°C$, the die surface temperature rises to a value close to the temperature of the work material, for

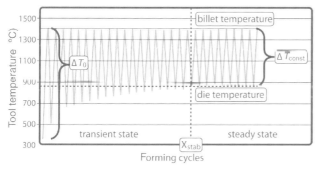

Figure 8.2 Schematic temperature evolution in steel thixoforming dies during sequenced processing in thixoforging and thixocasting without active temperature control.

example, 1400 °C. During the dwell time after forming and part ejection, the die temperature decreases to a value above the preheating temperature, until the next cycle starts. In the course of continuous forming, the lower die temperature increases until a steady state is reached, in which the die temperature is alternating between stable upper and lower boundaries. Two phases are distinguished on the time axis: a transient state characterized by severe thermal shocks in each forming step and a steady state in which thermal shocks are reduced compared with the onset of forming, while the average die temperature during operation is significantly higher than in the transient state. A process conducted in this way would require die systems able to operate at high temperatures while concurrently exhibiting outstanding thermal shock resistance. Since the upper thermal boundary is directly related to the work alloy, the only parameter to influence thermal process conditions is the die temperature. In order to reduce the transient state to a minimum and to obtain stable process conditions, the die temperature may be fixed by heating/cooling devices to a predefined value. The resulting die temperatures versus cycle times are depicted in Figure 8.3. Although thermal shocks on forming dies are high, the average die temperature is reduced compared with the aforementioned process layout. Hence die materials of excellent thermal shock resistance are required, whereas the demands on high temperature stability are decreased.

In an alternative approach, the die preheating temperatures may be increased to >1000 °C. This results in a significant decrease in thermal shock loads on forming dies but requires tool materials that can withstand these high operating temperatures. Moreover, a heating core has to be provided to establish the targeted die temperatures, in addition to a thermal insulation shell to protect the machinery environment from the heat generated in the die (Figure 8.4).

However, altering the die preheating temperature not only influences the demands on tool materials but also strongly affects the entire process layout. Low die temperatures lead to rapid solidification of the semi-solid slurry in the cavity, thereby decreasing the maximum flow length and increasing the minimum wall thickness. Hence this process layout is to be applied for manufacturing work pieces of bulk,

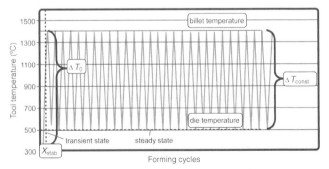

Figure 8.3 Schematic temperature evolution of steel thixoforming dies during sequenced processing in thixoforging and thixocasting with active cooling of the dies.

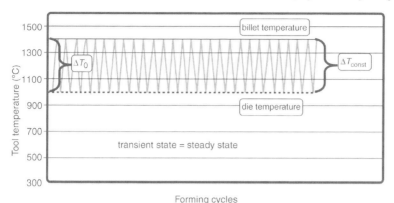

Figure 8.4 Schematic temperature evolution of steel thixoforming dies during sequenced processing in thixoforging and thixocasting using self-heating tools at high operating temperatures.

compact geometry by sequenced forming, that is, thixoforging and thixocasting. Owing to the short cycle time, the productivity of this process layout is high.

In contrast, increased die temperatures lead to retarded solidification and allow the production of rangy, thin-walled parts by increasing the flow length and decreasing the minimum wall thickness, preferentially in continuous forming methods such as thixoextrusion. However, the productivity of this process layout is reduced compared with the former concept.

Ceramic materials exhibit numerous beneficial properties in view of the demands on dies for the semi-solid processing of steel. The ionic/covalent bond character leads to high hardness, thermochemical stability and corrosion resistance against solid and molten metals up to temperatures noticeably above those of hot-working steels. Conversely, the resulting fracture toughness and thermal shock resistance are poor compared with metallic-bonded materials. One possible strategy to overcome these restraints and combine the advantageous properties of ceramics and metals is to deposit ceramic coatings of various thicknesses on metal substrates using the thermal spray technique or physical/chemical vapour deposition. Regarding the load profile as given in Figure 8.1, the mechanical loads are sustained by the substrate, whereas the main purpose of the coatings is to withstand corrosive and tribological attack. In the case of thin films applied by vapour deposition, the thermal gradient evolving within the coating is very small, hence the heat and thermal shocks introduced are uncritical for the coating but have also to be sustained by the substrate. Thermal spray coatings of greater thickness and defined porosity protect the substrate from thermal loads. However, residual porosity significantly reduces the load-bearing capacity of these coatings, which have to be adjusted carefully to the mechanical loads. Recent studies provide evidence that this porosity might be reduced successfully by hot isostatic pressing [4].

With respect to the maximum operating temperatures of common metallic substrate materials, this approach is confined to conventional process layouts of moderately preheated dies. Alternatively, bulk ceramic dies and die segments of high thermal shock resistance may be applied in this process layout. Apart from the strengths and limitations of different coatings and bulk ceramics regarding thixoforming process conditions, the characteristics of the respective manufacturing technologies have to be taken into consideration when designing tools for the semi-solid processing of steel parts of predefined geometry. That is, depending on the deposition technology, cavities and undercuts are difficult or impossible to coat, whereas processing of bulk ceramic green bodies is limited in size and geometric complexity. Moreover, the intensity of the separate loads acting on forming dies, as discussed above, varies locally inside the cavity as a function of work piece geometry. In order to account for these local load collectives and specific tool systems, a hybrid tool approach is proposed, in which individual tool solutions are developed for the targeted work piece geometry by combining the different die systems, where appropriate. An example of this hybrid concept is given in Figure 8.5, showing a thixoforging tool with a flat, physical vapour deposition (PVD)-coated upper swage, a plasma-enhanced chemical vapour deposition (PECVD)-coated liner and a bulk ceramic punch mounted on a steel plunger. Conversely, the demands on die materials to be applied in the concept of hot tools operating at temperatures >1000 °C, regarding high temperature strength and chemical stability in oxidizing atmospheres, are met exclusively by ceramic materials.

A test strategy (Figure 8.6) was established according to the load profile depicted above in order to examine the qualification of the different die materials and coating systems for the semi-solid processing of steels. Based on fundamental characterization methods typically applied for coatings and bulk ceramics, the resistance against thermomechanical, chemical and tribological loads was subsequently examined in

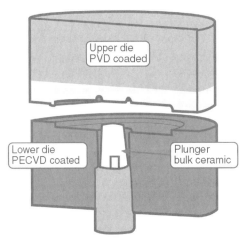

Figure 8.5 Hybrid tool concept.

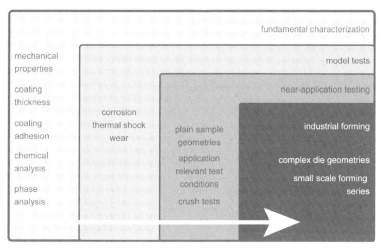

Figure 8.6 Testing strategy for evaluation and qualification of potential tool materials and systems.

model tests, which were custom-developed to reproduce the specific attack on dies during steel thixoforming separately [5, 6].

Further evaluations were undertaken using simple tool geometries in forming tests under defined experimental conditions. The tool systems tested in these experiments successfully were applied in small-scale forming series using geometrically complex dies under industrial conditions in forging presses and die-casting machines.

Forming experiments reveal the performance of the respective tool systems and die materials but also indicate the effect on form filling, shape accuracy and work piece quality. Owing to the pronounced temperature sensitivity of the rheological properties of the semi-solid slurry and the resulting effects on form filling behaviour and liquid phase separation (cf. Chapter 6), the different thermal die concepts, as proposed here, directly influence the work piece microstructure and the properties of the manufactured parts. Detailed descriptions of the characterization methods and model tests on PVD/PECVD-coated tools and bulk ceramic dies are given in the following sections. Experimental results on the load profile and the work piece quality obtained are presented and discussed.

8.2
Thin-Film PVD/PECVD Coating Concepts for Die Materials

Metallic alloys such as hot working steel and high-melting alloys, for example nickel- and molybdenum-based alloys, have been evaluated for application as die materials during semi-solid forming of steel [7–10]. The nickel steels are characterized by their low thermal conductivity and their high coefficient of thermal expansion (CTE) values. Consequently, they will be easily affected by strong thermal cycles and

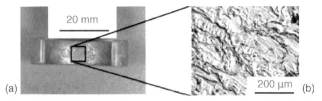

Figure 8.7 CuCrZr based alloy segment (a) and the surface (b) (after 10 produced parts).

soldering effects on the semi-solid steel. To avoid this, different tool materials have been tested in recent years regarding their applicability as die materials for steel thixoforming moulds. The processing of steel results in high surface loads, such as thermal, mechanical and tribological stresses. There are different concepts to address these challenges. One is to employ tool materials exhibiting high strength at working temperatures and low thermal conductivity (e.g. hot working steel).

Alternatively, die materials showing high thermal conductivity, such as copper- and molybdenum-based materials, may be applied. From this group of materials, the CuCrZr alloy 2.1293 with a high thermal conductivity ($320\,\mathrm{W\,mK^{-1}}$) was selected for forming experiments. After the production of 10 HS 6-5-2 steel parts using CuCrZr die material, the tool surface showed strong abrasion and erosion of the uncoated copper material (Figure 8.7). This degradation of the surface indicated that copper-based materials are not suitable for steel thixoforming.

Further investigations were concentrated on alloys exhibiting higher strength at elevated temperatures. Seeking metallic die material with low thermal conductivity, the hot working steel 1.2343 and the new developed 1.2999 from Deutsche Edelstahlwerke (Witten, Germany) were investigated. Also, a molybdenum-based alloy (TZM) from Plansee Holding (Reutte, Germany) with very high thermal conductivity was used. The mechanical and thermal properties are reported in Table 1.2. Pre-investigations such as corrosion tests in air and exposure to liquid steel showed that there is a need for tool protection for the uncoated metallic dies to withstand the chemical and mechanical wear associated with the semi-solid forming of steel [11].

Different nitride- and boride-based coatings were investigated for semi-solid processing of aluminium A356 and AA6082. Not only different coating systems were evaluated, but also sundries depositing technologies, such as magnetron sputter ion plating (MSIP)–PVD, arc ion plating (AIP)–PVD, PECVD and chemical vapour deposition (CVD) processes. Additionally, different interlayers were applied, for enhancing the adhesion to the top coat and also to support the top coating (Figure 8.8).

Starting with the material screening from standard titanium-based coatings such as TiN, TiAlN and TiB$_2$, the chromium nitride family offered very good durability in the forming tests. Especial the PVD-Cr$_{1-x}$Al$_x$N showed less adhesion to the semi-solid aluminium and high mechanical strength compared with all other coating types [11]. During the forming process in the contact zone between the semi-solid material and the coating, a small, dense Al$_2$O$_3$ and Cr$_2$O$_3$ layer is formed. Further investigations on a suitable coating material for the semi-solid forming of steel like

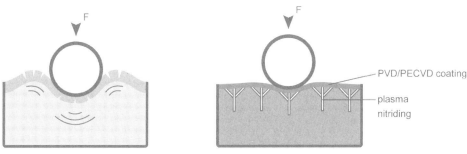

Figure 8.8 Coating concept for avoiding the softening of the steel substrate.

the corrosion tests were carried out at the GHI (Institute of Mineral Engineering) and showed that only the oxides withstand the corrosion attacks from the semi-solid steel. Also, thermocycle tests were carried out at the MCh (Institute of Materials Chemistry) using HS 6-5-2 as a counterpart material. These tests showed no counterpart interaction with the bulk material Al_2O_3, Al_2O_3/SiO_2 (mullite) and $MgAl_2O_4$. MgO- and Y_2O_3-stabilized ZrO_2 bulk materials were also tested and showed limited applicability for the thixoforming process. It is concluded that only oxide materials are suitable as potential coating systems. From the load profile shown in Figure 8.1, it is evident that the coatings need to protect the die's surface from abrasive and chemical wear. Oxidic ceramic coating materials often exhibit superior chemical resistance to other hard coatings [12].

8.3
PVD and PECVD Coating Technologies

There are different methods for the application of thin-film coatings. PVD and PECVD offer different advantages and disadvantages. Hence both technologies were investigated within the Collaborative Research Centre (SFB) 289.

PVD methods cover a range of thin-film deposition techniques, which include evaporation, laser ablation, vacuum arc-based deposition and many different modes of physical sputter deposition [13]. The PVD method used in the present work is *magnetron sputtering* (Figure 8.9), in which an electric field is applied between two electrodes immersed in an inert gas (argon) to create a plasma. One of these electrodes (the cathode) is the material source, the so-called target, and the chamber walls commonly act as the anode. The argon ions created in the plasma are attracted to the target (e.g. aluminium, zirconium) and sputter the latter, dislodging atoms [14]. These atoms are subsequently transferred in a controlled manner in an evacuated chamber to the substrate where the film growth proceeds atomistically. The typical operating pressure for magnetron sputtering ranges between 0.1 and 3 Pa, which means there is a very low probability of chemical reactions in the plasma. The deposition and properties of the films made by magnetron sputtering depend on a variety of parameters, such as the energy and direction of the incident particles, the

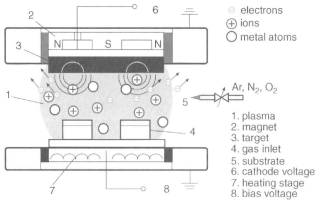

Figure 8.9 Schematic drawing of a magnetron sputtering PVD process.

temperature and composition of the substrate surface and the ion or electron bombardment [13]. Temperature and composition of the substrate surface have a fairly low impact on the arrival process, but can strongly influence the residence time or mobility of the atoms on the surface, which drastically affects structure evolution and the properties of the deposited films.

Magnetron sputtering methods are often used to synthesize compound materials by adding a reactive gas (O_2, N_2,) to the sputtering process. This process for which chemical reactions occur between the sputtered atoms and the reactive species at the substrate, on the chamber walls or at the target itself to create the compound is called *reactive sputtering* [13, 15]. As already mentioned, due to the low pressure used during the sputtering processes, reactions in the plasma are very unlikely. Nevertheless, after determination of the process window, this method permits reproducible fine control of the compound film stoichiometry and microstructure. Therefore, numerous industrial coaters are equipped with this technology.

CVD is a process in which the substrate is exposed to one or more volatile precursors which react chemically or decompose at the substrate surface to form a thin film. Often high substrate temperatures are employed to overcome the energetic barrier to synthesize the compound, which limits the range of suitable substrate materials. Therefore, plasma-assisted CVD processes have been developed [14] and were used here. The experimental setup used is shown in Figure 8.10 and discussed in a recent paper [17]. In the glow discharge, decomposition and activation of the volatile precursor species take place [14]. Hence chemical reactions are triggered at considerably lower temperatures than in thermal CVD processes. For example, α-alumina can be formed at about 1000 °C by CVD [16] whereas the synthesis temperature using PECVD was 550 °C [17].

Whatever the plasma-based deposition technique employed, the plasma–surface interaction during film growth defines the structure evolution and hence the properties of the coatings. Energy supplied by the arriving particles (neutrals, ions, clusters, etc.) and by heating the substrate is often considered as one of the main parameters affecting structure evolution [18].

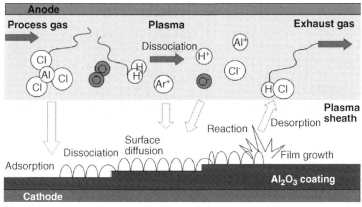

Figure 8.10 Schematic PECVD coating concept.

Among the most obvious differences between reactive magnetron sputtering and PECVD are the physical state of the primary source of the depositing species (which is solid for sputtering and gaseous for PECVD) [13, 14] and the collision probability (generally in magnetron sputtering collisionless transport is expected across the substrate sheath, whereas in PECVD collisions have to be taken into account) [17].

8.4
Multifunctional PVD Composites for Thixoforming Moulds

One of the challenging aspects of steel thixoforming is the development of a die concept which works up to a peak temperature of 1400 °C. The investigations focused on a PVD-coated metallic mould to withstand tribological loads such as thermal, chemical and mechanical interactions between the die surface and the forming material 100Cr6, HS6-5-2 or X210CrW12. Suitable tool material solutions are described in Chapter 1 (Table 1.2). These possible solutions seem to be compromises. Molybdenum and its alloys, for instance, suffer from their poor oxidation resistance in air at temperatures exceeding 600 °C. The following investigations focused on TZM and hot working tool steel as die material with a protective PVD coating on top, which seem to have the best combination of properties regarding the forming of steel. The requirements for the coating chemistry and architecture can be derived from the forming process and from the tool materials used. The challenge for a PVD protective coating of the TZM mould is to ensure chemical resistance against the semi-solid metal and oxygen. The degradation of the uncoated moulds results in the formation of volatile molybdenum oxides above 700 °C in air. There is also an internal demand in a sufficient bonding within the coating structure, resulting in a mismatch of the coefficient of thermal expansion and the Young's modulus. More than 90 different bulk ceramic materials were investigated during material screening for suitable coating materials, but only the Al_2O_3 and ZrO_2 bulk materials showed promising

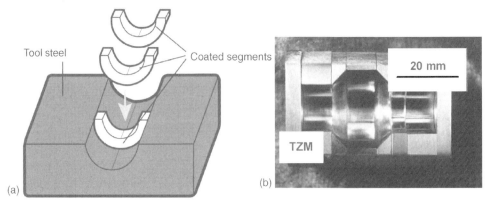

Figure 8.11 Schematic drawing of the segmented test die (a) and coated inserts (b).

behaviour during the corrosion tests [5]. It was concluded from these basic investigations that these materials are promising coating materials for tool protection.

The first investigations using PVD processes for thixoforming moulds for steel were carried out by Hornig [19]. During his investigations, he performed a material screening with different stable oxides deposited on hot working steel and TZM dies. To gain an understanding of which coating materials and coating concepts work for the forming process, different segments were coated and then tested at the Institute of Metal Forming (IBF), RWTH Aachen (Figure 8.11).

Hornig investigated the need for an interlayer and also the process parameters for the deposition of partially stabilized zirconia (PSZ)–ZrO_2, $MgAl_2O_4$ and Al_2O_3 (Table 8.1). This fundamental characterization combines the mechanical stability and the adhesion properties of the substrate–coating system. The depositions of the

Table 8.1 Deposition parameters for coated segments of the test die.

Process parameters	Etching phase	Interlayer TiN/CrN	Top layer		
			PSZ–ZrO_2–5%Y_2O_3	$MgAl_2O_4$	Al_2O_3
Process pressure (Pa)	1	0.75	1	1	1
Process time (s)	1200	600–1800	14 400	14 400	14 400
R.f. power (W)	100	—	—	—	—
R.f. frequency (MHz)	13.56	—	13.56	13.56	13.56
Ar:N_2 ratio (%)	—	33.4	—	—	—
Coating parameters					
Power supply	—	DC	r.f.	r.f.	r.f.
Power (W)	—	400	200	200	200
Deposition rate ($\mu m\,h^{-1}$)	—	6	0.3	0.125	0.125
Heating	—	Varied	Varied	Varied	Varied

oxidic top layers were performed using an r.f. power supply in a pure argon atmosphere. The targets were fully ceramic materials with different compositions, also mentioned in Table 8.1. The power supplied by the r.f. generator was kept constant (200 W) during deposition. The experiments show that the deposition rates were very low, compared with the DC deposition of nitrides (e.g. TiN). The structure was amorphous for the Al_2O_3 and $MgAl_2O_4$ films; only the ZrO_2-based coatings showed a crystalline structure [19].

The fundamental characterization of the deposited oxidic top coatings on the steel and TZM substrate showed a weak adhesion strength. The ceramic coatings were not able to follow the substrate deformation, which led to adhesive failure of the coating resulting from the low substrate stiffness. Therefore, further investigations were carried out using an interlayer as a support for the oxidic top layer coating. Figure 8.12 shows the impact of a TiN interlayer on the critical scratch load.

To verify the need for an interlayer in a real forming test, thixoforging experiments using a segmented test die (shown in Figure 8.11) were carried out later [20]. After the forming experiments, the segments were analysed using scanning electron microscopy (SEM) and energy-dispersive spectroscopy (EDS) analyses. The segments show the first signs of degeneration starting from the edges. These edges have the strongest impact of external loads due to their locations. Areas of onwelded material and plastic deformation can been seen in Figure 8.13a. Regarding the differences between coated and uncoated segments within the die, the coated TZM segments show almost no onwelded material and the inner area of the segments remained smooth and showed no changes. The best results concerning the edge stability were obtained for the bilayer coatings with a TiN interlayer and an oxidic top layer.

To increase the durability of the tools, further investigations were needed. The replacement of the amorphous top coating by applying a crystalline oxide top layer

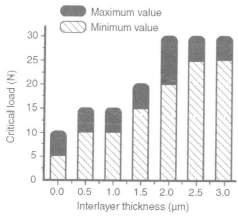

Figure 8.12 Influence of the TiN interlayer on the critical scratch load of Al_2O_3 coatings (thickness 1 µm) on TZM.

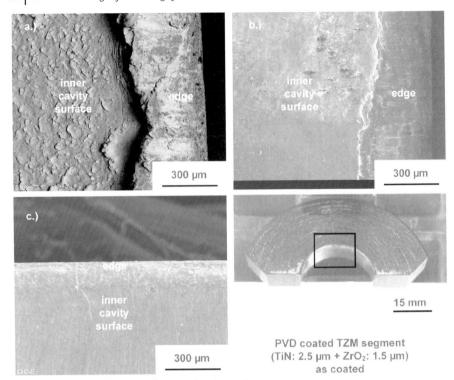

Figure 8.13 Performance of uncoated and coated segments after 10 produced parts (1.3333): (a) TZM uncoated; (b) coated with MgAl$_2$O$_4$; (c) coated TiN interlayer and ZrO$_2$ top layer; (d) location of the SEM images.

offers advantages such as higher chemical and thermal resistance, higher density and higher hardness. The oxides occur in different metastable crystallographic phases, with excellent properties, depending on the deposition temperature. Especially the γ-Al$_2$O$_3$ and tetragonal ZrO$_2$ (t-ZrO$_2$) phases offer, compared with other phases, high strength, toughness and high thermal stability [21, 22]. Also, there is a need to enhance the stability and support for the interlayer towards higher thermal stability and more support of the top layer. TiAlN offers, compared with TiN, higher thermal stability and high hardness by adding aluminium to the cubic TiN [23]. For further development, the TiAlN enhanced the support for the oxidic top layer.

8.4.1
Deposition Process Development of Crystalline PVD Al$_2$O$_3$

Different PVD technologies are available to generate Al$_2$O$_3$ thin-film coatings. One technology is the r.f. sputtering of ceramic target materials [19, 24]. However, the low deposition rate (only 0.12–0.48 µm h^{-1} in the case of alumina) [19, 24] and the high expenditure are disadvantages of these processes and restricted industrial applications [25]. The major problem with the common DC sputtering technology is the so-called target poisoning depending on the oxygen ratio. The chemical reaction takes place not only on the substrate, but also on the chamber walls and on the target. When a certain oxygen partial pressure is exceeded, the target is totally poisoned and the deposition rate decreases. On the other hand, a low oxygen partial pressure leads to the deposition of metallic, conductive material. For this reason, the partial pressure has to be controlled closely. Another problem is the formation of an insulating layer on the shielding and the targets, which leads to potential displacement and an increasing tendency for arcing, respectively. This makes the reactive DC sputtering unattractive for industrial applications. To solve these problems, Schiller *et al.* used pulsed power supplies [27, 28]. A further beneficial effect of the pulse technology is the increase in the plasma density [28]. In 1998, Belkind *et al.* examined the influence of different pulse parameters on the reactive sputtering of alumina by applying the unipolar pulse technique [29]. The critical frequency where arcing occurs depends on cathode current, pressure and reverse time (duration of the positive discharging time) and lies in the range 1–80 kHz. It increases with decreasing pressure and reverse time.

8.4.1.1 Experimental Details for the Development of γ-Al$_2$O$_3$

The development of crystalline γ-Al$_2$O$_3$ for application in tool inserts was done using a Model Z400 laboratory sputtering PVD coating machine (Leybold Heraeus, Hanau, Germany). For deposition of the bilayer system (TiAlN/γ-Al$_2$O$_3$, a Melec SPIK 2000 A pulsed power supply was used. For the target material, an aluminium metal target of 99.2% purity and a titanium–aluminium metal target (50:50 at.%) with a purity of 99.9% was used. Prior to deposition, the vacuum chamber is evacuated to a base pressure of 7.8×10^{-5} mbar. The process gas argon and the reactive gases nitrogen and oxygen possess a purity of 99.999%. The Ar, N$_2$ and O$_2$ flow rates are controlled independently by mass flow controllers. For the TiAlN interlayer the N$_2$ concentration was kept constant at 30%. while the oxygen flow was varied between 2 and 16%. The variation of the deposition parameters is reported in Table 8.2. The substrates were also precleaned by r.f. sputtering for 30 min using the parameters listed in Table 8.2.

8.4.1.2 Results and Discussion

The challenge for the development of crystalline Al$_2$O$_3$ coatings lies in avoiding the formation of a dielectric reactive layer at the targets. A strong O$_2$ flow-dependent hysteresis effect is observed and needs to be stabilized within a small process window. At a significant oxygen flow rate, the effect of target poisoning is observed by a drop in voltage and deposition rate (coating time = constant) (see Figure 8.14). This effect marks the transition region of the target's surface from a metallic (zone 1) to a crystalline (zone 2) and finally to an amorphous zone (zone 3). By forming a nanocrystalline structure, an increase in mechanical properties and hardness was achieved for an oxygen flow rate of 2.25 sccm (Figure 8.14) [30].

To investigate the high-temperature stability during the thixoforming process, high-temperature X-ray diffraction (XRD) measurements were performed with the

Table 8.2 Deposition parameters for TiAlN/γ-Al$_2$O$_3$.

Process parameters	Etching phase	Interlayer (Ti$_{0.5}$Al$_{0.5}$)N	Top layer Al$_2$O$_3$
Process pressure (Pa)	1	0.5	0.2–0.5
Process time (s)	1800	3600	3600
R.f. power (W)	100	—	—
R.f. frequency (MHz)	13.56	—	—
N$_2$, O$_2$ concentration (%)	—	30	2–16
Pulse parameters			
Pulse mode/pulse power (W)	—	Unipolar/500	Bipolar/800
Pulse sequence ($t_{on-}/t_{off-}/t_{on+}/t_{off+}$) (µs)	—	5/25	5–13/5–13/5–13/5–13
M.f. (kHz)	—	33	19.2–50
Pulse voltage (V)	—	637	280–360
Pulse current (A)	—	0.8	0.9–2

developed coatings. These investigations showed the stability of the γ-Al$_2$O$_3$ up to 1000 °C [25]. To enhance the weak adhesion of crystalline Al$_2$O$_3$ (10 N scratch load without an interlayer), the developed TiAlN–γ-Al$_2$O$_3$ system offers a critical scratch load up to 70 N on the hot working steel substrates. Figure 8.15 shows the deposited γ-Al$_2$O$_3$ without an interlayer (Figure 8.15a) and with the TiAlN interlayer (Figure 8.15b).

8.4.1.3 Summary of the Development of the TiAlN–γ-Al$_2$O$_3$ Bilayer System

By using pulsed power supplies, it is possible to deposit fine crystalline γ-Al$_2$O$_3$ with good mechanical and tribological properties compared with an uncoated material. The coated samples are analysed by common thin-film techniques such as nanoindentation and XRD. The deposition of γ-Al$_2$O$_3$ is limited to a very low deposition

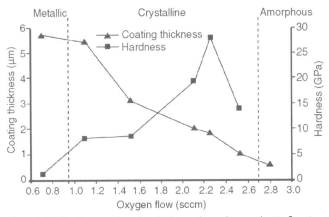

Figure 8.14 Hardness and coating thickness depending on the O$_2$ flow (coating time = constant).

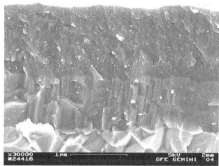

Figure 8.15 Monolayer γ-Al₂O₃ (a) and TiAlN/γ-Al₂O₃ bilayer (b).

pressure (0.2 Pa) and a high target power density (78 W cm^{-2}). The deposited coatings offer a high critical scratch load (70 N), very high hardness (28 GPa) and high Young's modulus (351.8 GPa). After the promising results of this material screening, this type of coating is being further developed for coating on industrial-sized coating units to coat tool inserts for application and forming experiments.

8.4.1.4 Upscaling γ-Al₂O₃ from a Laboratory-scale Unit to an Industrial Coating Unit

The upscaling of a deposition process developed on a laboratory coating unit to an industrial coating unit still remains a challenge. Especially the transfer of oxidic coatings is difficult, because of their hysteresis behaviour during deposition. There are different adjustable parameters, such as pumping speed of the turbopump [the Z400 laboratory coater has a 5.5 times higher pumping capacity than the CemeCon (Würselen, Germany) CC800/9 SinOx industrial unit] and the power density is twice as high in the Z400. Furthermore, the gas inlet in the laboratory coater is located near the substrate, whereas in the CC800/9 SinOx unit it is attached next to the cathodes. These parameters strongly influence the coating behaviour and the coating properties.

8.4.1.5 Experimental Details for the Development of γ-Al₂O₃ on an Industrial Coating Unit

Deposition is performed using a four cathode setup in a two-by-two dual cathode arrangement on a CemeCon CC800/9-SinOx industrial coating unit. The coatings are produced by co-sputtering of two TiAl targets and two Al targets. The size of the TiAl targets is 88 × 500 mm. The Al targets have a purity of 99.2%. For the Ti₅₀Al₅₀ targets the purity is about 99.5% for the titanium and of 99.9% for the cylindrical aluminium inserts within the sputter track. During deposition, the samples are moved in a planetary motion to ensure a constant film thickness distribution at the substrates. Before deposition, the samples are ion etched in an argon atmosphere. For this substrate cleaning process, an m.f. power source is used. The coatings are deposited using a MELEC pulse power supply (SPIK 2000A). For the (Ti$_{1-x}$Al$_x$)N interlayer a total pressure of 500 mPa by reactive sputtering in a mixed atmosphere of argon and nitrogen is used. The deposition of the Al₂O₃ top layer is performed by

Table 8.3 Process parameters for coating deposition.

Process parameters	Etching phase	Interlayer $(Ti_{1-x}Al_x)N$	Top coat γ-(Al_2O_3)
Heating power (kW)	16	16	16
Process step time (s)	3600	8500	8000
Total pressure (mPa)	Flow controlled	500	Flow controlled
M.f. voltage (250 kHz) (V)	650	—	—
Bias voltage (V)	—	80 (DC)	25 (pulsed m.f.)
Nitrogen flow (mln)	—	60	—
Oxygen flow (mln)	—	—	varied
Argon flow (mln)	150	Pressure controlled	250
Target power density (W cm^{-2})	—	13.7	7.95
Anode voltage (V)	—	70	—

using two aluminium targets (target size 88 × 500 mm) using the dual cathode arrangement (power 3500 W, pulse frequency 18.5 kHz). The process is flow controlled: the argon is kept at 250 sccm while the oxygen flow is varied. The oxygen partial pressure was controlled by adjusting a certain voltage at the cathodes. The process parameters are listed in Table 8.3.

8.4.1.6 Results of the Deposition of γ-Al$_2$O$_3$ on an Industrial Coating Unit

The properties of the deposited coatings are strongly dependent on the working point in the hysteresis (Figure 8.16). To obtain the right coating properties for the dies, different working points are examined. Two of these working points presented as an example are compared using different thin-film analysis equipment such as phase analysis (XRD), mechanical properties (nanoindentation) and SEM. One point is chosen near the metallic sputtering mode at a cathode voltage of 560 V, and the second one in the lower area at 460 V near the fully poisoned region. With an increase in oxygen partial pressure, the deposition rate dropped from 40 nm min^{-1} at 560 V to 18 nm min^{-1} at 460 V.

Further investigations by using the XRD patterns in these working points show a strong increase of the γ-phase from 460 to 560 V. At 560 V, all important γ-peaks can be identified. For the γ-Al$_2$O$_3$ the JCPDS card 10-425, for the substrate the JCPDS card 25-1047 and for the cubic TiAlN interlayer the JCPDS card 38-1420 for TiN and the JCPDS card 25-1495 for AlN are used [31]. These results were compared with those for the γ-Al$_2$O$_3$ coating of the laboratory coater (Figure 8.17).

This strong difference in the crystallinity can also be seen in the SEM images (Figure 8.16). Comparing the different working points 460 and 560 V on the industrial coating unit and the operation point of 310 V on the laboratory coater, only 560 and 310 V show all 100% intensity peaks of γ-Al$_2$O$_3$, the $\langle 400 \rangle$ and $\langle 440 \rangle$. With an increase in crystallinity of the deposited coating, the mechanical properties also changed massively. The hardness and Young's modulus are obtained by using nanoindentation. The hardness increased from 11.2 ± 1.1 GPa at a deposition voltage

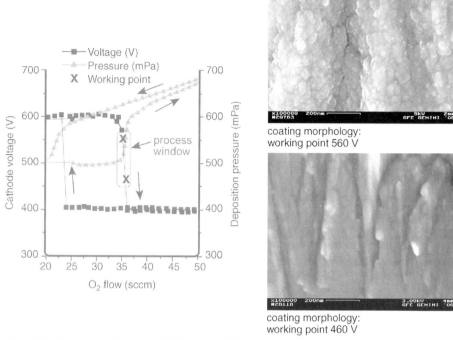

coating morphology:
working point 560 V

coating morphology:
working point 460 V

Figure 8.16 Hysteresis behaviour and SEM image of the two working points.

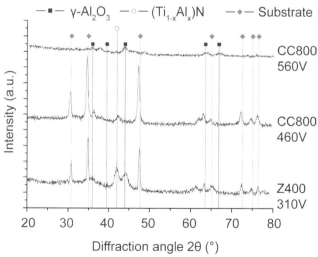

Figure 8.17 X-ray diffraction pattern of the different deposited coatings on the laboratory and industrial coating unit.

of 460 V to 27.07 ± 1.7 GPa at 560 V. The Young's modulus also increased from 246.7 ± 24.3 GPa at 460 V to 368.6 ± 10.16 GPa at 560 V. A result of these investigations is that the deposition process window is small and for the properties of the coatings it is necessary to operate in this window. Modern plasma diagnostic equipment can be used to control the deposition process. First investigations using optical emission spectroscopy (OES) have shown the ability to control this small process window [31].

8.4.2
Deposition Process Development of Crystalline PVD ZrO_2

The advantage of the ZrO_2-based coatings compared with the Al_2O_3 coatings is high chemical inertness against liquid steel. These results are confirmed by the material screening of Meyer-Rau [5]. Zirconium oxide occurs in three stable crystallographic phases, depending on the temperature. The monoclinic phase is stable up to 1170 °C, the tetragonal phase is stable within the range 1170–2370 °C and at higher temperatures the cubic phase is stable [32]. The thermal and mechanical properties depend strongly on the phase structure [34]. In comparison with the other phases, the tetragonal phase of ZrO_2 offers high strength and toughness values. Various methods have been investigated to stabilize the promising high-temperature tetragonal phase at room temperature. The most common technology used is to add dopants such as Y_2O_3, MgO and CaO to suppress the tetragonal-to-monoclinic transformation.

8.4.2.1 Experimental Details for the Development of t-ZrO_2
The investigations on the deposition of crystalline ZrO_2 were carried out on a Leybold Heraeus Z400 laboratory sputtering PVD coating machine for the development of a suitable deposition process window and a multilayer structure using tungsten as an interlayer with a high thermal conductivity. For deposition, a zirconium metal target with a purity of 99.5% and a tungsten metal target with a purity of a 99.9% were used. The argon and oxygen flow rates were independently controlled by mass flow controllers. The O_2 gas concentration was determined by the ratio of O_2 mass flow rate to the total mass flow rate (Ar and O_2). As a pulsed power supply, an Eifler bipolar pulse generator was used. The variations of the deposition parameters are reported in Table 8.4. The substrates were heated to up to 300 °C. The substrates were also precleaned by r.f. sputtering using the parameters listed in Table 8.4.

8.4.2.2 Results and Discussion
By using pulsed power supplies, it is possible to deposit fine crystalline ZrO_2 with different mechanical and tribological properties. The coated samples show different properties that depend strongly on the chosen process working points. The variation of the deposition parameters such as deposition pressure and oxygen flow rate allow different phases of ZrO_2 to be synthesized. It is also possible to stabilize t-ZrO_2 at room temperature without the use of dopands such as Y_2O_3, CaO or MgO. The deposition parameters of these samples are presented in Table 8.5.

Table 8.4 Deposition parameters for the multilayer system W–ZrO$_2$.

Process parameters	Etching phase	Interlayer tungsten	Top layer ZrO$_2$
Process pressure (Pa)	1	0.5	0.3–1
Process time (s)	1800	3600	7200
R.f. power (W)	100	—	—
R.f. frequency (MHz)	13.56	—	—
N$_2$, O$_2$ concentration (%)	—	—	2–16
Pulse parameters			
Pulse mode/pulse power (W)	—	DC/400	Unipolar/500
Pulse sequence (t$_{on-}$/t$_{off-}$) (μs)	—	5/25	3.56/2.67
M.f. (kHz)	—	0	33
Pulse voltage (V)	—	Power controlled	980–1022
Pulse current (A)	—	—	0.5

Table 8.5 Properties of four representative coatings.

Property	Metallic ZrO$_2$	Monoclinic ZrO$_2$	Tetragonal + monoclinic ZrO$_2$	Tetragonal ZrO$_2$
Depositing pressure (Pa)	1	1	0.5	0.3
Coating thickness (μm)	9.9	3.8	8.3	6
O$_2$ (sccm)	1.76	3.96	2.07	1.7
O$_2$/Ar (%)	2	2	5	7
Phase composition	Metallic	Monoclinic	Tetragonal + monoclinic	Tetragonal

To study the effect of the different crystalline phase on the properties of the films, four characteristic samples were selected. By changing the O$_2$ to argon ratio, a drop in the deposition rate in nanometres per minute can be observed. The effect of the oxygen flow rate on the coating thickness is shown in Figure 8.18 This behaviour can be explained by the hysteresis effect of the deposition process [33]. With increasing oxygen flow rate, the coating structure changed from metallic to crystalline and finally to amorphous (Figure 8.19). The phase composition of the deposited samples was determined using XRD as described previously. By changing the oxygen to argon ratio, the experiments revealed that with a deposition pressure of 1 and 0.5 Pa only a small amount of tetragonal phase can be deposited. For the coating, the development of t-ZrO$_2$ coatings with very small crystallite sizes is needed to stabilize this phase [34]. Calculations using the Debye–Scherer equitation show that the domain size for the t-ZrO$_2$ coatings is in the range 4–5 nm.

Figure 8.19 shows the cross-sections of the investigated samples. The cross-section show in Figure 8.19c indicates a very fine-grained morphology for the tetragonal phase compared with the other three SEM images.

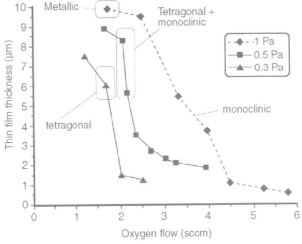

Figure 8.18 Effect of oxygen flow rate on the film thickness.

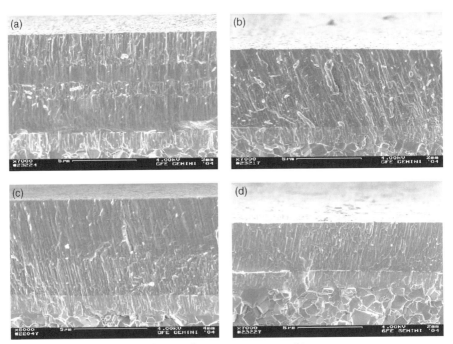

Figure 8.19 Cross-section of (a) metallic zirconium structure, (b) tetragonal + monoclinic ZrO_2 structure, (c) tetragonal ZrO_2 structure and (d) monoclinic ZrO_2 structure.

Table 8.6 Mechanical properties of the deposited coatings.

Property	Metallic ZrO$_2$	Monoclinic ZrO$_2$	Tetragonal + monoclinic ZrO$_2$	Tetragonal ZrO$_2$
Hardness, H_{Univ} (GPa)	16.5	5.7	15.1	21.4
Young's modulus (GPa)	238.7	146.8	241.9	292.7

The mechanical properties of all samples are given in Table 8.6. The t-ZrO$_2$ exhibits the highest hardness and Young's modulus of all deposited ZrO$_2$ modifications.

8.4.2.3 Summary for the Development of t-ZrO$_2$

The investigations have shown that it is possible to find a process window where stabilized t-ZrO$_2$ can be synthesized without the use of dopands such as Y$_2$O$_3$, CaO or MgO. The t-ZrO$_2$ exhibits the highest hardness and Young's modulus of all deposited ZrO$_2$ modifications. The deposition of t-ZrO$_2$ is limited to a narrow process window. Further developments such as upscaling to an industrial coating unit are needed to apply these promising coatings to tool inserts.

8.4.2.4 Experimental Details on the Development of t-ZrO$_2$ on an Industrial Coating Unit

For upscaling, a co-sputtering setup with two TiAl targets and two zirconium targets was used. The size of all targets was 88×500 mm. The purity of the zirconium targets was 99.2%. For the Ti$_{50}$Al$_{50}$ targets, the purity was about 99.5% for the titanium and 99.9% for the cylindrical aluminium inserts within the sputter track. During deposition, the samples were moved in a planetary motion to ensure a constant film thickness distribution at the substrates. Prior to deposition, the samples were ion etched in an argon atmosphere. For the etching process, an m.f. power source was used. The coatings were deposited by using a MELEC pulse power supply (SPIK 2000 A). For the TiAlZr N interlayer a total pressure of 500 mPa by reactive sputtering in a mixed atmosphere of argon and nitrogen was used. To enhance adhesion, the coating process started first with a pure TiAlN interlayer and after a coating time 1000 s zirconium was added to the process. Pre-examination showed that the adhesions with a TiAlZrN interlayer suffer in comparison with a graded one. The deposition of a ZrO$_2$ top layer was performed by using two zirconium targets (purity 99.2%; target size 88×500 mm) and the dual cathode arrangement (power 3500 W, pulse frequency 18.5 kHz). The process was flow controlled: the argon flow was kept constant at 250 sccm while the oxygen flow was varied. The oxygen partial pressure was set by adjusting a certain voltage at the cathodes. The process parameters are listed in Table 8.7.

8.4.2.5 Results of the Deposition of t-ZrO$_2$ on an Industrial Coating Unit

The investigations of the deposition process of the ZrO$_2$ top layer showed nearly the same behaviour as the γ-Al$_2$O$_3$ deposition. The properties of the deposited coatings

Table 8.7 Process parameters for the coating deposition.

Process parameters	Etching phase	Interlayer TiAlZr	Top coat t-(ZrO$_2$)
Heating power (kW)	16	16	16
Process step time (s)	3600	5000	6000
Total pressure (mPa)	Flow controlled	500	Pressure controlled
m.f. voltage (250 kHz) (V)	650	—	
Bias voltage (V)	—	80 (DC)	25–40 (pulsed m.f.)
Nitrogen flow (sccm)	—	60	—
Oxygen flow (sccm)	—	—	Varied
Argon flow (sccm)	150	Pressure controlled	250
Target power density (W cm^{-2})	—	13.7	7.95
Anode voltage (V)	—	70	—

are strongly dependent on the working point in the hysteresis (Figure 8.20). Different points were investigated: two deposited coatings with 650 and 600 V cathode voltages are presented here (Figure 8.20).

The properties of the coated samples from the laboratory coater and the industrial coating unit are compared in Table 8.8. The two working points show

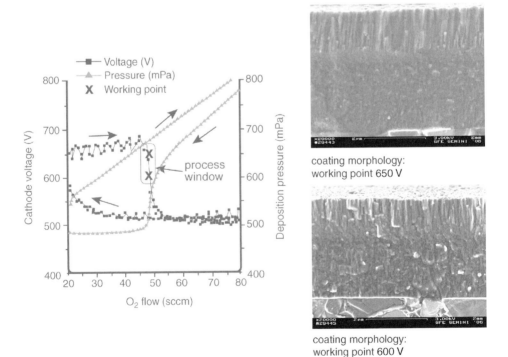

coating morphology:
working point 650 V

coating morphology:
working point 600 V

Figure 8.20 ZrO$_2$ hysteresis and two deposited coatings.

Table 8.8 Mechanical and phase properties of the deposited coatings.

Property	Sample name			
	Monoclinic ZrO₂	Tetragonal ZrO₂	ID 650V	ID 600V
Coating unit	Z400	Z400	CC800/9	CC800/9
Phase type	Monoclinic	Tetragonal	Monoclinic + tetragonal	Monoclinic + tetragonal
Young's modulus (GPa)	146.76 ± 13.62	292.7 ± 33.2	213.3 ± 22.6	294.6 ± 23.8
Hardness, H_{Univ} (GPa)	5.72 ± 1.09	21.35 ± 3.44	8.06 ± 1.5	13.9 ± 1.9

that with a decrease in the cathode voltage (increase in O_2 partial pressure), the hardness and Young's modulus increase. The amount of tetragonal phase also increases.

The investigations showed that it is not possible to deposit only tetragonal phase with this deposition parameter setup. Recent investigations showed that a small grain size below 6 nm is needed to stabilize only the tetragonal phase [37]. By increasing the bias voltage to 40 V, the ion bombardment on the substrate also increases, so it can be seen that only the tetragonal phase is growing on the substrate. These deposition parameters are taken into account for the coating of the forming dies.

8.4.3
Model and Near Application Tests

To test the behaviour of the different coated samples, model tests were developed to study by isolating single process loads, that is, wear, corrosion and thermal shock resistance. The results obtained by these tests led to the identification of different degradation and failure mechanisms that are related to the specific loads applied in each test [5, 35, 36]. After assembling the data obtained in model tests for each coating system, the prediction of material behaviour during thixoforming process is difficult. Practical fast and cheap forming experiments and setups are needed. The high-temperature compression tests combine mechanical, thermal and corrosive attack in one test setup. For the investigation, both die materials 1.2999 and TZM and both coating systems γ-Al_2O_3 and t-ZrO_2 were selected. The experiments were carried out at the IBF (Institute of Metal Forming), conducted using a servo hydraulic testing machine (Carl Schenck, Darmstadt, Germany), a maximum testing load of 640 kN and a velocity of 600 mm s^{-1}. The testing machine was equipped with a tube furnace for compression tools and an induction heater for steel billets. The compression tests were carried out under an inert gas atmosphere. The testing tool plate was fixed on the lower end of the top plunger. Billets were placed on a ceramic compression plate

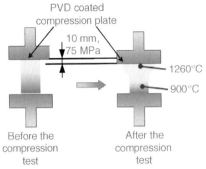

Figure 8.21 Experimental setup for the compression tests.

situated on the movable plunger. The top plunger with the coated tool plate was heated by the tube furnace to a temperature of $\sim$400 °C. The setup was comparable to the tool temperature in the current semi-solid forging process. The steel billet was heated up to the semi-solid range. The testing procedure is illustrated in Figure 8.21. The steel billet was heated up to 1350 °C within 202 s and compressed at $\sim$1260 °C. The billet was compressed at a velocity of 10 mm s^{-1} up until a displacement of 10 mm was reached. The contact duration of hot billets with the tool was 1 s per test cycle. The procedure was repeated 50 times for the each coated plate. The stress on the surface of the compression plates was 75 MPa.

The PVD-coated compression plates do not show any visible signs of coating failure or rupture at the surface. X-ray grazing incidence measurements show that the deposited phase structures γ-Al$_2$O$_3$ and also the high-temperature stable t-ZrO$_2$ phase are still present after forming operations [40]. No traces of phase transformation from γ-Al$_2$O$_3$ to α-Al$_2$O$_3$ or tetragonal to monoclinic ZrO$_2$ are observed, despite the high surfaces temperatures. With respect to adhesion towards forming materials, t-ZrO$_2$ displays the least built-up of iron, which can be gathered from the EDS data shown in Figure 8.22 after 50 compression cycles.

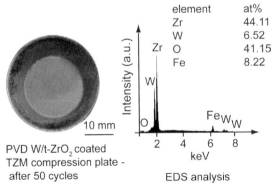

Figure 8.22 Compression plate after 50 forming cycles.

The EDS graph clearly shows the zirconium reflex of the coating and the tungsten reflex of the interlayer. Traces of iron can also be detected. The Al_2O_3-coated plates show more iron build-up than the ZrO_2-coated plates. However, both ceramic PVD coatings show no visible signs of serve erosion or abrasion. These investigations have shown that both coating candidates are suitable for industrial forming operations.

8.4.4
Applicability of PVD-Coated Dies in Steel Thixoforming

Thixoforging, -casting and -extrusion dies are coated with application of the parameters determined on the above-mentioned samples. Prior to deposition, the tool substrates are hardened to 54 HRC and subsequently cleaned ultrasonically with industrial cleaning equipment containing surfactant solutions and rinsed with deionized water. Finally, the dies are dried in hot air. During deposition, the dies are moved in a planetary motion to ensure an even film distribution. Figure 8.23 shows (a) a coated thixoextrusion die and (b) three coated inserts for the thixocasting process. For the bilayer coating system, the TiAlN interlayer is 5.2 and 2.3 µm for the γ-Al_2O_3 top coat.

The forming experiments on the different coated tools were carried out at the IBF and GI (see also Chapter 9). The coated parts show good performance during the thixoforming tests. The different coated parts were now analysed for phase stability with EDS, SEM and XRD. For the extrusion dies, the coatings showed excellent performance so far in the first experiments [38] (Figure 8.24).

The coated samples were also investigated in the thixoforging process. Figure 8.25 shows the results of the forming experiments carried out at the IBF with the hybrid tool concept. A schematic drawing of the concept is shown in Figure 8.5. The upper dies are coated with γ-Al_2O_3 and t-ZrO_2 PVD coatings. For the examination, a γ-Al_2O_3 coated insert was analysed after 33 forming cycles using X210CrW12 as forming material.

The SEM and EDS investigations showed that there are several traces of iron in the centre of the coated tool after the forming experiments. The coating is still present here; this can be seen by the aluminium and oxygen peaks. Also, the coating remains stable around the edges. Small pieces of iron can be detected here. No traces of iron can be seen in the inner area.

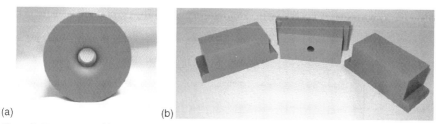

(a)   (b)

Figure 8.23 PVD coated hot working steel and TZM dies with TiAlN/γ-Al_2O_3.

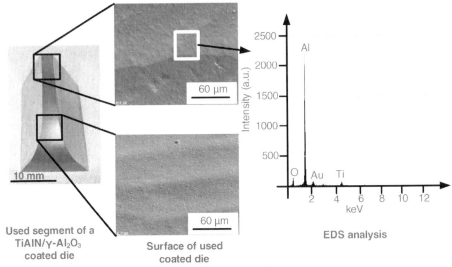

Figure 8.24 SEM and EDS analysis of the used coated extrusion segment.

8.4.5
Perspectives for the Application of PVD-Coated Dies in Steel Thixoforming

Competitive solutions for tools have to provide high mechanical strength at high temperatures, good resistance against thermal shock and the ability to withstand the attack of the semi-solid material. The thixoforming experiments showed promising results for the developed tooling coating concepts and materials. Both systems,

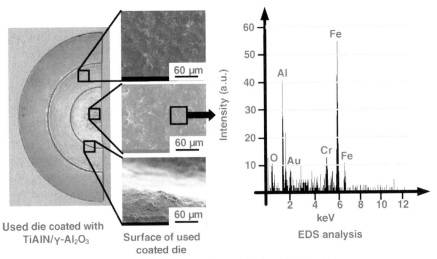

Figure 8.25 Coated tool after 33 forming cycles and SEM and EDS analysis.

γ-Al$_2$O$_3$ and t-ZrO$_2$, worked well in the forming experiments. A suitable interlayer concept was developed to provide good adhesion in order to support the oxidic PVD top layer. The combination of oxide coatings with a TiAlN interlayer showed good results after SEM and EDS evaluation. A comparison between uncoated and coated dies confirmed the stability of the oxide coating system and its potential for technological application in the semi-solid forming process.

8.5
Developing Al$_2$O$_3$ PECVD Coatings for Thixoforming Moulds

Alumina (Al$_2$O$_3$) coatings are known to exhibit advantageous mechanical properties and also chemical inertness and oxidation resistance at high temperatures [39]. Thin alumina films act as diffusion barriers for Ni, Al, Ti and Ta even at 1100 °C during annealing for 100 h [40] and it has been shown that this material is a suitable candidate for coatings used in wear and hot gas protection applications [41, 42]. Therefore, in this work, we investigated alumina coatings for protecting die surfaces during the semi-solid processing of steel [43]. Despite the availability of tooling concepts for the semi-solid processing of Al-based alloys [44], the increase in the tool lifetime is an almost unaddressed challenge for the semi-solid processing of steel [43]. The degradation of the die is enabled through the interaction between the die and the casting alloy at temperatures >1300 °C [45] and may result in the formation of intermetallic phases, chemical reaction products, thermal fatigue cracking and wear by erosion and corrosion [46, 47]. Alumina, exhibiting the previously discussed properties, is therefore a promising candidate to fulfil the requirements of semi-solid processing of steel. Alumina as a polymorphous material exists in several modifications (α-, κ-, γ-, δ-, θ-Al$_2$O$_3$). The alumina polymorphs presenting the highest protective performances are the thermodynamically stable α-alumina and the metastable κ-alumina. Various deposition techniques including CVD [42] and PECVD methods have also been employed to synthesize alumina coatings at relatively low temperatures. Lin *et al.* reported the growth of amorphous alumina for deposition temperatures ranging between 200 and 600 °C [48], while Täschner *et al.* reported a mixture of α- and γ-alumina and even phase pure α-alumina, utilizing an unipolar pulsed DC plasma at a substrate temperature of 650 °C [49].

Although the reduction of the deposition temperature allowed by these latter methods is significant, it is not sufficient for coatings on steel since tempering of steel tools occurs at approximately 550 °C [50]. The only published studies reporting deposition of α alumina at temperatures permitting deposition on steel tools are based on homoepitaxial [51] and localized epitaxial [52] growth on chromium templates. However, the reported deposition rates (1 nm min^{-1}) of both of these processes are not compatible with the requirements of the die coating industry. The major aim of our work within the SFB 289 project was to design a process to perform the deposition of α-alumina at substrate temperatures allowing deposition on steel dies, namely at 550 °C. To achieve that goal, a bipolar DC PECVD method has been developed. The systematic study of the relationship between the process conditions and coating constitution allowed the determination of a phase diagram for alumina as a function of substrate temperature and electrical power injected into the plasma. Special attention has been devoted to the mechanical properties of the coatings. Their behaviour has been correlated with coating constitution and microstructure. The latter can be porous. This porosity evolution can be understood by considering different mechanisms, including the formation of residual chlorine bubbles in the coatings. Finally, characterization of the adhesion, the tribological behaviour and the thermal shock resistance has been performed in order to evaluate the performances of the coatings in 'close to real' conditions.

8.5.1
Experimental Procedure

Alumina thin films were deposited in a bipolar pulsed PECVD system. The plasma was generated by means of a pulse switch unit (SPIK 1000A, Melec), which was connected to a DC generator (MDX, Advanced Energy). Two stainless-steel electrodes with a diameter of 160 mm were biased with a negative peak voltage (V_P) ranging from -720 to -900 V. A schematic drawing of the experimental setup is shown in Figure 8.26.

The pulse unit was operated in the bipolar symmetric mode with a rectangular voltage pulse form. The negative and positive pulse times were adjusted to 50 and 30 μs, respectively, and the time between pulses was set to 5 μs (Figure 8.27).

The ion irradiation period of the growing film corresponds to the negative pulse, which is followed by a positive period in which the film is under electron bombardment and, consequently, the insulating layer is discharged. The discharge current and voltage measurements were performed with a current probe (Tektronix TM P6003) and a high-impedance voltage probe (Elditest GE 8115), respectively. The substrate material was tempered hot working steel X38CrMoV5-1. Prior to deposition, the

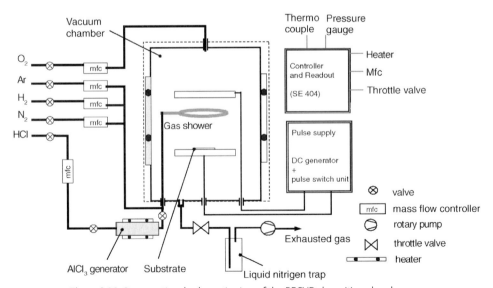

Figure 8.26 Cross-sectional schematic view of the PECVD deposition chamber.

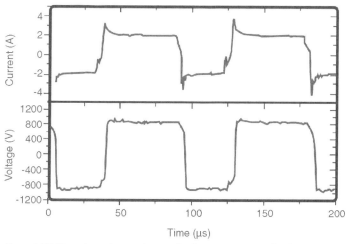

Figure 8.27 Time-dependent evolution of substrate current and voltage [53].

substrates were mirror polished, ultrasonically cleaned in acetone and methanol and positioned on the cathode, which served as substrate holder. The substrates were then heated to the deposition temperature, T_s, (500–600 °C) in an Ar-H$_2$ plasma. T_s was monitored by means of a Ni-Cr-Ni thermocouple attached to the substrate holder. The deposition was carried out in an AlCl$_3$–O$_2$–H$_2$–Ar mixture at a total pressure of 175 Pa and a total gas flow rate of 22.3 slh (standard litres per hour). Except for AlCl$_3$, all gases had a purity of 99.999%. AlCl$_3$ was generated *in situ* by flowing HCl over Al chips of purity >99.99% at a temperature in the range 500–550 °C. At this temperature, the reaction is complete and the AlCl$_3$ flow can be controlled by means of the HCl flow. The precursor content (AlCl$_3$ + O$_2$), P_C, and the precursor ratio (AlCl$_3$/O$_2$), P_R, were varied keeping both the total flow rate and the pressure constant by adjusting the Ar and H$_2$ flows. The experimental conditions are summarized in Table 8.9.

The morphology and chemical composition of the deposited films were measured by SEM and EDS, respectively. The crystallographic structure was determined by grazing incidence X-ray diffraction (GI-XRD) using 900 W Cu Kα radiation with a Siemens D500 diffractometer, which was operated at a grazing incidence of 3°. Planar thin foils were examined in a FEI Tecnai G^2 TF 20 Ultra-Twin field emission gun transmission electron microscope operated at 200 keV. Hardness and elastic modulus were measured by nanoindentation using a triboindenter apparatus from Hysitron. The tip was calibrated against fused quartz according to the method described by Oliver and Pharr [54]. We assumed a Poisson's ratio of 0.3 and each sample was characterized by at least 20 indents at a maximum penetration depth of <5% of the film thickness. Scratch tests were used to evaluate the adhesion of the films by measuring the critical load. The tribological properties of the coatings were evaluated using a CSM Instrument tribometer operated in ball-on-disk configuration and at a linear speed of 5 cm s^{-1}. A 100Cr6 steel ball was loaded on to the coating surface with a normal load of 2 N and the steady-state value of the friction coefficient was evaluated after a sliding distance of 200 m. The tests were performed at 25 °C and 80% humidity. Finally, thermal shock resistance was evaluated by pressing an M2 steel surface against the

Table 8.9 Experimental conditions during the deposition experiments: peak voltage (V_P), deposition temperature (T_s), pressure, precursors ratio (P_R) and precursor content (P_C).

	Peak voltage (V)	Temperature (°C)	Pressure (Pa)	$AlCl_3/O_2$ ratio	$AlCl_3 + O_2$ content (%)
Set of experiments 1: variation of V_P and T_s	−720 to −905	500–600	175	1	2
Set of experiments 2: variation of P_c	−900	600	175	1.15	0.8–5.3
Set of experiments 3: variation of P_R and T_s	−900	540–600	175	0.5–1.15	2–3.45

coated sample at temperatures of $1170 \pm 30\,°C$ for 5 s. The sample was then cooled under an air stream for 15 s. This procedure was repeated 1000 times. After this treatment, a cross-section of the sample was evaluated by SEM and EDS.

8.5.2
Results and Discussion

8.5.2.1 Chemical Composition and Constitution
The influence of the power density (P_D) on the films' constitution was studied by varying the peak voltage from −720 to −905 V (set of experiments 1, Table 8.9). P_D was calculated by multiplying the measured current by the voltage, which was then normalized with respect to the cathode surface area. The influence of P_D on the chemical composition of the films was evaluated by EDS. No dependence of the Al/O ratio which was measured to be 0.59 ± 0.02 at.% was observed. Nevertheless, it was found that all films contain chlorine at levels from 0.9 ± 0.01 to 2.3 ± 0.01 at.% depending on their constitution. This will be discussed later. The phase-formation data from 12 experiments were compiled into one diagram correlating T_S and P_D with the film constitution (Figure 8.28).

The data in the diagram can be grouped into three regions according to the constitution of the films, namely region I, phase pure γ-alumina; region II, mixture of γ- and α-alumina; and region III, phase pure α-alumina. In this diagram, it is clearly observed that, on increasing the energy supplied to the growing films (thermal or through particle bombardment), α-alumina is preferentially grown. It is important to note that by increasing P_D to $6.6\,W\,cm^{-2}$ and T_S at about $560\,°C$ it is possible to grow phase pure α-alumina.

According to theory, the mobility and dwell time of adsorbed particles on the surface are important for the film constitution. These parameters are mainly controlled by the energy provided to the growing film. In PECVD processes, this energy can be provided not only thermally by substrate heating as in conventional CVD processes, but also by collision of adsorbed particles with the ions bombarding the surface. Due to the latter phenomena, the adsorbed particles received additional energy which led to higher mobility and therefore to modification of the film

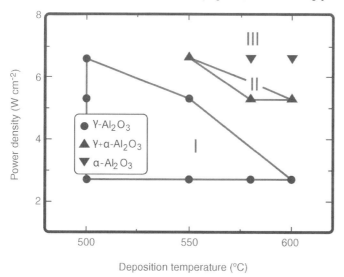

Figure 8.28 Phase-formation diagram of alumina films deposited by PECVD [44].

structure [55]. Based on the results presented here, it is proposed that both ion flux and ion energy at the substrate surface play an important role in the formation of α-alumina. Thus, a significant reduction in the deposition temperature can be obtained by enhancing the surface mobility of the deposited atoms by increasing the electrical power density at the substrate, which leads to an increase in both ion flux and ion energy at the substrate surface [44].

Other process parameters than the peak voltage can be controlled in order to tune the ion bombardment, for example, by varying the precursor content, P_C, as has been shown using the second set of experiments (see Table 8.9). In order to quantify the ion bombardment, we determined the normalized ions flux (NIF) according to a method suggested by Kester and Messier [56]. The NIF is a function of the ratio between the measured ion flux and measured deposition flux. The contribution of secondary electrons to the ion flux is assumed to be negligible and the sticking coefficient of the deposited flux is assumed to be 1. The modification of P_C leads to significant modifications of the deposition rate (a) and, therefore, of the NIF. This is presented in Figure 8.29, where one can observe a significant increase in a as a function of P_C.

Conversely, the NIF decreases from 480 to 50, meaning a decrease in the energy supplied to the growing film by ion bombardment. As expected, the constitution of the deposited films has a strong dependence on NIF, as shown in Figure 8.29. As the NIF is increased from 50 to 100, phase pure γ-alumina is formed, whereas at NIF ≥140, phase pure α-alumina is observed (Figure 8.30).

Again, the importance of the ion bombardment on the phase constitution of alumina films is clearly demonstrated and is explained by considering the additional energy supplied to the growing films. For the discharge conditions employed within this set of experiments, we have estimated, based on a simple model put forward previously, that the number of collisions in the cathode sheath is approximately 16 [17], with an average

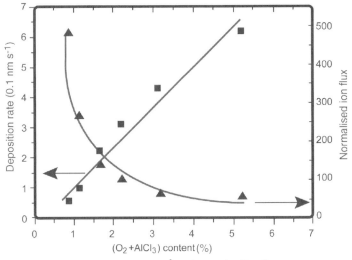

Figure 8.29 Deposition rate (0.1 nm s^{-1}) and normalized ion flux as a function of the precursor concentration (AlCl$_3$ + O$_2$) [57].

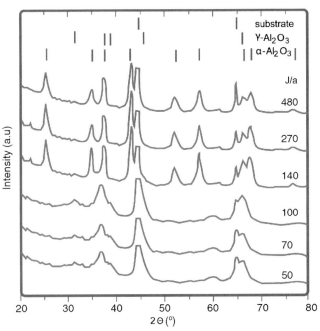

Figure 8.30 Evolution of the crystallographic structure of alumina films with increasing normalized ion flux values for films deposited at 600 °C [57].

ion energy of approximately 100 eV [53]. It is reasonable to assume that 100 eV ions bombarding the growing film surface may help diffusion of adsorbed species.

8.5.2.2 Film Morphology and Porosity

The morphology of the films was evaluated by SEM and transmission electron microscopy (TEM). Three pore populations were observed and are referred to as micro-, meso- and nanopores. Micropores with a diameter in the sub-micrometre range are clearly visible in the SEM images presented in Figure 8.31.

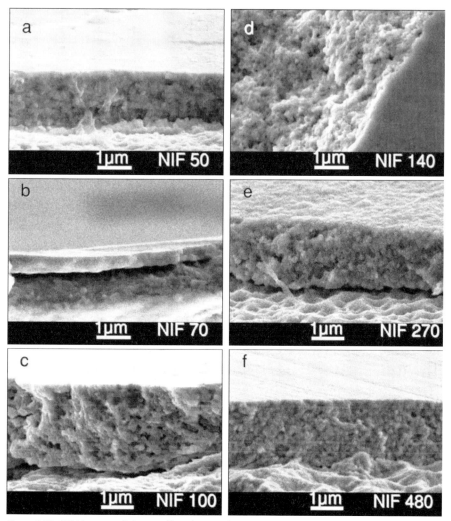

Figure 8.31 SEM images of alumina films deposited at normalized ion fluxes of (a) 50, (b) 70, (c) 100, (d) 140, (e) 270 and (f) 480 [61].

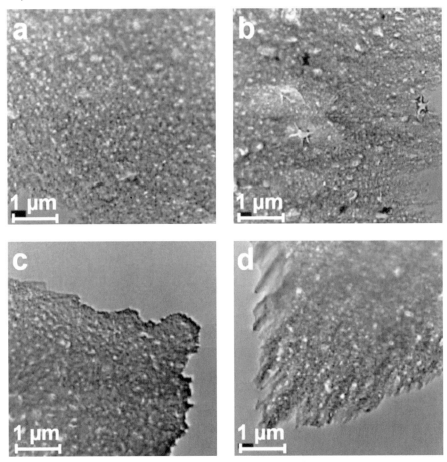

Figure 8.32 Bright-field TEM images of alumina films deposited at normalized ion flux values of (a) 50, (b) 100, (c) 140 and (d) 480 [64].

Mesopores are those pores with diameters varying between approximately 10 and 30 nm observable on bright-field TEM images (Figure 8.32).

Finally, in Figure 8.33, **the** bright-field TEM images show pores belonging to the third population, the nanopores, which have diameters of about 5 nm or smaller.

Several mechanisms could explain the formation of these porous structures. The first is the production of point defects due to ion bombardment and subsequent agglomeration at grain boundaries, dislocations and free surfaces resulting in the formation of pores and voids. A second mechanism which may contribute to the observed porous structure is void formation parallel to the growth direction due to self-shadowing in competitive growth. Finally, the formation of the porous structure under ion bombardment may also be understood based on ion bombardment-induced bubble formation [58, 59]. The incorporation of Ar and N in reactive bias-sputtered TiN films [58] and the incorporation of Cl during PECVD growth of alumina have been

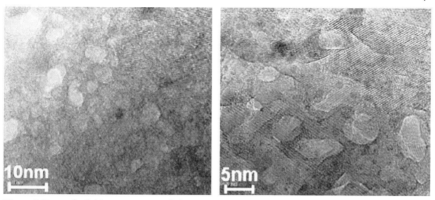

Figure 8.33 Bright-field TEM images of alumina films showing typical distribution of nanopores. The film was deposited at a normalized ion flux of 50 [60].

reported [53, 60]. Hence it is conceivable here that during growth with Cl$_2$ gas, the supersaturated alumina matrix precipitates the incorporated Cl, which may then form bubbles that give rise to the observed porosity. Incomplete decomposition of the AlCl$_3$ molecule in the plasma may be one of the origins of the incorporation of Cl into the films, as has been proposed based on optical emission measurements [53].

8.5.2.3 Mechanical Properties

The mechanical properties of the coatings were determined by nanoindentation. Increasing the NIF from 50 to 100 caused a decrease in the elastic modulus from 227 ± 7 to 152 ± 8 GPa for γ-alumina. A further increase in the NIF to 140 resulted in the growth of α-alumina with an elastic modulus of 220 ± 5 GPa. A detrimental decrease in the elastic modulus to 179 ± 7 GPa was observed with a final increase in the NIF to 480. Hence the elastic modulus values of both the α- and γ-alumina films decrease with increase in NIF. The highest elastic-modulus values were measured for the γ- and α-alumina films grown at minimal NIF, which is necessary for the formation of the corresponding phase. This behaviour may be understood by considering the formation of porous microstructure observed by TEM, which may influence the mechanical properties of the films.

An empirical correlation between measured elastic modulus and density values of bulk alumina samples was developed by Knudsen [61] based on a proposal by Spriggs [62]. The latter has been compared with literature reports [63–65] that discuss the correlation between elastic properties and density, and strong agreement has been observed [60]. The porosity of the bulk samples can be calculated by using the following expression:

$$E = E_0 \exp(-bP) \tag{8.1}$$

where E is the elastic modulus of the porous body, E_0 is the elastic modulus of a non-porous body, b is an empirical constant and P is the pore volume fraction of the body. Equation 8.1 was compared with literature reports that discuss the correlation

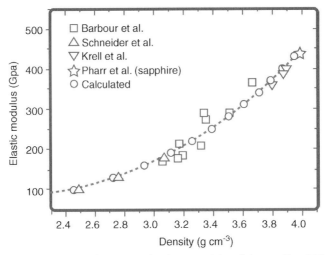

Figure 8.34 Effect of density on the elastic modulus of alumina films [57].

between density and elastic modulus data for thin films [63–65]. Very good agreement between the previously published data for thin films and Equation 8.1 can be observed in Figure 8.34.

Assuming that the alumina thin films discussed here can be represented by Equation 8.1 with $E_0 = 441$ GPa [56] and $b = 3.98$ g cm^{-3} (which is the theoretical density of non-porous alumina), the film density was estimated based on the elastic modulus values measured by nanoindentation. The evolution of the density as a function of NIF is presented in Figure 8.35.

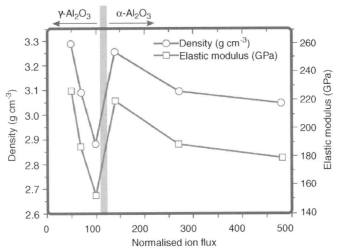

Figure 8.35 Calculated film density and elastic modulus plotted as a function of normalized ion flux [61].

For all NIF values, it appears that the estimated densities and the corresponding elastic modulus are lower than the theoretical values. It is reasonable to assume that the pore formation is responsible for this behaviour.

Performance-related tests were used in order to study material properties by isolating individual process loads such as adhesion to the substrate, friction and thermal shock resistance. The results obtained in these experiments led to the identification of degradation and failure mechanisms that are related to the specific loads applied in each test. For these tests, the alumina films deposited on tempered hot working steel at different AlCl$_3$/O$_2$ ratios (P_R) and substrate temperature were chosen (set of experiments 3, Table 8.9).

Adhesion is one of the most important properties of a coating since inadequate adhesion may result in delamination and hence in the loss of all desired functionality of the surface coating. Adhesion was evaluated by using so-called scratch test measurements. The resulting scratch traces are presented in Figure 8.36.

It has been observed that the critical load (F_C) for γ-alumina films is about 20 N and is independent of T_S. For α-alumina films, F_C varies strongly with T_S. At 600 °C, $F_C = 10$ N; when T_S is decreased to 560 °C, F_C increases to 20 N. For a given phase (γ- or α-alumina), no influence of P_R was observed. In order to enhance the adhesive properties, alumina films were deposited on previously nitrided steel. This process enables F_C to be increased to 50 ± 5 N independently of the deposition conditions.

Another important parameter which can affect the lifetime of a tool is friction arising during manufacturing processes. Figure 8.37 shows the friction coefficient of the steel ball against alumina coatings plotted versus T_S and P_R.

As a general observation, the average friction coefficient values increase with decreasing P_R and T_S. The resistance of the as-deposited films to stresses caused by thermal shock treatment was investigated for films deposited at $T_S = 560$ °C. The

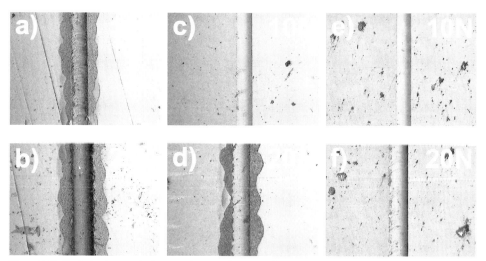

Figure 8.36 Optical micrograph showing the scratch traces of alumina films deposited at an AlCl$_3$/O$_2$ ratio of 0.5 and at temperatures of (a, b) 600 °C, (c, d) 580 °C and (e, f) 560 °C [66].

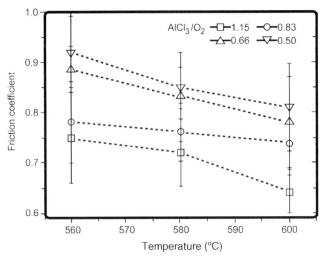

Figure 8.37 Friction coefficient measured during high-temperature tribological tests of the deposited alumina films [56].

evaluation of the coatings after the treatment by cross-sectional SEM and EDS analysis was aimed at identifying possible adhesive or cohesive failure and also signs of possible interdiffusion taking place through cracks or other defects in the films. The alumina films show very good resistance to thermal shock; as can be seen in Figure 8.38, no crack formation or erosion of the films was observed. Moreover, after contact with steel, the film surface was smooth without indication of damage or delamination. Possible diffusion of the counterpart material into the substrate was not observed based on EDS experiments.

The results of these performance-related tests confirmed the applicability of the alumina films for die protection during semi-solid processing of steel.

8.5.3
Conclusions

Alumina thin films were deposited on tempered hot working steel by PECVD in an $AlCl_3$–O_2–H_2–Ar mixture. The constitution and mechanical properties of the films were found to be strongly dependent on the deposition conditions such as substrate temperature, gas mixture and peak voltage. It has been shown that α-alumina films can be grown at substrate temperatures as low as $560\,^{\circ}C$. Finally, the constitution, morphology, impurity incorporation and elastic properties of the alumina thin films were found to depend on the normalized ion flux. The changes in structure and properties may be understood by considering surface and bulk diffusion-related mechanisms. The formation of porous structures is observed. The formation of the porosity may be understood by considering vacancy generation, migration and subsequent clustering and/or the incorporation of Cl during growth, migration and bubble formation. The elastic properties of the films were strongly affected by the formation of the pores in the structure and were intimately linked to the normalized ion flux.

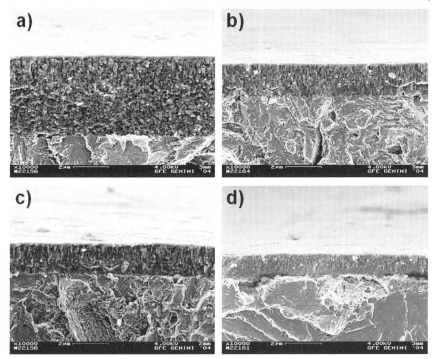

Figure 8.38 Cross-sectional micrographs of the alumina films after thermal shock experiments. The films were deposited at 560 °C and at AlCl$_3$/O$_2$ ratios of (a) 1.15, (b) 0.83, (c) 0.66 and (d) 0.5 [69].

Performance-related tests such as adhesion, high-temperature tribological and thermal shock evaluations were performed to evaluate the behaviour of coatings in a 'close to real' environment. The results underline that alumina thin films deposited by PECVD are very promising candidates for die protection during semi-solid processing of steel.

8.6
Bulk Ceramic Forming Tools

8.6.1
Candidate Die Materials

Ceramic materials are renowned for their outstanding high-temperature properties, notable strength, hardness, wear and corrosion resistance, leading to their application in a wide range of industrial processes. Owing to the high production costs of ceramics compared with metals, these materials are typically utilized at locations where metallic parts do not withstand the operating conditions, for example, process temperatures above 800 °C in oxidizing atmospheres and/or in contact with chemically aggressive media or high frictional loads.

In ferrous metallurgy, ceramics are mainly applied as refractory materials in blast furnaces, converters, tundishes and so on, providing a means to melt ores and metals and to handle the respective liquids in steelmaking. The main function of these refractories is to withstand the aggressive melts and slags. However, despite the sophisticated material solutions available today, the refractory parts are consumed during operation. Moreover, refractories contain a defined amount of residual porosity to improve thermal shock resistance.

Dense engineering ceramics are applied in metallurgy as boron nitride (BN)-based side dams for thin strip steel casting [67], owing to its excellent corrosion resistance and low friction coefficient. Oxide ceramics, for example alumina (Al$_2$O$_3$), zirconium oxide (ZrO$_2$) and aluminium titanate (Al$_2$TiO$_5$), are used for protective tubes for thermocouples and various items in direct contact with ferrous melts.

The demands on materials properties in these applications are lower compared with the load profile acting on thixoforming dies. Since refractories are not part of load-bearing structures and merely have to sustain their own weight, the required properties are high corrosion resistance, low thermal conductivity and a specific compressive strength. Likewise, hexagonal boron nitride, being a solid lubricant, exhibits relatively low strength and is commonly regarded as a functional rather than a structural material. Oxide ceramics show reasonable strength values and excellent corrosion and wear resistance, but thermal shock resistance is poor. Hence one of the main prerequisites for successful application of these ceramics is to establish thermally stable operating conditions, which is detrimental to the process conditions of sequenced metal forming techniques, that is, thixoforging and thixocasting.

Carbides and nitrides, for example, silicon carbide and silicon nitride, although exhibiting excellent thermal shock resistance among dense ceramics and strength levels superior to oxide ceramics, are not chemically stable in contact with iron at elevated and high temperatures [68, 69], which is why they are not considered appropriate for application in ferrous metallurgy.

With semi-solid processing of high-melting alloys being based on experience and findings obtained in light metal thixoforming, the principal approach at the beginning of interest in steel thixoforming was merely to transfer existing knowledge to this new group of work materials. Regarding tool development, this implied corrosion resistance being regarded as a key property for the selection of die materials. Thus, with the target of identifying suitable tool materials for the semi-solid processing of steel, Beyer [70] and Behrens *et al.* [71] independently carried out screenings of ceramics with a view to chemical interaction with steels at high temperatures. Thermochemical investigations revealed that a number of oxide ceramics (Al$_2$O$_3$, ZrO$_2$, etc.) are stable in the presence of typical steel grades at temperatures of interest, whereas most of the carbides and nitrides considered show significant reaction with ferrous alloys, in particular silicon nitride and silicon carbide [74]. This is confirmed by steel melt corrosion tests performed by Beyer, who observed severe weight loss for non-oxide ceramics after testing, whereas Al$_2$O$_3$- and ZrO$_2$-based ceramics showed notably higher corrosion resistance [70]. Further investigations included the wetting of potential ceramic die materials by liquid steel. According to the determined wetting angle, Behrens *et al.* [71] classified Al$_2$O$_3$, ZrO$_2$, spinel $MgAl_2O_4$ and SiAlONs as suitable for steel thixoforming. Several Si_3N_4-based ceramics, aluminium titanate (Al_2TiO_5) and numerous mixed oxides are categorized as suitable to a limited extent; SiC, TiN and cordierite are considered unsuitable.

Since the first forming trials using ceramic dies and die segments revealed thermal shock acting on the die surface to be critical, corrosion resistance was replaced by thermal shock resistance as a selection criterion for the semi-solid processing of high-melting alloys. Hence silicon carbide and silicon nitride came back into focus due to their excellent strength, thermal shock resistance and fracture toughness, despite their lower ranking in corrosion resistance and wetting angle screenings. Consequently, in recent work on semi-solid processing of steel, silicon nitride was selected exclusively as tool material [7, 32, 70–73]. However, only a few data are available on the actual performance of Si_3N_4 dies. Beyer [70] observed rupture of externally prestressed, axial-symmetric Si_3N_4 liners after only a few forming cycles. Failure was attributed to tensile stresses generated by the armouring on the one hand and insufficient thermal shock resistance and fracture toughness on the other. Analysis of die surfaces after application yielded evidence of corrosive attack, with deterioration of the surface quality [74].

Behrens and co-workers [7, 75] used Si_3N_4 plunger tops for steel thixoforming and regularly observed rupture of the ceramic die parts due to shrinking of the steel part on the ceramic. Neither analysis of the stresses arising in the ceramic die part nor analysis of the die surface after testing was reported. Other workers, although reporting the application of Si_3N_4 die parts, indicated that stress resistance was unsatisfactory in steel thixoforming experiments, without giving any details on tool performance [73, 76].

8.6.2
Ceramic Die Concepts for Steel Thixoforming

In a first approach to the selection of ceramic die materials for the semi-solid processing of metals, the tribochemical attack on forming dies is the most critical load, since the service life of metal forming dies is typically limited by the maximum tolerated shape inaccuracy. However, with the envisaged thixoforming of high-melting alloys using ceramic dies, the thermal process loads proved to be decisive for material selection. As discussed in Section 8.1, the upper thermal process boundary is given by the work alloy, whereas the die preheating temperature determining the lower thermal process boundary may be varied. Any change in thermal process conditions unequivocally affects the entire process layout, owing to the pronounced temperature dependence of the liquid phase content and thus the rheological properties of the semi-solid steels. Furthermore, the microstructure of the work pieces and the resulting mechanical properties are affected. A schematic diagram of the interconnection of thermal boundary conditions, process parameters and work piece quality is depicted in Figure 8.39.

With reference to the above-mentioned process analysis, bulk ceramic tools for the semi-solid processing of steels were applied following two different die concepts: (i) conventionally heated tools at working temperatures of <500 °C and (ii) self-heating tools operating at >1000 °C. The characteristics of the different approaches are given in Table 8.10.

Table 8.10 Characteristics of bulk ceramic die concepts for steel thixoforming.

Die concept	Conventionally preheated dies	Self-heating dies
Forming method	Thixocasting, thixoforging, thixoextrusion	
Boundary conditions		
Process-derived		
• High primary material temperature		
• Chemical attack by aggressive ferrous liquid phase		
• Tribological attack by solid particles		
Equipment-derived		
• Die temperature ~200–400 °C	• Die temperature >1000 °C	
• Rapid solidification of the semi-solid slurry	• Retarded solidification of the semi-solid slurry	
• Short cycle times	• Increased cycle times	
• Short flow lengths	• Long flow paths	
• High forming forces required	• Low forming forces required	
Tool demands		
• Long-term temperature stability at ~400–500 °C	• High temperature stability up to 1200–1400 °C	
• Short-term temperature stability up to primary material temperature	• Thermal shock resistance required: $\Delta T \approx$ 200–600 K	
• Thermal shock resistance required: $\Delta T \approx 1000$ K	• Excellent long-term corrosion resistance	
• Good corrosion resistance	• Sufficient wear resistance	
• Excellent wear resistance		

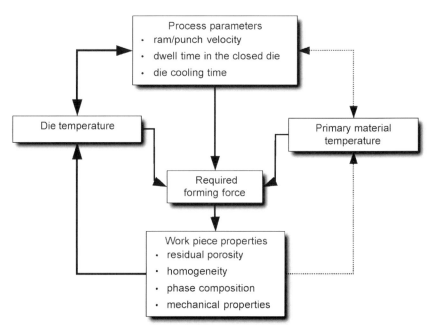

Figure 8.39 Schematic diagram of the interaction between thermal boundary conditions, process parameters and work piece quality.

In the following sections, these concepts are introduced and the general load profile as given in Section 8.1 is modified with respect to the die materials selected for each concept. Adapted testing schemes are derived according to Table 8.10 in order to identify the specific strengths and limits in model tests, near-application tests and small-scale forming series. Experimental results are reflected on the predefined load profile, forming test results and work piece quality obtained. An evaluation of the applicability of bulk ceramic dies for the semi-solid processing of steels is given in concluding remarks.

8.6.3
Conventionally Heated Tools

8.6.3.1 Characteristic Load Profile and Die Material Selection

The selection of suitable die materials has to be carried out initially regarding the short-term effects, namely mechanical and thermal loads during forming. When using conventionally heated forming tools operating at 500 °C, the characteristic load profile essentially consists of severe thermal shock on the die surface in contact with the partially liquid alloy. Hence the most critical die material property derived from this load profile is thermal shock resistance, in conjunction with high strength levels.

Silicon carbide (SiC) and silicon nitride (Si_3N_4) meet those demands on die material properties and are commonly used as structural materials in high-temperature applications. Silicon carbide exhibits a high thermal conductivity, which is why it is often used for heat exchangers, whereas silicon nitride shows very high thermal shock resistance among dense ceramic materials. Silicon nitride ceramics typically consist of two modifications: α-Si_3N_4, typically being present in the form of equiaxed grains, and β-Si_3N_4, in the form of elongated grains of needle-like appearance, owing to significant differences in the c-axis length of the respective unit cells. Controlling the α/β ratio and the grain morphology allows for tailoring of the microstructure and mechanical properties of Si_3N_4. The inherent high thermal conductivity of silicon nitride further promotes thermal shock resistance. Moreover, mechanical strength and fracture toughness are high compared with other engineering ceramics, being in the region of 800 MPa at 1000 °C and 5 MPa $m^{0.5}$, respectively.

Hence the chemical interaction of SiC and Si_3N_4 with semi-solid steels under thixoforming conditions is decisive for material selection. Both ceramics react with oxygen in ambient air to form a dense, superficial SiO_2 layer acting as a diffusion barrier for further reaction, thereby passivating the surface. Owing to the poor sinterability of these materials, sintering aids are required to promote liquid phase formation during densification, with the respective elements to be taken into account when considering oxidation and chemical interaction with steels. These sintering additives are known to affect drastically the oxidation behaviour of the parent ceramic by the formation of ternary or quaternary phases facilitating oxygen diffusion, inhibiting in turn the formation of a passivating oxide layer [68]. In the case of Si_3N_4, the release of gaseous nitrogen after decomposition further aggravates this effect by causing pores in the reaction layer. Some workers have also reported an enrichment of the sintering additive elements in the contact zone [77, 78]. Thus, critical application temperatures for SiC and Si_3N_4 under oxidizing conditions are determined by the eutectic temperature of the respective chemical composition, being typically in the range 1200–1500 °C for commercially available Si_3N_4 ceramics. Owing to the better sinterability and hence reduced amount of sintering additives required for densification, the critical temperatures for SiC are in the range 1700–1800 °C [79].

The oxidation mechanisms of SiC and Si_3N_4 are strongly dependent on the oxygen partial pressure. Whereas at high oxygen partial pressures passive corrosion with formation of SiO_2 occurs, at lower partial pressures gaseous SiO is formed [81, 82], leading to linear weight loss with time by evaporation. Oxygen partial pressures in non-encapsulated steel thixoforming tool setups vary from close to 1 bar prior to billet insertion to estimated values of $<10^{-24}$ bar in the contact zone between the semi-solid slurry and the die surface during form filling. Prediction of phase reaction is difficult due to the diversity of the process conditions; concurrently, measuring oxygen pressures in the contact zone is extremely difficult since sensors have to be in direct contact with molten and solid steel under forming forces.

When considering the chemical interaction between silicon carbide and silicon nitride in contact with iron at high temperatures, SiC is less stable than Si_3N_4. Silicon nitride decomposes in the presence of iron at temperatures >1250 °C and reacts with iron to form iron silicides [80] accompanied by release of gaseous nitrogen. Silicon carbide shows similar behaviour, but reaction starts already at 850 °C [81, 82]. Moreover, instead of nitrogen gas, carbon is released, being readily incorporated in the iron structure. This in consequence leads to an increase in the carbon content of the steel alloy, which is undesired in terms of mechanical properties of the thixoformed parts.

Hence silicon nitride was selected as the die material for the semi-solid processing of steel. Apart from forming trials to investigate the resistance of Si_3N_4 dies to thermomechanically induced stresses, the tribochemical attack on the die surface during forming has to be investigated carefully to ensure sufficient shape accuracy and service life.

8.6.3.2 Corrosion Experiments and Forming Series

Fully dense silicon nitride ceramics with Y_2O_3 and Al_2O_3 added as sintering additives were applied in corrosion experiments and small-scale forming tests. A typical microstructure in the as-received state is shown in Figure 8.40. The use of standard grades of typical composition allows for comparison of the present results with the findings of other workers in similar applications. Steel-grade 100Cr6 was used for thixoforging experiments; for details on composition and thermophysical properties, see Chapter 3.

Following the general scheme given in Figure 8.6, model tests were developed in order to isolate specific loads from the load profile (cf. Section 8.1). Apart from typical melt corrosion tests performed to examine the corrosion resistance against non-ferrous and ferrous liquids at temperatures up to 1650 °C, adapted corrosion tests were used to reproduce characteristic conditions during semi-solid forming [6]. Equipping the melt corrosion test with a linear unit allowed for cyclic testing by repeated rapid immersion of the sample in the melt. Thereby, test

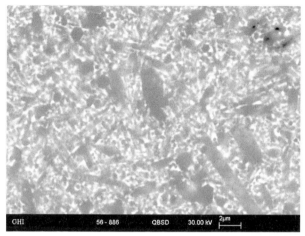

Figure 8.40 SEM image depicting a typical microstructure of gas pressure sintered silicon nitride (GPSN) consisting of elongated β-Si$_3$N$_4$ grains in a matrix of fine, equiaxed α-Si$_3$N$_4$, grains, surrounded by a glassy grain boundary phase containing large amounts of sintering additives Al$_2$O$_3$ and Y$_2$O$_3$.

samples were subjected to thermal shock similar to the conditions present during application.

Likewise, the behaviour of potential die materials in contact with solid and semi-solid steel was investigated using a tailor-made contact corrosion test (Figure 8.41). The test setup permits the investigation of chemical interaction between diffusion couples under static load in different atmospheres. For steel contact corrosion tests, the standard test temperature was 1300 °C, in air or under flowing argon. Selecting a test temperature below or within the melt interval of the respective alloy additionally permits evaluation of corrosion effects in contact with metals in the solid or semi-solid state.

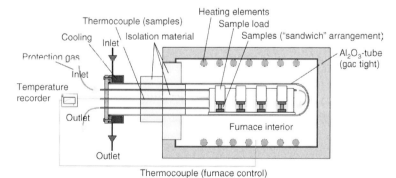

Figure 8.41 Schematic diagram of the contact corrosion test setup developed for the investigation of chemical interactions of ceramics with solid and semi-solid metals.

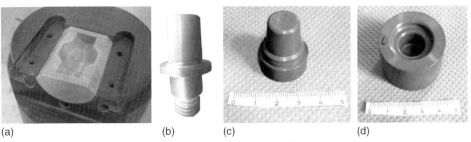

(a) (b) (c) (d)

Figure 8.42 Si$_3$N$_4$ thixoforging dies. (a) Segmented half-shell;
(b-d) plunger tops of different geometry.

Analysis of corrosion samples was carried out by careful preparation of cross-sections of the samples for SEM. Element distributions and phase compositions were determined using EDS and XRD of the reaction products.

In the course of the collaborate research project, an extensive screening of nearly 100 potential ceramic die materials regarding their chemical interaction with aluminium, copper and steel alloys targeted for thixoforming was carried out [5, 86]. With reaction mechanisms of the respective material pairings being investigated, the most promising die materials regarding corrosion resistance for the different applications were identified. Mechanical strength and thermal shock resistance were further investigated in near-application crush tests. Steel billets were partially remelted and positioned on plane circular samples of potential die materials. Subsequently, a hydraulic press exerted a predefined mechanical pressure on the billets. The sample surface was inspected macroscopically after each test cycle. Fifty cycles without damage were set as the target value for successful testing of the respective samples. Small-scale thixoforging series were carried out using different types of ceramic dies and inserts in order to evaluate the performance of silicon nitride in true steel thixoforming conditions. First experiments focused on several Si$_3$N$_4$ die segments of C-ring shape that were mounted in a hot-working steel tool frame (Figure 8.42a). Preheated steel billets were inserted in the fully assembled tool in the form of a half-shell, followed by a plane punch pressing the semi-solid steel into the cavity. Additional forming experiments were performed using Si$_3$N$_4$ plunger tops of different geometry (Figure 8.42b-d) in the frame of the hybrid tool concept established within the collaborate research centre. Plunger tops are necessarily made of ceramic materials since these die parts inherently are subjected to the highest thermal process loads during immersion into the semi-solid steel.

After testing, the ceramic dies and die segments were inspected visually. Optical microscopy and SEM were used for fractographic determination of crack origins on the fracture surface. Wear and corrosive attack on the die surface were examined using SEM and EDS analysis of cross-sections.

8.6.3.3 Corrosion Mechanisms and Forming Performance

The model tests applied to investigate the chemical interaction of candidate die materials with semi-solid steels revealed strong differences in applicability and test results. Chemical attack in melt corrosion tests was significantly higher than in contact corrosion, exacerbating analysis of corrosion samples and identification of reaction mechanisms. This is aggravated by the necessity to use casting powder to prevent oxidation of the steel melt, in particular when applying cyclic immersion of the samples. Chemical interaction of ceramic samples with the steel melt is overlaid by reaction with the chemically aggressive, low-viscosity slag formed, acting as a separating agent between the test sample and the liquid steel. Experimental results hence allow for a rapid qualitative ranking of potential die materials, whereas phase formation between Si_3N_4 and steel in the semi-solid state was examined by contact corrosion tests. Significant differences in chemical interaction were observed depending on the atmosphere present during testing. Under oxidizing conditions, superficial oxidation of silicon nitride occurs, independently of the steel contact, which was confirmed by EDS analysis of lateral parts of the Si_3N_4 corrosion samples. Silicon/oxygen ratios of approximately 1:2 lead to the conclusion that Si_3N_4 decomposes by releasing gaseous nitrogen and reacting with the surrounding oxygen to form SiO_2:

$$Si_3N_4 + 3O_2 \rightarrow 3SiO_2 + 2N_2 \tag{8.2}$$

Nitrogen gas is either removed from the contact zone through cracks and channels in the reaction layer or is entrapped in pores of different size within this layer (Figure 8.43). Concurrently, the grain boundary phase of near-surface Si_3N_4, consisting mainly of the sintering additives Y_2O_3 and Al_2O_3, is mobilized and segregates in the contact zone in the form of a precipitation seam containing high amounts of yttrium and aluminium. Both effects, release of nitrogen gas and mobilization of the grain boundary phase, lead to depletion of the contact zone from intergranular phase and the formation of considerable porosity.

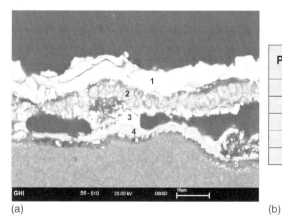

Point	Element (at.%)				
	Si	Al	Fe	Cr	O
1	-	-	37.4	0.5	62.0
2	5.8	1.2	3.1	20.4	69.1
3	1.2	-	31.6	1.9	65.4
4	14.2		19.1	2.0	64.8
5	30.0	-	4.1	-	65.9

(a)　　　　　　　　　　　　　　　　(b)

Figure 8.43 (a) SEM image of the contact zone of Si_3N_4/X210CrW12 contact corrosion at 1330 °C/2 h in air. (b) Results of EDS analysis.

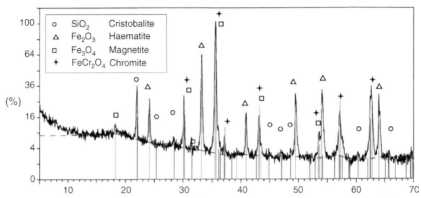

Figure 8.44 XRD spectra of the contact zone of Si_3N_4/X210CrW12 contact corrosion at 1300 °C/2 h.

The above-mentioned reactions are superimposed by chemical interaction with steel and scale in the areas in contact to steel. On top of the SiO_2 layer a thick scale layer is observed under oxidizing conditions, consisting of ferrous oxides with varying amounts of chromium in solid solution. XRD analysis of this layer yields magnetite, haematite and chromite as major constituents in addition to cristobalite (Figure 8.44).

Enclosed in this layer metallic precipitations are detected that exhibit a high amount of steel alloying elements, for example, chromium, manganese and tungsten, significantly exceeding the element content of the respective steel grade applied. Thus, alloying elements are mobilized in the bulk of the steel and diffuse to the contact zone. The entire scale layer is permeated by channels, cracks and pores resulting from the nitrogen gas released during decomposition of Si_3N_4 trying to escape from the contact zone.

As expected, chemical interaction of Si_3N_4 and steel in an argon atmosphere differs from the behaviour in air: depletion of the grain boundary phase in the upper Si_3N_4 layers is less pronounced. Likewise, precipitations containing a high amount of the sintering additive elements Y and Al are observed only locally in the contact zone. On top of the Si_3N_4 ceramic a metallic layer is detected, consisting mainly of iron and silicon. Embedded in this layer metallic precipitations are found, showing high amounts of steel alloying elements, and also large pores that are ascribed to the nitrogen gas emerging during decomposition of Si_3N_4. Although this gas can readily escape from the contact zone through the powdery and brittle scale layer in the case of oxidizing conditions, the lack of such a layer in an argon atmosphere leads to entrapment of the gas in the contact zone, eventually resulting in very large cavities formed due to the gas pressure. While more pronounced in an argon atmosphere, interdiffusion of silicon and iron is observed independently of the experimental parameters.

Examination of silicon nitride die parts after application in small-scale forming series revealed that under real thixoforming conditions a combination of the aforementioned mechanisms observed in model tests in air and in an argon atmosphere is present. Although the inner die surfaces showed a mirror finish

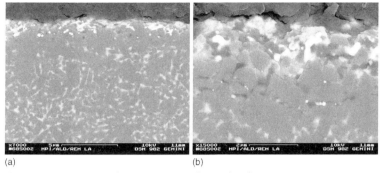

(a)   (b)

Figure 8.45 SEM images of the contact zone of Si$_3$N$_4$ thixoforging dies (below) after small scale steel thixoforming series. (a) Overview; (b) magnified view.

after application (cf. Figure 8.46), a reaction layer of 1–3 µm thickness was detected by SEM analysis of the respective cross-sections (Figure 8.45) after approximately 50 forming cycles. The contact zone shows similar characteristics to those observed in contact corrosion experiments: (i) depletion of intergranular phase in the contact zone and (ii) enrichment of steel alloying elements.

Regarding the short-term effects, rupture of Si$_3$N$_4$ die parts occurred spontaneously in crush tests and small-scale forming series, predominantly after only a few forming cycles, if the strength of the material was exceeded locally (Figure 8.46).

According to analysis of the particular process conditions and fractographic examination of fracture surfaces, three main failure causes are distinguished, ordered by the observed frequency:

- continuously superimposed thermally induced stresses, resulting from the repeated introduction of heat on the die surface in contact with the semi-solid steel during forming cycles;

Figure 8.46 Si$_3$N$_4$ thixoforming dies after application in thixoforging experiments. Typical failure effects.

- shrinking of the solidifying steel part on the ceramic die parts, inducing compressive loads that cause tensile loads at distant locations in the bulk;
- mechanically induced stresses arising due to insufficient fixation of the ceramic die parts in metallic tool frames.

It is noteworthy that in many cases the integrity of the Si_3N_4 dies was maintained even after rupture if the tool frame successfully fixed the fracture surfaces. This enabled the operator to continue forming without any decrease in work piece quality until a predefined number of forming cycles was accomplished. In general, the surface quality of the as-formed steel parts was excellent in the case of Si_3N_4 dies [35, 85]. This is attributed to the poor wetting of Si_3N_4 by liquid steel, facilitating part ejection and inhibiting welding of steel residues on the die surface.

8.6.3.4 Applicability of Silicon Nitride Dies in Steel Thixoforming

The experimental results using bulk Si_3N_4 forming dies for the semi-solid processing of steels are discussed in terms of the performance related to the short- and long-term effects as defined in the general load profile (cf. Figure 8.1). In view of the short-term effects comprising mechanical and thermal loads, the results show an evident sensitivity of Si_3N_4 die parts regarding thermally and mechanically induced stress states evolving during the entire course of forming, including preheating, billet insertion, forming, solidification stage, part ejection and die opening time [85]. Failure causes may be categorized according to (i) process loads, (ii) die construction and (iii) process parameters.

The first category comprises the required forming forces and thermal loads exerted by the semi-solid slurry in contact with the die surface. Experimental results indicate that thermally induced stresses are decisive in comparison with the mechanical forces which are reduced compared with conventional forging and casting by taking advantage of the rheological effect of thixotropy. The second category includes construction details that are essential when using ceramic forming dies, that is, the fixation of die parts. Clamping or joining of ceramics to metallic tool parts is difficult due to the distinct mismatch in thermal expansion. This is further aggravated by residual stresses in the solder layer after cooling from high-temperature brazing, which may contribute to the evolution of critical stress distributions inside the ceramic part when overlaid by the above-mentioned process-induced stresses. If rotational-symmetric dies are used, prestressing of the dies by shrink-fitting of (hard) metal armouring rings may be beneficial. However, the compressive stresses induced prior to application may cause critical tensile stresses in die parts protruding into the cavity. Moreover, armouring stresses rapidly decrease with increasing die temperature and are eliminated at shrinking temperature, reducing the efficiency of this approach.

The third category includes the effects of various process parameters on stresses arising in the ceramic dies, namely dwell time of the shaped steel part in the die and deviant loads, such as insufficiently preheated steel billets inserted in the cavity. The most crucial aspect among these is the on-shrinking of steel on ceramic dies during solidification of the formed part. The tensile stresses generated beneath the compressed parts of the die are of high magnitude, in most cases significantly exceeding the material's strength even for high-strength ceramics such as Si_3N_4. Consequently, the dwell time of the steel part in the die has to be exactly adjusted to the minimum solidification time required for ejection without any deterioration in shape.

Concerning the long-term performance of Si_3N_4 steel thixoforming dies, the tribochemical wear is of crucial importance for the determination of service life. Experimental results show substantial agreement of the corrosive effects observed in model tests, both in air and an argon atmosphere, and thixoforging trials. This indicates that the oxygen partial pressure in the die during thixoforming varies strongly, since oxidation of Si_3N_4 is observed in addition to metallic phases containing high amounts of iron and silicon, suggesting the formation of Fe(Si) solid solutions and iron silicides.

After thixoforming experiments, the reaction products are found on the die surface in the form of a discontinuous tribochemical layer, which develops according to the following sequence:

1. superficial heating of the Si_3N_4 ceramic to temperatures >1000 °C, accompanied by softening of the glassy grain boundary phase;
2. relative enrichment of alloying elements (Mn, Cr, V) on the steel side due to process-induced segregation;
3. out-squeezing of the ceramic glassy phase by compression and shear forces;
4. redox reactions between glassy phase, Si_3N_4 and steel/steel melt;
5. formation of a tribochemical layer as a result of the above-mentioned redox reactions;
6. delamination, fragmentation and shearing of this layer;
7. iteration of the sequence with the subsequent forming cycle.

This sequence and the corresponding effects are confirmed by observations of other workers on Si_3N_4 forming dies for the semi-solid processing of steel [70]. With regard to tool performance, the behaviour of the resulting tribochemical layer on the die surface is decisive. The physico-chemical properties of this layer can be controlled only by the amount and composition of ceramic sintering additives, since other parameters of influence cannot be altered, such as the chemical composition of the steel grade and forming temperatures. The use of alternative sintering aids for Si_3N_4 ceramics on the target for promoting the formation of refractory reaction layers to inhibit diffusion from constituents to the contact zone in both the ceramic and the steel seems to be an appropriate measure to improve the corrosion resistance of Si_3N_4 dies. However, apart from chemical interaction with iron as the main constituent of steel, the alloying elements have to be taken into account. Previous work on the corrosion resistance of various ceramics with metals (Al, Cu, steel) aimed at thixoforming showed that the reaction between sintering additive elements and steel alloying elements might be decisive for corrosive attack rather than reaction with iron [86]. Thus, with the target of tailoring the grain boundary phase of Si_3N_4 ceramics for application as forming dies in steel thixoforming, the entire chemical composition of the respective alloy has to be considered. Changing the work alloy may therefore lead to significantly reduced service life of the dies. The development of tailored ceramic materials has not been reported hitherto, presumably owing to the dependence on the steel grades intended for thixoforming and the extensive effort for the development of specific Si_3N_4 grades. Since the reaction layer thickness on Si_3N_4 dies after thixoforming experiments is small and due to the competing reaction mechanisms, in conjunction with the difficulties in determining exact oxygen partial pressures evolving in the contact zone, prediction of corrosion rates is not feasible. Hence the corrosion resistance of Si_3N_4 dies has to be investigated in long-term forming series under genuine thixoforming conditions.

8.6.4
Self-heating Dies

8.6.4.1 Die Materials and Tool Characteristics
The load profile acting on self-heating dies operating at temperatures >1000 °C differs significantly from that of conventionally heated dies. Thermal shock is drastically reduced and may even be eliminated from the load category of short-term effects in near-isothermal forming. Consequently, thermally induced stresses are decreased, whereas the demands on high-temperature stability are increased. This in turn allows for the application of thermal shock-sensitive materials, thereby considerably expanding the range of potential ceramic die materials. Mechanical loads are of the same magnitude as in conventional process layouts, possibly slightly reduced due to the high temperature facilitating material flow in the cavity. Regarding the long-term effects, corrosive attack is intensified due to the increased temperature in the contact zone.

Hence material selection is carried out first according to corrosion resistance instead of thermal shock resistance. Oxide ceramics such as alumina (Al_2O_3) and zirconium oxide (ZrO_2) are known to exhibit excellent corrosion resistance against ferrous melts and slags, which is confirmed by their application as primary phases in numerous refractories for steelmaking [68, 69]. Both ceramics are also widely used for wear protection in the textile industry and for lubricant-free antifriction bearings [68]. Although the strength levels of oxide ceramics typically are lower than those of silicon nitride or silicon carbide, these materials are successfully used as structural parts, if critical tensile stresses are prevented or reduced to uncritical values. Based on experimental results of extensive screening of the corrosion resistance of ceramic materials in contact with solid, semi-solid and molten steel, alumina was selected as the tool material for self-heating dies due to its higher ratio of thermal shock resistance to corrosion resistance in comparison with zirconium oxide. This allowed for a broader thermal process window for die construction, since no experience is available concerning high-temperature dies in steel thixoforming. Tool construction thus has to meet the following demands derived from the process analysis given in Section 8.1:

1. The inner surface of the die shaping the steel must be heatable to a minimum temperature that is less than 150 K lower than the solidus temperature of the steel grade to be formed, in the case of steel X210CrW12 being 1250 °C.
2. The die surface must be corrosion, infiltration and wear resistant in contact with semi-solid steel at operating temperatures.

3. The temperature at the outer shell of the die, in contact with the metallic tool frame, must not exceed 500 °C, desirably 300 °C, to ensure a long service life of the surrounding tool frame.
4. The material between the hot die and the outer tool frame must exhibit sufficient strength to bear the process forces, but also has to sustain the severe temperature gradient present after die preheating.

The first demand is met by a huge number of dense and porous ceramic and refractory materials, whereas the second demand considerably narrows the range of potential die materials. The third and fourth demands reveal the crucial point of self-heating dies operating at high temperature, since the tool frame has to be protected from the heat generated in the die. The low thermal conductivity of thermal insulation materials required hereto is typically realized by porosity, thereby affecting mechanical strength. Thus, in a first approach to this concept, a multi-shell tool construction was envisaged, in which the demands are fulfilled separately by the individual shells (Figure 8.47). The inner forming die (A) is surrounded by a heating core (B) that is resistance heated to the operating temperature. The inner parts are fixed into an insulation shell (C) that is covered by a metallic tool frame (D). In an advanced layout the heating core may be replaced by an induction coil to enhance heating velocity.

This die concept is predestined for, although not limited to, thixoextrusion, since this forming method exhibits relatively stable thermal process parameters in comparison with sequenced forming methods, that is, thixoforging and thixocasting. The applicability for the latter processes was evaluated by additional prototypes in order to demonstrate the transferability of this concept to alternative forming methods.

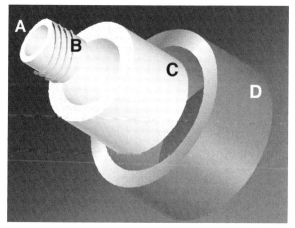

Figure 8.47 Schematic tool construction for self-heating ceramic dies for the semi-solid processing of high melting point alloys. (A) shaping die; (B) heating core; (C) insulation body; (D) tool frame.

8.6.4.2 **Tool Design and Construction**

The shaping surface of ceramic die parts (Figure 8.47A) is manufactured using state-of-the art ceramic machining technologies, for example, CNC-controlled green machining and finishing using diamond tools. These technologies allow for accurate shaping of complex freeform surfaces in the green state and a mirror-like surface finish after hard machining. If sintering shrinkage and process control are monitored precisely, the as-sintered parts are ready to use without additional surface finish, thereby significantly reducing production costs.

Various heating systems may be applied to obtain the desired temperatures in or near to the two-phase field of the work alloy: Resistance heating of ceramic parts may be realized either by Kanthal heating coils or molybdenum disilicide ($MoSi_2$) heating elements, depending on the required temperature range. In a more sophisticated tool construction inductive heating may be applied, either by indirect tool preheating via inductive coupling of metal parts inserted into the die prior to forming or by direct heating via coupling elements incorporated into the ceramic die parts.

The most critical aspect in the construction of self-heating steel thixoforming dies is their implementation in tool frames of existing forming machines such as forging presses, high-pressure die casting machines and extruders. While the major stresses during forming, evolving due to thermal gradients and the applied forming forces, have to be sustained by the inner die, the isolation body surrounding the inner parts and containing the heating core has to fix the inner cavity in its position and tolerate the temperature gradient to the outer metallic tool frame. Construction of this shell is dependent on the target application (thixoforging, thixocasting, thixoextrusion) and may be accomplished using a broad variety of available thermal insulation materials, including high-temperature insulation wool, vacuum-formed refractory fibre parts, lightweight insulating firebricks and refractory castables in order of decreasing thermal insulation efficiency and increasing load-bearing capacity. Additionally, the power supply has to be connected to the heating core. This implies that cables are laid through the metallic tool frames and the insulation body to the heating elements. Prototypes of self-heating ceramic tools for different applications are introduced and discussed in the following.

As stated above, thixoextrusion is characterized by comparatively stable process conditions due to semi-continuous feeding techniques (e.g. billet-on-billet extrusion). Thus, the thermal profile during forming is relatively stable and dynamic forces are negligible in relation to cyclic forging and casting methods. Moreover, extrusion forces are not acting in the same direction as the critical stresses in the die. Tool construction is hence focused on self-supporting ceramic recipients that are merely centred in a heating unit rather than being part of the load-bearing structure. One possible tool construction is illustrated in Figure 8.48, showing a prototype consisting of an alumina recipient and a silicon nitride die of complex geometry. This tool setup is capable of providing die temperatures of >1500 °C.

In contrast to thixoextrusion, thixoforging tools are subjected to dynamic loads and higher peak stresses during forming. Additionally, critical stresses may arise in the same direction of the applied forming forces, thereby further aggravating mechanical

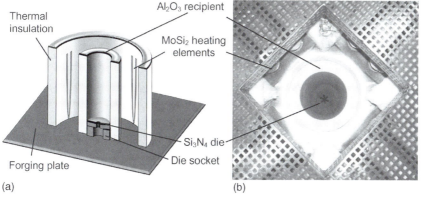

Figure 8.48 Self-heating ceramic thixoextrusion die.
(a) Schematic cross section; (b) view into recipient.

impacts. Thus, the inner forming die has to be supported by the isolation shell made of a low-porosity refractory castable. A tool prototype for self-heating ceramic thixoforging dies is presented in Figure 8.49. The inner alumina die is surrounded by a Kanthal A1 heating core and heated to 1000 °C prior to forming. Temperatures of the steel forging table were less than 100 °C.

Similarly to thixoforging tools, self-heating ceramic thixocasting moulds have to withstand cyclic mechanical loads, although the process forces are applied indirectly on the tool surface. However, special attention has to be paid to the construction of the gate system in order to obtain a smooth material flow into the cavity. Figure 8.50 shows a self-heating thixocasting mould consisting of alumina plates containing a geometrically complex freeform-shaped cavity. The insulation body surrounding the alumina moulds is made of lightweight insulating firebricks.

These prototypes reveal the variety of possible tool setups and prove the principle applicability of the underlying tool concept for semi-solid metal forming.

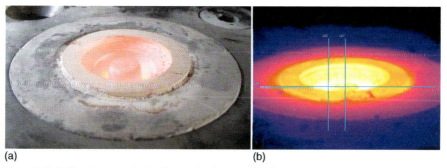

Figure 8.49 Self-heating ceramic thixoforging die. (a) Heated tool;
(b) IR image showing temperatures of 1000 °C in the centre
(yellow) and <200 °C outside the insulation layer (blue).

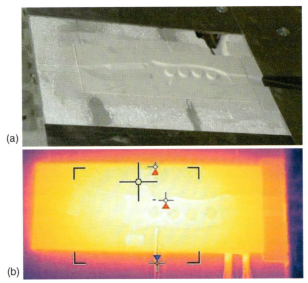

(a)

(b)

Figure 8.50 Self-heating ceramic thixocasting mould (a) and IR image of tool at operating temperature (b), showing temperatures of 1000 °C in the centre (yellow) and 400 °C outside the insulation layer (blue).

8.6.4.3 Forming Performance and Wear Resistance

Results of thixoforming experiments using the ceramic tools introduced above confirmed the beneficial influence of highly preheated tools on the form filling behaviour of the semi-solid slurry, requiring very low forming pressures. As observed in forming experiments using conventionally preheated ceramic dies (cf. Chaper 10), excellent surface quality of the as-formed parts is obtained. Details of process parameters and work piece quality of the respective forming experiments are given in Chapter 11.

With regard to tool performance, results indicate that thermal shock is effectively reduced to a minimum, allowing for the application of thermal shock-sensitive but highly corrosion resistant oxide ceramics. However, the applied prototypes revealed the weak points of tool construction. Rupture of ceramic die parts occurred due to insufficient mechanical strength of the insulation shells and unintended process loads such as canted extrusion discs. Moreover, heating cores of the prototypes did not provide a fully homogeneous temperature distribution throughout the parts of the die in contact with the ceramic surface. Hence, despite the active heating of tools, thermally induced stresses may also contribute to critical stress states evolving in the ceramic parts. Owing to the complex tool construction and the high operating temperatures, determination of the causes of fracture of prototypes is difficult.

Regarding wear and corrosion resistance of the alumina dies and moulds, the tool surfaces after forming experiments macroscopically yielded no evidence of wear or corrosive attack. This is ascribed to the high corrosion resistance and hardness of this

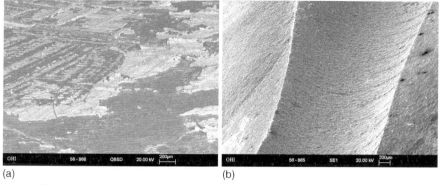

Figure 8.51 SEM images of alumina die surfaces after steel
thixoextrusion experiments. (a) Die inlet; (b) Die land.

ceramic material. Adhesion of steel on the die surfaces was observed sporadically and
only locally, owing to the poor wetting of alumina by steel. This advantageous
behaviour promotes the aforementioned high surface quality of formed parts.

Microscopic analysis of the dies carried out using SEM and EDS substantiates
these findings. Alumina die surfaces applied in steel thixoextrusion experiments, as
depicted exemplarily in Figure 8.51, appear to be unaffected by tribochemical attack
of the semi-solid steel slurry. This is demonstrated by the machining grooves
originating from green machining of the die still being visible after application,
thereby acting as indicators of tribological attack (Figure 8.51a).

Although the accumulated contact time of alumina dies with semi-solid steel was
short in small-scale forming series performed using self-heating ceramic dies, the
findings are consistent with investigations on the chemical interaction of alumina
ceramics with steel alloys in the temperature range of interest for steel thixoforming [5]. Thus, the potential of oxide ceramics in semi-solid metal forming with regard
to improved tool life is clearly discernible.

8.6.4.4 Applicability of Self-heating Ceramic Dies for Steel Thixoforming

The die concept of self-heating ceramic dies for the semi-solid processing of steel has
been realized and tested by the construction of prototypes for thixocasting, thixoforging and thixoextrusion. The intended beneficial effects on the flow behaviour of the
semi-solid slurries, reduction of process forces and geometric complexity of work
pieces have been accomplished in small-scale forming series.

Regarding the thermomechanically induced stresses, denoted short term effects
according to the general load profile depicted in Figure 8.1, self-heating alumina dies
showed sufficient resistance in near-isothermal steel thixoextrusion experiments. A
detailed analysis of the temperature-dependent mechanical strength of the alumina
dies in steel thixoextrusion is given elsewhere [2]. Regardless of the excellent wear and
corrosion resistance of thixoextrusion dies observed herein, a prediction on long-
term performance of such tools is not feasible on the basis of current experiments,
since contact times with the semi-solid slurry were too short.

The technical drawbacks of the existing prototypes are directly related to tool construction and may be addressed by a straight bottom-up tool design. This includes finite-element calculations on thermomechanical stresses arising during preheating and service. The critical aspects of this concept have been identified and should be addressed in future work. Owing to the severe impact on the process layout itself, the influence on process parameters and work piece quality has to be monitored carefully, as discussed in Chapter 11.

Concerning general applicability, the vast perspective of this tool concept should be pointed out, in particular with a view to tapping the full potential of the advantages offered by forming metals in the semi-solid state, and even higher melting alloys than steel.

8.6.5
Perspectives for the Application of Bulk Ceramic Dies in Steel Thixoforming

Two different ceramic die concepts for the semi-solid processing of steel are introduced, considering the process demands on die properties and the specific demands of distinct ceramic die materials on their operating conditions. The characteristic strengths and limitations of both die concepts are identified on the basis of adapted model tests, their performance in small-scale forming series and the work piece quality of thixoformed steel parts. Regarding the application of conventionally heated Si_3N_4 dies operating at temperatures <500 °C, the stress states evolving during service have to be evaluated carefully prior to tool manufacturing to prevent rupture. Preferentially, this is carried out by finite element analysis of the intended die design, taking into account the different stress origins depicted in Section 8.6.3.3. This will unquestionably lead to geometric limitations in die design and restrictions of certain process parameters, that is, die closing time and total cycle time, and thus result in individual design guidelines derived from those calculations. Furthermore, long-term forming series are required to understand completely the tribochemical interaction of Si_3N_4 ceramics with semi-solid steels under the multifaceted thixoforming process conditions. Results obtained in such experiments will contribute to the improvement of engineering ceramics and refractories based on Si_3N_4 in metallurgical applications.

The novel concept of high-temperature dies for steel thixoforming working at die temperatures >1000 °C was validated on the example of steel thixoextrusion. It was shown that the notably increased die temperatures not only allowed the application of thermal shock-sensitive ceramics of high corrosion and wear resistance, but also significantly improved work piece quality in view of the geometric complexity of the formed parts and phase homogeneity in comparison with previous studies by other workers [87–89] on steel thixoextrusion. Further work has to be carried out on the appropriate design of such tools for target applications. The principal results of tool construction and process parameters may be transferred to innovative metal forming technologies, such as isothermal forging and superplastic forming of other high-melting alloys that are difficult to shape using conventional forming methods.

Thus, two die concepts have been developed that provide useful means for successful semi-solid processing of steels. However, the narrow process window inherent to steel thixoforming and the necessity for evaluating process parameters and designing specific tools for each intended work piece geometry make die construction a challenging task. In this respect, the latter concept of high-temperature dies is advantageous in terms of process stability owing to the comparatively stable defined operation parameters.

8.7
Conclusions and Perspectives

The work on PVD process technology focused on the improvement of tool life in semi-solid metal forming by developing an adapted deposition process for oxide coatings on hot working steel and TZM substrates. Two different candidates were identified as suitable coating materials, γ-Al_2O_3 and t-ZrO_2, using tailored model tests, owing to the high hardness of alumina in the relevant temperature range and less adhesion of the semi-solid steel forming material on zirconium oxide. The PVD coatings γ-Al_2O_3 and t-ZrO_2 were confirmed as suitable coating materials. In order to improve adhesion to the substrate, a TiAlN interlayer was provided to increase the coating performance. Regarding thin-film synthesis, the objectives of the study were to reduce the deposition temperature to values below the annealing temperatures of the substrate materials and to determine parameters for reliable and reproducible thin-film properties. This was achieved by a pulsed PVD process with tight and fast process control to prevent target poisoning. Coating experiments showed that substrates with surface features up to an aspect ratio of 1:2 can be coated successfully with a constant coating thickness. The applicability of PVD-coated hot working steel and TZM dies was demonstrated in small-scale steel thixoforming (thixoforging, -casting and -extrusion) series. The long-term stability of γ-Al_2O_3 and t-ZrO_2 coatings has to be evaluated in industrial forming series. Future work will focus on enhancement of the PVD process on the target for coating substrates of complex surface morphology.

The aim of the PECVD process technology development was to ensure a reduction in the deposition temperatures for α-Al_2O_3 at substrate temperatures allowing deposition on steel dies, namely at 550 °C. To achieve that goal, a bipolar DC PECVD method was developed. The constitution and mechanical properties of the films were found to be strongly dependent on the deposition conditions such as substrate temperature, gas mixture and peak voltage. The elastic properties of the films were strongly affected by the formation of the pores in the structure and were intimately linked to the normalized ion flux. Performance-related tests such as adhesion, high-temperature tribological and thermal shock evaluations were performed to evaluate the behaviour of coatings in a 'close to real' environment. The results underline that alumina thin films deposited by PECVD are very promising candidates for die protection during semi-solid processing of steel.

The excellent high-temperature properties inherent to ceramics make them candidates for die materials for high-temperature metal forming, for example, semi-solid processing of steel, in a first approach. Although ceramic coatings on metal substrates are one possible strategy to benefit from these properties, this implies that tools need to be cooled during operation. Consequently, the process layout is determined by the thermal boundary conditions, limiting the applicability of bulk ceramics to those exhibiting the highest available thermal shock resistance and fracture toughness, that is, silicon nitride. The suitability of silicon nitride has been demonstrated, the operating limits identified and design guidelines derived. Conversely, this process layout leads to restrictions on work piece quality and geometric complexity of produced parts due to rapid solidification.

Hence an alternative bulk ceramic tool concept based on the separation of forming and solidification has been introduced, increasing freedom of shape and allowing for the application of highly corrosion-resistant but thermal shock-sensitive ceramics. The viability of this tool concept has been demonstrated on the example of thixoextrusion. Forming results and work piece quality are highly encouraging, opening new perspectives for future metal forming processes.

List of Symbols and Abbreviations

a	deposition rate [nm s^{-1}]
AIP	arc ion plating
Al$_2$O$_3$	aluminium oxide
Al$_2$TiO$_5$	aluminium titanate
b	empirical constant [g cm^{-3}]
CrAlN	chromium aluminium nitride
CTE	coefficient of thermal expansion
CVD	chemical vapour deposition
DC	direct current
E	elastic modulus of a porous body [GPa]
E_0	elastic modulus of a non-porous body [GPa]
EDS	energy-dispersive spectroscopy
F_C	critical load [N]
GPSN	gas pressure sintered silicon nitride
HPDC	high-pressure die casting
HSS	high-speed-steel
JCPDS	Joint Committee on Powder Diffraction Standards
m.f.	middle frequency
MSIP	magnetron sputter ion plating
NIF	normalized ion flux
OES	optical emission spectroscopy
P	pore volume fraction [g cm^{-3}]
P_C	precursor content [%]
P_D	power density [W cm^{-2}]
PECVD	plasma-enhanced chemical vapour deposition
P_R	precursor ratio

PSZ	partially stabilized zirconia
PVD	physical vapour deposition
r.f.	radiofrequency
sccm	standard cubic centimeters per minute
SEM	scanning electron microscopy
Si_3N_4	silicon nitride
SiC	silicon carbide
TEM	transmission electron microscopy
TiAlN	titanium aluminium nitride
TiN	titanium nitride
T_s	deposition temperature [°C]
TZM	titanium–zirconium–molybdenum (molybdenum-based alloy)
V_P	peak voltage [V]
XRD	X-ray diffraction
Y_2O_3	yttrium oxide
ZrO_2	zirconium oxide

References

1 Baur, J. (2000) Anlagen für das thixoschmieden. *Wt Werkstattstechnik*, **90** (10) 441–445.

2 Münstermann, S. (2007) Development of ceramic die concepts for steel thixoforming. PhD Thesis, RWTH Aachen University, Shaker, Aachen.

3 Omar, M.Z., Palmiere, E.J., Howe, A.A., Atkinson, H.V. and Kapranos, P. (2005) Thixoforming of a high performance HP9/4/30 steel. *Materials Science and Engineering A*, **395**, 53–61.

4 Bobzin, K., Lugscheider, E., Maes, M. and Abdel-Samad, A. (2005) The influence of hot isostatic pressing on plasma sprayed coatings properties. *Surface and Coatings Technology*, **201**, 1224–1227.

5 Meyer-Rau, S. (2002) Corrosive interactions between ceramics and semi-solid aluminium, copper and steel alloys. PhD Thesis, RWTH Aachen University, Mainz Publishers, Aachen.

6 Meyer-Rau, S. and Telle, R. (2005) Testing strategies for corrosive interactions of ceramics with semi-solid and molten metal alloys. *Journal of the European Ceramic Society*, **25**, 1049–1055.

7 Behrens, A., Haller, B., Desery, C., Stegner, F., Schober, R., Wagemann, A., Wohlfahrt, D. and Lengsdorf, C. (2005) Thixoformingtechnologie für hochschmelzende Stahlwerkstoffe, PZH Produktionstechnisches Zentrum, Garbsen, Abschlussbericht zum BMBF Verbundprojektes Thixostahl, Report Nr FKZ 02 PD 2120-02PD 2125.

8 Bleck, W. (2004) THIXOCOMP – Improvement of Steels and Tools for Thixoforming of Complex Structural Parts. Institut für Eisenhüttenkunde, RWTH Aachen University, Abschlussbericht zum EU-Projekt Nr G5RD-CT-2001-00488.

9 Koesling, D., Tinius, C., Cremer, R., Hirt, G., Morjan, U., Nohn, B., Wittstamm, T. and Witulski, T. (1999) Untersuchungen zum Thixoforming als Alternative zur Herstellung komplex geformter Stahlschmiedeteile. EFU Gesellschaft für Ur-/Umformtechnik mbH, Simmerath, Abschlussbericht der Studiengesellschaft Stahlanwendung Projekt P322.

10 Hartmann, D., Morjan, U., Nohn, B., Zimmer, M., Hoffmann, A. and Braun, R.

(2002) Erprobung metallischer Werkzeugkonzepte für das Thixoforming von Stahl, EFU Gesellschaft für Ur-/Umformtechnik mbH, Simmerath, Abschlussbericht der Studiengesellschaft Stahlanwendung Projekt P483.

11 Lugscheider, E., Bobzin, K., Maes, M. and Immich, P. (2005) High performance $(Cr_{1-x},Al_x)N$-PVD coatings for thixoforming of Al-MMC, Proceedings of the 5th THE Coatings Conference, Kallithea, Greece, pp. 169–175.

12 Lugscheider, E. and Hornig, Th. (2000) Multifunktionale Schichten zum Thixoforming von Stahl – Eine Herausforderung für neue Werkstoffkonzepte, Schutzschichten für Hochtemperatur-Anwendungen, Workshop, pp. 12.1–12.10.

13 Rossnagel, S.M. (2003) Thin film deposition with physical vapor deposition and related technologies. *Journal of Vacuum Science and Technology A – Vacuum Surfaces and Films*, **21**, S74–S87.

14 Orhing, M. (1992) *The Materials Science of Thin Films*, Academic Press, San Diego, CA.

15 Berg, S. and Nyberg, T. (2005) Fundamental understanding and modeling of reactive sputtering processes. *Thin Solid Films*, **476**, 215–230.

16 Müller, J. (2004) Abscheidekinetik und Transporteigenschaften von CVD-Schichten aus alpha-Al_2O_3 als Diffusionsbarriere auf Nickelbasislegierungen. Doctoral Thesis, RWTH Aachen University.

17 Kyrylov, O., Kurapov, D. and Schneider, J.M. (2005) Effect of ion irradiation during deposition on the structure of alumina thin films grown by plasma assisted chemical vapour deposition. *Applied Physics A – Materials Science and Processing*, **80**, 1657–1660.

18 Westwood, W.D. (2003) *Sputter Deposition*, AVS Publications, New York.

19 Hornig, T. (2002) Entwicklung von Werkstoffverbunden für den Einsatz in Thixoformingwerkzeugen für die Aluminium- und Stahlverarbeitung. Doctoral Thesis, RWTH Aachen University.

20 Lugscheider, E., Hornig, Th., Kopp, R., Kallweit, J. and Möller, T. (2001) Oxide PVD-coatings for use on dies for semi-solid metal (SSM-) forming of steel. *Advanced Engineering Materials*, **3**, 998–1001.

21 Gao, O., Meng, L.J., Santos, M.P., Teixeira, V. and Andritschky, M. (2000) Characterisation of ZrO_2 films prepared by rf reactive sputtering at different O_2 concentrations in their sputtering gas. *Vacuum*, **56**, 143–148.

22 Aita, C.R., Wiggins, M.D., Whig, R. and Scanlan, C.M. (1996) Thermodynamics of tetragonal zirconia formation in a nanolaminated film. *Journal of Applied Physics*, **79**, 1176–1178.

23 Tönshoff, K., Mohlfeld, A., Leyendecker, T., Fuß, H.G., Erkens, G., Wenke, R., Cselle, T. and Schwenck, M. (1997) Wear mechanisms of $(Ti_{1-x},Al_x)N$ coatings in dry drilling. *Surface and Coatings Technology*, **94–95**, 603–609.

24 Salama, C.A.T. (1970) RF sputtered aluminium oxide films on silicon. *Journal of the Electrochemical Society*, **117**, 913–917.

25 Bobzin, K., Lugscheider, E., Knotek, O., Maes, M., Immich, P. and Piñero, C. (2005) Development of multilayer TiAlN + γ-Al_2O_3 coatings for difficult machining operations. *Materials Research Society Symposium*, **890**, 27–32.

26 Frieser, R.G. (1966) Phase changes in thin reactively sputtered alumina films. *Journal of the Electrochemical Society*, **113** (4), 357–360.

27 Schiller, S., Goedicke, K., Reschke, J., Kirchhoff, V., Schneider, S. and Milde, F. (1993) Pulsed magnetron sputter technology. *Surface and Coatings Technology*, **41**, 331–337.

28 Schiller, S., Goedicke, K., Kirchhoff, V. and Kopte, T. (1995) Pulsed technology – a new era of magnetron sputtering, Society of Vacuum Coaters, 38th Annual Technical Conference Proceedings, pp. 293–297.

29 Belkind, A., Freilich, A. and Scholl, R. (1998) Pulse duration effects in pulse-power reactive sputtering of Al₂O₃. *Surface and Coatings Technology*, **108–109**, 558–563.

30 Bobzin, K., Lugscheider, E., Maes, M. and Immich, P. (2006) Alumina PVD tool coatings for the use in semi solid metal forming of steel. *Journal of Solid State Phenomena*, **7**, 704–707.

31 Bobzin, K., Lugscheider, E., Maes, M. and Piñero, C. (2006) Relation of hardness and oxygen flow of H₂O₃ coatings deposited by reactive bipolar pulsed magnetron sputtering. *Thin Solid Films*, **494**, 255–262.

32 Wong, M.S., Chia, W.J., Yashar, P., Schneider, J.M., Sproul, W.D. and Barnett, S.A. (1996) High-rate reactive DC magnetron sputtering of ZrOₓ coatings. *Surface and Coatings Technology*, **86–87**, 381–387.

33 Koski, K., Hölsä, J. and Juliet, P. (1999) Properties of zirconium oxide thin films deposited by pulsed reactive magnetron sputtering. *Surface and Coatings Technology*, **120–121**, 303–312.

34 Schofield, M.A., Aita, C.R., Rice, P.M. and Gajdardziska-Josifovska, M. (1998) Transmission electron microscopy study of zirconia–alumina nanolaminates grown by reactive sputter deposition. Part I: zirconia nanocrystallite growth morphology. *Thin Solid Films*, **326**, 106–116.

35 Kopp, R., Shimahara, H., Schneider, J., Kurapov, D., Telle, R., Münstermann, S., Lugscheider, E., Bobzin, K. and Maes, M. (2004) Characterization of steel thixoforming tool materials by high temperature compression tests. *Steel Research International*, **75**, 569–576.

36 Lugscheider, E., Baerwulf, S., Bobzin, K. and Hornig, T. (2000) PVD-hard coatings protecting the surface of thixoforming tools. *Advanced Engineering Materials*, **2**, 33–37.

37 Lugscheider, E., Bobzin, K., Maes, M. and Immich, P. (2006) Developing PVD zirconium-oxide coatings for use of thixoforming of steel, Proceedings of the International Tooling Conference, pp. 449–456.

38 Knauf, F., Hirt, G., Immich, P. and Bobzin, K. Analysis and benchmark of the influence of process variables for thixoextrusion of steel, ESAFORM 2007. Proceedings of the 10th ESAFORM Conference on Materials Forming, pp. 1173–1178.

39 Kramer, B.M. and Judd, P.K. (1985) Computational design of wear coatings. *Journal of Vacuum Science and Technology A*, **3**, 2439–2444.

40 Müller, J. and Neuschütz, D. (2003) Efficiency of α-alumina as diffusion barrier between bond coat and bulk material of gas turbine blades. *Vacuum*, **71**, 247–251.

41 Shaw, L. and Abbaschian, R. (1995) Al₂O₃ coatings as diffusion barriers deposited from particulate-containing sol–gel solutions. *Journal of the American Ceramic Society*, **78**, 3376–3382.

42 Funk, R., Schachner, C., Triquet, C., Kornmann, M. and Lux, B. (1976) Coating of cemented carbide cutting tools with alumina by chemical vapor deposition. *Journal of the Electrochemical Society*, **123**, 285–289.

43 Kyrylov, O., Cremer, R. and Neuschütz, D. (2002) Influence of thin coatings deposited by PECVD on wear and corrosion resistance of moulds for semi-solid processing, Proceedings of the 6th International Tooling Conference, pp. 863–869.

44 Kyrylov, O. (2003) Abscheidung und Charakterisierung von PECVD-Aluminiumoxidschichten. PhD Thesis, RWTH Aachen University.

45 Bleck, W. and Püttgen, W. (2004) Fundamentals of thixoforming processes. *Steel Research International*, **75**, 531–536.

46 Quirmbach, P., Telle, R., Mertens, H.P. and Kopp, R. (1995) Rheocasting – Eine Herausforderung an keramische Formwerkzeuge. *Keramische Zeitschrift*, **47**, 891–897.

47 Pierre, D., Peronnet, M., Bosselt, F., Viala, J.C. and Bouix, J. (2002) Chemical interaction between mild steel and liquid Mg–Si alloys. *Materials Science and Engineering B*, **94**, 186–195.

48 Lin, C.H., Wang, L. and Hon, M.H. (1997) Preparation and characterization of aluminum oxide films by plasma enhanced chemical vapor deposition. *Surface and Coatings Technology*, **90**, 102–106.

49 Täschner, C., Ljungberg, B., Alfredsson, V., Endler, I. and Leonhardt, A. (1998) Deposition of hard crystalline Al$_2$O$_3$ coatings by bipolar pulsed DC PACVD. *Surface and Coatings Technology*, **108–109**, 257–264.

50 Böhler Edelstahlhandbuch. Böhler Edelstahl, Kapfenberg, Germany (1998).

51 Maeda, T., Yoshimoto, M., Ohnishi, T., Lee, G.H. and Koinuma, H. (1997) Orientation-defined molecular layer epitaxy of α-Al$_2$O$_3$ thin films. *Journal of Crystal Growth*, **177**, 95–101.

52 Jin, P., Nakao, S., Wang, S.X. and Wang, L.M. (2003) Localized epitaxial growth of α-Al$_2$O$_3$ thin films on Cr$_2$O$_3$ template by sputter deposition at low substrate temperature. *Applied Physics Letters*, **82**, 1024–1026.

53 Kurapov, D. (2005) Structure evolution, properties and application of alumina thin films deposited by PECVD. PhD Thesis, RWTH Aachen University.

54 Oliver, W.C. and Pharr, G.M. (1992) An improved technique for determining hardness and elastic modulus using load and displacement sensing indentation experiments. *Journal of Materials Research*, **7**, 1564–1583.

55 Nastasi, M., Mayer, W.J. and Hirvonen, J.K. (1996) *Ion–Solid Interactions*, Cambridge University Press, Cambridge.

56 Kester, D.J. and Messier, R. (1992) Phase control of cubic boron nitride thin films. *Journal of Applied Physics*, **72**, 504–513.

57 Kurapov, D., Reiss, J., Trinh, D.H., Hultmann, L. and Schneider, J.M. (2007)

Influence of the normalized ion flux on the constitution of alumina films deposited by plasma-assisted chemical vapor deposition. *Journal of Vacuum Science and Technology A*, **25**, 831–836.

58 Hultman, L., Sundgren, J.E., Markert, L.C. and Greene, J.E. (1989) Ar and excess N incorporation in epitaxial TiN films grown by reactive bias sputtering in mixed Ar/N$_2$ and pure N$_2$ discharges. *Journal of Vacuum Science and Technology A*, **7**, 1187–1193.

59 Catania, P., Rox, R.A. and Cuomo, J. (1993) Phase formation and microstructure changes in tantalum thin films induced by bias sputtering. *Journal of Applied Physics*, **74**, 1008–1014.

60 Täschner, C.H., Ljungberg, B., Endler, I. and Leonhardt, A. (1999) Deposition of hard crystalline Al$_2$O$_3$ coatings by pulsed DC PACVD. *Surface and Coatings Technology*, **116–119**, 891–897.

61 Knudsen, F.P. (1962) Effect of porosity on Young's Modulus of alumina. *Journal of the American Ceramic Society*, **45**, 94–95.

62 Spriggs, R.M. (1961) Expression for effect of porosity on elastic modulus of polycrystalline refractory materials, particularly aluminum oxide. *Journal of the American Ceramic Society*, **44**, 628–629.

63 Barbour, J.C., Knapp, J.A., Follstaedt, D.M., Mayer, T.M., Minor, K.G. and Linam, D.L. (2000) The mechanical properties of alumina films formed by plasma deposition and by ion irradiation of sapphire. *Nuclear Instruments and Methods in Physics Research Section B*, **166–167**, 140–147.

64 Krell, A. and Schädlich, S. (2001) Nanoindentation hardness of submicrometer alumina ceramics. *Materials Science and Engineering A*, **307**, 172–181.

65 Schneider, J.M., Larsson, K., Lu, J., Olsson, B. and Hjövarssopn, B. (2002) Role of hydrogen for the elastic properties of alumina thin films. *Applied Physics Letters*, **80**, 1144–1146.

66 Kurapov, D. and Schneider, J.M. (2004) Adhesion and thermal shock resistance of Al_2O_3 thin films deposited by PACVD for die protection in semi-solid processing of steel. *Steel Research International*, **75**, 577.

67 Lesniak, C. and Schmalzried, C. (2007) Non-oxide ceramics for high temperature metallurgical processes. In: Heinrich, J.G and Aneziris, C. (Eds.) Proc. 10th ECerS Conf., Göller-Verlag, Baden-Baden, 2007, 2206–2211, ISBN 3-87264-022-4.

68 Salmang, H., Scholze, H. and Telle, R. (2006) *Keramik*, Neuauflage, Springer, Berlin.

69 Routschka, G. (ed.) (2004) *Pocket Manual of Refractory Materials – Basics, Structure, Properties,* 2nd edn, Vulkan Verlag, Essen.

70 Beyer, C. (2005) Entwicklung prozessoptimierter Werkzeugtechnologien anhand eines seriennahen Zielbauteiles beim Thixoschmieden von Stahl. PhD Thesis, RWTH Aachen University.

71 Behrens, B.-A., Haller, B., Fischer, D. and Schober, R. (2004) Investigations on steel grades and tool materials for thixoforging, Proceedings of the 8th International Conference on Semi-solid Processing of Alloys and Composites (S2P), Limassol, 21–23 September 2004, TMS Publications, Warrendale, PA.

72 Kapranos, P. (2005) Thixoforming of high melting point alloys. *Journal of the Japan Foundry Engineering Society*, **77**, 518–525.

73 Rassili, A., Adam, L., Fischer, D., Robelet, M., Demurger, J., Cucatto, A., Klemm, H., Walkin, B., Karlsson, M. and Flüss, A. (2004) Improvement of materials and tools for thixoforming of steels, Proceedings of the 8th International Conference on Semi-solid Processing of Alloys and Composites (S2P), Limassol, 21–23 September 2004, TMS Publications, Warrendale, PA.

74 Lobert, M. and Telle, R. (2004) Investigation of tribological attack of Si_3N_4 dies during thixoforging of steel, Proceedings of the 47th International Colloquium on Refractories, Aachen, pp. 218–221.

75 Behrens, B.A., Haller, B. and Fischer, D. (2004) Thixoforming of steel using ceramic tool materials. *Steel Research International*, **75** (8/9), 561–568.

76 Behrens, B.A., Fischer, D., Haller, B., Rassili, A., Pierret, J.C., Klemm, H., Studinski, A., Walkin, B., Karlsson, M., Robelet, m., natale, l. and alpini, f. (2006) series Production of Thixoformed Steel Parts. *Solid State Phenomena*, **116–117**, 686–689.

77 Biswas, S.K., Mukerji, J. and Das, P.K. (1994) Oxidation of silicon nitride sintered with yttria and magnesia containing nitrogen rich liquid. *Key Engineering Materials*, **89–91**, 271–274.

78 Singhal, S.C. (1977) Oxidation of silicon nitride and related materials, Proceedings of the NATO Advanced Study Institute on Nitrogen Ceramics, 16–27 August 1976, University of Kent, Canterbury, Riley, F.L. (ed.), Noordhoff International Publishing, Leyden, pp. 607–626.

79 Nickel, K.G., Fu, Z. and Quirmbach, P. (1993) High-Temperature Oxidation and Corrosion of Engineering Ceramics. *Journal of Engineering for Gas Turbines and Power*, **115**, 76–82.

80 Grieveson, P. (1977) Some considerations of the thermodynamics of gas–solid reactions, Proceedings of the NATO Advanced Study Institute on Nitrogen Ceramics, 16–27 August 1976, University of Kent, Canterbury, Riley, F.L. (ed.), Noordhoff International Publishing, Leyden, pp. 153–174.

81 Schiepers, R.C.J., van Beek, F.J.J., van Loo, F.J.J. and de With, G. (1993) The interaction between SiC and Ni, Fe, (Fe,Ni) and steel: morphology and kinetics. *Journal of the European Ceramic Society*, **11**, 211–218.

82 Schiepers, R.C.J., van Loo, F.J.J. and de With, G. (1988) Reactions between α-silicon carbide ceramic and nickel or iron. *Journal of the American Ceramic Society*, **71**, C284–C287.

83 Meyer-Rau, S., Münstermann, S. and Telle, R. (2002) Study on the behaviour of

ceramic materials in contact with solid and molten aluminium alloy, Stahl und Eisen Special: refractories for ironmaking, foundries and non-iron metallurgy, Proceedings of the 45th International Colloquium on Refractories, Aachen, pp. 20–24.

84 Münstermann, S. and Telle, R. (2004) A ceramic tool concept for the thixoforming of steel, Proceedings of 8th International Conference on Semi-solid Processing of Alloys and Composites (S2P), Limassol, 21–23 September 2004, TMS Publications, Warrendale, PA.

85 Telle, R., Münstermann, S. and Beyer, C. (2006) Design, construction and performance of Si_3N_4 tool parts in steel thixoforming processes. *Solid State Phenomena*, **116–117**, 690–695.

86 Guo, Y., Long, S. and Telle, R. (2005) Effect of dopants on the corrosion behaviour of zirconia by steel at high temperature. *Key Engineering Materials*, **280–283**, 999–1004.

87 Miwa, K. and Kawamura, S. (2000) Semi-solid extrusion forming process of stainless steel, Proceedings of 6th International Conference on Semi-solid Processing of Alloys and Composites (S2P), Turin, 27–29 September 2000, pp. 279–281.

88 Rouff, C., Bigot, R., Favier, V. and Robelet, M. (2002) Characterization of thixoforging steel during extrusion tests, Proceedings of the 7th International Conference on Semi-solid Processing of Alloys and Composites (S2P), Tsukuba, A-56, pp. 355–360.

89 Sugiyama, S., Li, J.Y. and Yanagimoto, J. (2004) Semi-solid state extrusion of plain carbon steel, Proceedings of the 8th International Conference on Semi-solid Processing of Alloys and Composites (S2P), TMS Publications, Warrendale, PA.

Part Four
Forming of Semi-solid Metals

Thixoforming: Semi-solid Metal Processing. Edited by G. Hirt and R. Kopp
Copyright © 2009 WILEY-VCH Verlag GmbH & Co. KGaA, Weinheim
ISBN: 978-3-527-32204-6

9
Rheocasting of Aluminium Alloys and Thixocasting of Steels[*]

Matthias Bünck, Fabian Küthe, and Andreas Bührig-Polaczek

9.1
Casting of Semi-solid Slurries

The technological and economic potential of innovative materials and their processing is increasingly in the focus of technology-oriented companies. The demands on cast parts for the automotive and aerospace industries, but also in wide areas of the mechanical engineering sector, are still rising. The demand for lighter structures with higher mechanical properties has to be combined with stringent economic and ecological aspects. These apparent inconsistent requirements have to be tackled with the development of innovative manufacturing and material concepts.

Overall, three forming operations have been investigated to produce parts in the collaborative research centre. Forging processes are performed by using a hydraulic forging press and the thixocasting is done in a cold chamber high-pressure die-casting machine.

Thixocasting is a process where an ingot billet is squeezed into a closed die by a shot piston, comparable to high-pressure die casting. The process is mainly performed on conventional real-time controlled die-casting machines where the shot chamber system is adapted to the semi-solid billet insert [1–4].

There are four main aspects for the thixocasting route. First, the velocity of the slurry is significantly faster than in thixoforging but lower than in conventional high-pressure die casting. A normal piston velocity is about 0.3–1 m s^{-1} for a part thickness of 5 mm. Together with the flow property of the semi-solid material, this leads to a laminar die filling. By the closed flow front air enclosed in the die cavity is evenly conducted through venting channels and the division plane.

Second, the liquid phase can range from 30% to a maximum of 60%. The use of a faceplate within the gating system reduces the entry and inclusion of damaging surface oxides from the reheated billet.

[*]A List of Symbols and Abbreviations can be found at the end of this chapter.

Third, the shot weight is greater than the weight of the part because of the material of the biscuit, gating system and the overflows. The use of multiple moulds is as usual for thixocasting as for the high-pressure die casting (HPDC) process.

Fourth, the semi-solid metal (SSM) casting processes seem very useful for filigree and complex geometries. Complexity includes undercuts formed by core pullers and also long flow lengths enabled by a high shot velocity and thick- and thin-walled sections in one part.

9.2
SSM Casting Processes

The SSM casting processes are usually divided into three main types: thixocasting, rheocasting and thixomoulding (Figures 9.1 and 9.2).

The thixocasting process is usually divided into three steps: pre-material billet production, reheating into the semi-solid state and forming.

Depending on the alloy, several methods for producing a pre-material with a globular microstructure have been developed. During reheating of the pre-material billets, very close control of the process is needed to ensure the correct solid/liquid fraction and a homogeneous temperature distribution of the billet. After reheating, the semi-solid billet is placed in the modified shot chamber of an HPDC machine or a squeeze casting machine and formed [5, 6].

In contrast to thixocasting, the rheocasting processes deal with integration of the pre-material production and the subsequent preparation of a semi-solid slurry with thixotropic properties [7]. An important development in the field of rheocasting was the so-called 'new rheocasting' introduced by UBE, Japan. With this technology, the generation of a globular microstructure is not caused by agitation during solidification but directly through forced nucleation by cooling into the semi-solid state. Melt just above its liquidus temperature is poured into a permanent steel mould, of a defined geometry, where it spontaneously cools into the semi-solid state. At the same time, nucleation of the material takes place according to the heterogeneous nucleation theory and through controlled cooling the desired globular microstructure is produced.

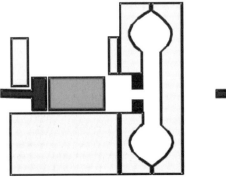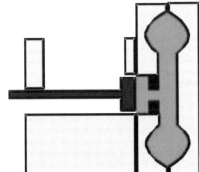

Figure 9.1 Layout of a thixocasting process.

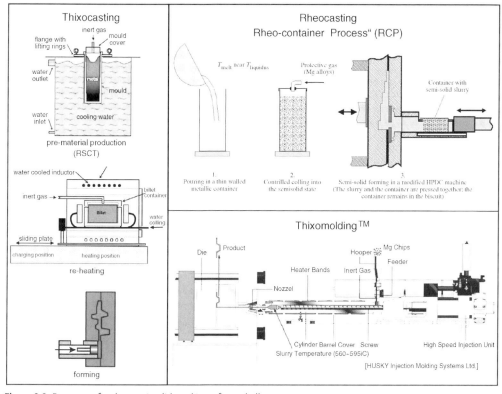

Figure 9.2 Processes for the semi-solid working of metal alloys.

When the material reaches the desired temperature, the mould is lightly inductively heated, to melt the material in contact with the mould wall and thus facilitate the extraction of the billet.

The newly developed rheo-container process (RCP) (Figure 9.3) for light metal alloys should solve the problem of transferring the semi-solid billet from the preparation mould to the shot sleeve of a modified HPDC machine [9]. The melt is directly poured into a disposable container. When the metal reaches the desired thixotropic conditions, the slurry and container are pressed in the HPDC machine. The metal flows into the cavity of the mould while the container remains in the biscuit. Because of this, the use of specially designed moulds for the cooling of the melt and eventual reheating of a pre-material is not necessary.

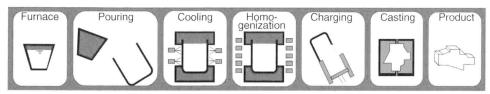

Figure 9.3 New rheocasting process (RCP) [8].

Another interesting method is the so-called cooling channel process [10], which will be presented in Section 9.4.1. The melt is poured on to an inclined channel just above liquidus and caught in an isolated ceramic container. The metal flow induces the heterogeneous nucleation of solid particles and the retarded cooling in the container promotes globular growth of the solid particles. After reaching the desired solid fraction, the slurry is tilted in the shot sleeve of the HPDC machine and formed.

More complex conditions exist during the one-step process of semi-solid injection moulding or thixomoulding (Figure 9.2) where feedstock particulates of a specific nature are subjected to the simultaneous influence of the heat supplied from an external source and shear force imposed by the injection screw in a machine very similar to those used for the injection of thermoplastics [11]. This process is especially effective for processing magnesium alloys.

9.2.1
Developmental History of the Part Liquid Aluminium Processing

Numerous companies have utilized the advantages of the processing of semi-solid materials for developing SSM technology since the 1990s. Table 9.1 gives the most important manufacturers of pre-material, users and international large academic facilities [6].

Table 9.1 Relevant manufacturer and R&D facilities for Al thixocasting.

Pre-material	Producer	R&D[1]
• Alcoa (Alumax) (Canada)	• AEMP (USA)	• Aachen RWTH (Germany)
• Algroup Alusuisse (Switzerland)	• Algroup Alusuisse (CH)	• Brunel University (UK)
• Hertwich Engineering (Austria)	• Aluthix (The Netherlands)	• ETH Zürich (Switzerland)
• NorskHydro [F&E] (Norway)	• Audi [F&E[2]] (Germany)	• FhG Bremen (Germany)
• Northwest (USA)	• Bühler (Switzerland)	• INPG Grenoble (France)
• Ormet (USA)	• DaimlerChrysler [F&E[2]] (Germany)	• LK Ranshofen (Austria)
• Pechiney (France)	• Form Cast (USA)	• MIT (USA)
• SAG Thixalloy (Austria)	• Honda Engineering (Japan)	• Osaka University (Japan)
	• Hot Metal Molding (USA)	• Pusan University (South Korea)
	• Madison Kipp (USA)	• SINTEF (Norway)
	• SAG (Austria) (Germany)	• Ancona University (Italy)
	• Stampal (Italy)	• Sheffield University (UK)
	• UBE Industries (Japan)	• Stuttgart University (Germany)
	• Volkswagen [F&E[2]] (Germany)	• WPI (USA)
	• Weber & Broutin (Italy)	

[1] Research & Development.
[2] Forschung & Entwidclung.

Two Italian companies gained the first experience with use of the innovative thixoforming technology in the test series or batch production of aluminium components in 1992–93.

The processing of semi-solid materials has been an increasingly favoured topic of worldwide research activities, with a considerable increase in the production of thixocasting components [12]. Attention has been devoted to the production of mechanically highly loaded and particularly pressure-tight and weldable safety components for the automotive industry [13–15], and the lot sizes of some parts is over 2 million pieces per year [16]. The component weight varies between 10 g and 10 kg [17]. Examples of thixocasting parts from the series and from the F&E in Europe and the USA show that the method is suitable particularly for the production of complex and highly resilient components (Figure 9.4). The following application classes can be distinguished:

Thixotec [Bosch AG]   Injection pump [Bühler]   Wheel hub [GF]

Spaceframe knot Audi A8 [SAG Thixalloy]   Doorframe part Audi A3 [Audi AG]   Brake master cylinder [EFU]

Connecting flange hydraulic accumulator [VW AG]   Steering arm for DC W220 [Alcan Singen GmbH]   Belt redirector for DC [SAG Thixoalloy]

Figure 9.4 Examples of aluminium thixocast applications.

- components with tightness requirements (pressure connections, brake master cylinders);
- thick-walled, highly loaded components;
- thin-walled components (structure components with high strength and elongation);
- components of special materials (materials with particulate reinforcement).

Aluminium-based alloys have been almost exclusively used as the material. In series applications, the preferred alloys are mainly A356 (AlSi7Mg0.3) and A357 (AlSi7Mg0.6) due to their advantageous mechanical characteristics and good processability [5]. On the other hand, the wrought alloys 2024, 5083, 6012, 6082 and 7075 offered high potential due to their high strengths. The tendency for hot cracks and microporosities induced by solidification can be significantly lowered in comparison with conventional casting processes [18]. The thixocasting of aluminium-based materials can show excellent mechanical qualities; Figure 9.5 presents the strengths obtainable for a T6 heat treatment.

Since the end of the 1990s, the industrial use of the thixoforming has declined steadily. The reasons for this change can be found in both the economic and technological framework conditions. A main reason is that most applications need specially made, expensive pre-material. The necessity for a larger number of single-coil heating devices leads to an increase in investment and operating costs. The longer lifetime of the dies and the several seconds shorter cycle time do not compensate for these costs completely. A solution would have been the extended use of rheoforming plants, which were not widely available at that time. The smallest deviations in the temperature of the billet were an important process-specific disadvantage because they influence the flow behaviour considerably with a direct change of the liquid phase content.

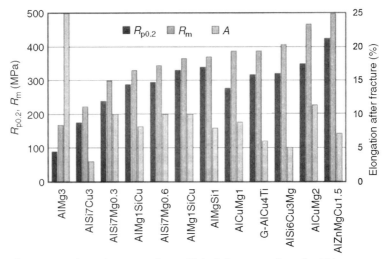

Figure 9.5 Mechanical properties for established aluminium alloys after T6 heat treatment [19].

For these reasons, the same production rates as with conventional methods still cannot be reached for the relatively young thixocasting process. Finally, the choice of safety parts in the driving work area and other sensitive components is important. The high mechanical requirements and the small fault sizes have led to high scrap rates. The boundary conditions for use turned out to be increasingly difficult since there were a growing number of methods with similar possible applications. The Vacural pressure die-casting, the Poral method and the Cobapress method were alternatives in the chassis suspension, and also the squeeze casting method and a modern ingot casting [20]. Thixoforming has become the production alternative for very qualitative aluminium components, [21] which deliver good quality [22] but is more expensive than alternative methods. In this situation, these competing methods have fully exploited their potential, and thixoforming has only seldom been adopted. Thixalloy Components GmbH, as a motorcar sub-supplier of high-strength castings, decoration parts, solderable parts and also structural parts, is currently a successful company with annual sales of more than 4 million euros.

At that time, particle amplified and also over-eutectic alloys also offered further material classes. It was known that an unwanted separation of particles and metal melt could occur due to the density difference between the metal and particles during processing in the normal die-casting process. Because of the high content of solid phase in processing in the semi-solid area, this phenomenon is a considerably restricted. First examinations were carried out on AlSi-based alloys with SiC or TiB_2 particle additions [1, 5]. The wear-resistant alloys AlSi17Cu4Mg and AlSi25CuMgNi also could be processed by thixocasting [23, 24] and show comparably high wear resistances with respect to a metal–matrix composites (MMCs).

Until now, the conventional die casting of high-melting point is practicable only for some brass-based materials. The two standardized copper–zinc pressure die-cast alloys CuZn37Pb and CuZn15Si4 show a liquid interval between 890 and 910 °C and 830 and 900 °C, respectively. The high pouring temperatures, however, generally do not allow very long form stand times of approximately 10 000 parts.

In first examinations, it was reported that the mechanical qualities of thixocast model components of the copper bronze CDA905 are similar to those of conventionally cast parts [25, 26].

The development of copper pressure die-casting has attracted greater attention in the newer literature for the electrical industry despite the load on the tools. To improve the degree of effectiveness and the load alternation behaviour, aluminium pressure die-casting rotors and cages could be replaced with copper pressure die-casting of energy-saving three-phase motors [27]. Examinations have already been carried out successfully of die-casting copper-based alloys in the semi-solid state [28]. Different sample components were produced for the determination of properties [29, 30].

The wide alloying range of modern steel materials makes possible broad variations of the quality profile through which numerous industrial applications become possible in combination with adapted heat treatment strategies. Steel is still therefore one of the s most frequently used structural materials worldwide [31]. Steel materials offer an often equal or sometimes even higher lightweight construction potential than light metal materials (aluminium/magnesium) or synthetic materials due to their outstanding mechanical qualities despite the high specific gravity of between 7.6 and 8.2 g cm^{-3}. The exploitation of this potential by the continuing development of innovative production technologies for the production of complex, high-performance and weight-optimized steel components could lead to an expansion of their competitiveness of the foundry industry.

The thixocasting process with steel materials offers unusual possibilities due to its great design flexibility and the net-shaped qualities of the components in combination with the productivity of a pressure die-casting method.

At the beginning of the 1980s, examinations of the pressure die casting of alloyed steels were subject to different basis tests. In addition to the complex warming of suitable pre-materials, the processing of the steel materials at modified presses or in die-casting machines showed a variety of difficulties.

Major emphasis was put on austenitic and other high-alloyed steels (X5CrNi18-10/X2CrNil9-11/HS6-5-2) because these showed a globulitic formation of the solid phase due to the conventional production route [stress-induced and melt-activated (SIMA) route, powder metallurgy] during reheating [32]. The extensive examinations for the pre-material, its reheating and the subsequent design to model parts clarified that the steel materials are in principle usable for thixocasting. The complex remelting behaviour of carbon steels led to considerable problems with respect to reproducible warming, although high-speed steels due to their content of eutectic phase (HS 6-5-2-1, HS 12-1-4-5) were consequently often examined. In the course of the examinations, it became clear that even commercially available semi-finished material showed globulitic formation of solid particles during reheating [33].

The first components of more complex geometry showed satisfactory mechanical qualities in addition to a good illustration precision and contour acuity (Figure 9.6). In 1999, a practically relevant project at the Studiengesellschaft für Stahlanwendung started with the aim of producing a motorcar connecting rod by thixo lateral extrusion with the use of a modified multi-piston press. The mechanical quality demanded could not, however, be achieved, which could be explained by the rapid failure of the tool materials and also method-specific difficulties.

Pump cover with biscuit   Connecting rod, steel C70, overflow

Figure 9.6 Components of projects P322 and P443.

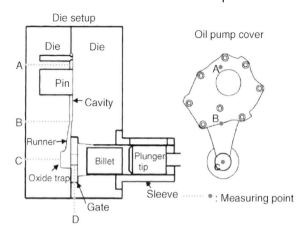

Figure 9.7 Oil pump cover for Honda cars.

In a Japanese thixocasting project, moulding methods, optimal billet heating speeds and copper die materials were established for an iron thixocasting process. The manufacturing method was studied with an oil pump cover (Figure 9.7). An improvement in the volume efficiency of the oil pump was confirmed due to the stiffness improvement and clearance reduction effects.

9.3
Thixocasting of Steel Alloys

The multitude of steel materials with their wide range of mechanical properties, relatively low costs combined with the advantages of the productive high-pressure die-casting process open up numerous areas of application. This chapter takes a closer look at the current work in the field of thixocasting of steel alloys.

9.3.1
Inductive Heating for Steels: Adaptation and Process Control

Already in the processing of aluminium alloys it was found that complete form filling and good component characteristics depend on the globulitic microstructure.

With regard to the melting behaviour of steel materials, high requirements are placed on the reheating process. First, the warming must be fast enough that coarsening of the solid phase is reliably avoided. Since steel billets have almost three times the energy content of an Al billet, the nominal power must be greater than at least 15 kW kg^{-1}. Otherwise, the industrial user is forced to invest in 10 or more individual heating devices to use the short cycle times of a pressure die-casting machine when required. Second, precision of the reheating process is required because for some annealing steels a variation of only 15 K can makes a difference of 30–60% in the content of liquid phase. A few degrees difference in temperature decides whether the reheating matches the chosen tool and process window. Third, the warming must be controlled sufficiently exactly that the homogeneity of the liquid phase varies by only a few percent about the bolt length and the bolt crosscut. This homogeneity can be quantified with axial and radial temperature measurements. Since the heat conductivity of the steels is lower than that of aluminium, more time must be allowed for the temperature compensation. Furthermore, the radiation losses have to be considered and without countermeasures lead to a solid billet skin of some millimetres thickness, with which the oxide restraint system of the Deutsche Gesellschaft für Materialforschung (DGM) would be rapidly overtaxed. Fourth, good reproducibility is imperative for constant, high component quality. Close control and regulation, high-quality components and correctly selected materials in the direct surroundings of the billet are prerequisites. The avoidance of oxidation is an essential point for steel billets since no closed oxide film develops as on a semi-solid aluminium billet and the oxidation would steadily and rapidly progress. Inductive heating best fulfils the qualification profile for many materials and for this reason is frequently used [34].

The heating device from Elotherm (Figure 9.8) works with frequencies between 750 and 950 Hz according to the oscillating circuit principle. The plant consists of a water-cooled, cylindrical coil with 29 meanders embedded in an isolating mass.

The oscillating circuit parameters are included after every constructive change in the form of a calibrating measurement in neutral to determine the dependence on the output power. During the subsequent heating the computer accesses this file.

The method of measurement of the working energy was implemented by Elotherm and integrates the real energy entry with consideration of the oscillating circuit loss resistance RV0. The energy entry is determined in discrete timing steps with a reference specification, the so-called energy set point curve that has been adapted previously to the material and its geometry.

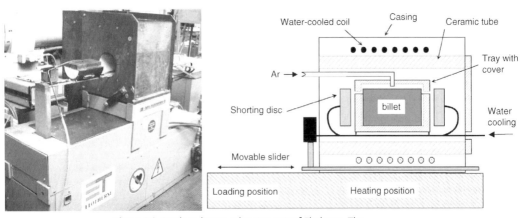

Figure 9.8 Single inductive reheating unit of Elotherm. The horizontal layout of the billet tray allows wide variations of the liquid fraction.

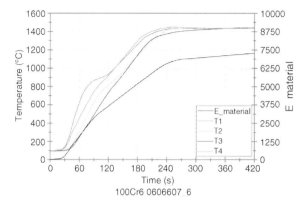

100Cr6 0606607_6

Figure 9.9 Middle temperature courses of the steels 100Cr6, 42CrMo4 and X210CrW12. Change of the electromagnetic qualities at ~780 °C. As in the case of aluminium, the temperature course is very flat according to admission to the SSM region and hardly allows the assignment of the liquid phase content. The end of the dissolution of the chromium carbides becomes clear in the hold step. The energy set point courses show characteristic differences in the transition in the hold step (around 300 s) and in the power.

With the help of some heating trials, an energy set point curve is successful. This curve consists in principle of four periods: start-up, fast heating, reduced ferritic phase and homogenization phase (Figure 9.9). Following the short start-up phase, the performance can be increased rapidly to achieve the maximum converter system voltage because of the good coupling of the steel to the coil in the ferromagnetic area (for 1.7225: 70 s).

After the Curie temperature in the outer billet shell has been exceeded, the power must be reduced slightly (300 s) in phase three. With this power, the bolt is heated to the semi-solid range since both the amount for the heating of the billet and the share of the melting heat must be found. The time for switching over to the homogenization phase has to be determined when the bolt reaches its desired liquid phase content but also largely keeps its form. A temperature compensation is realized by heat conduction in this hold step (Figures 9.10). The middle billet temperature is constant and merely the overheating of the outer shells is used to warm the inside and bring the liquid state contents into line. The tray drives out automatically for all alloys after 420 s.

In the early years, the pre-material billet had the dimensions ϕ 55 × 160 mm and a shot weight of 3.5 kg. The billet was moved into the coil recumbently in a tray. Due to the high energy content of the high-carbon semi-solid steel billet at temperatures around 1300 °C and also the occurrence of high-temperature corrosion, new solutions for the tray had to be developed. The new framework of the tray was constructed of magnetically soft stainless-steel sheets to protect the reheating unit from melting. The billet holder was lined with a specifically designed ceramic insulation insert, which actually carries the semi-solid billet (Figure 9.11). Additionally, a cover with an integrated inert gas supply was designed.

It turned out to be disadvantageous that a scale had formed up to the triggering of the shot by oxidation and cooling at the bolt surface and that the liquid state content was not completely homogeneous.

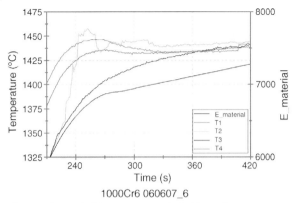

1000Cr6 060607_6

Figure 9.10 Detail of the transition in the hold step: Temperature compensation in the 100Cr6 billet by heat conduction. The overheating of the outer shells (T1, T4) delivers energy into the bolt inside (T2, T3). The liquid state contents adjust strongly since the difference in temperature is 9 K at the end of heating.

In an improved solution, complete encapsulation of the billet in a tight-fitting high-temperature isolation unit was applied. In addition, the reheated billet should not be removed from the encapsulation shell. This was achieved by keeping the semi-solid billet in the heating and transportation container and driving it out by the shot piston itself [shot-through cartridge (STC)].

To be able to use the same moulds as built for Al alloys, the billet diameter was increased to 76 mm, which made possible a maximum shot weight of 6.05 kg at a length of 170 mm.

For isolation, vacuum-formed parts have been produced consisting of 48% Al_2O_3 and 52% SiO_2, which provide a sufficient supporting effect for the more than 5 kg weight of the billet, a good isolation behaviour and a short time resistance for temperatures over 1490 °C (Figure 9.12). At the density of 320 kg m^{-3} used here, the piston needs ~115 N to compress the material from the diameter 76 mm to the shooting piston diameter of 85 mm radially.

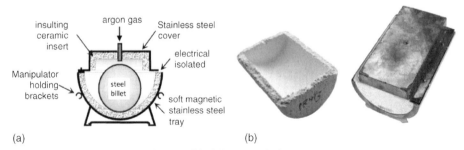

(a)

(b)

Figure 9.11 Schematic drawing of the billet tray including cover and billet (a) and manufactured ceramic insulation inserts in the initial state (b).

Figure 9.12 Components of cartridge system, cartridge in shot chamber.

The isolated billet has a protection tube that ensures sure handling from the reheating device to the shot chamber and an exact positioning in the shot chamber axially. A crack-resistant ceramic pipe made of aluminium titanate was used which results in an almost ideal quality profile. The measurements confirmed the excellent isolation effect of the concept with a temperature loss of less than 8 K in 60 s.

For the cartridges, a new charging zone for the shot chamber has to be designed. For precise threading into the pressure part of the shot chamber, first a piston of length 320 mm was used. Because of increased wear, the length of the high-strength CuCoNiBe piston was increased to 580 mm from an alloying last. This piston generates a sharp edge to separate the isolation material. This isolation material protects the piston from excessive heat entry and reliably prevents excessive wear.

9.3.2
Casting Technology and Systems

9.3.2.1 Systems Engineering for Thixo- and Rheoprocesses
To be able to characterize semi-solid materials during and after thixoforming, a locking force of 725 tons in a machine from Bühler AG is used (Figure 9.13). The real-time shot control permits exact programming of the piston velocity (0.02–8 m s^{-1}), movement and pressure during the fractions of a second that are needed to fill a mould cavity. Depending on thickness or geometry, the metal velocity can be increased or reduced at predetermined points and after complete filling a post-pressure can be applied to compact the material. The dies can be tempered with thermal oil, using three independent heater cooler units from Thermobiel with two circuits each up to a temperature of 350 °C. A Fondarex system to reduce the presence of air in the mould cavity completes the peripheries of the HPDC machine.

9.3.2.2 Process-adapted Shot Chamber System
A process-adapted shot chamber has been designed and constructed in close cooperation with Ortmann Druckgiesstechnik GmbH. In the preparation for the construction, an approximate simulation of the thermal loads occurring during the thixocasting process of semi-solid steels was performed with the numerical

Figure 9.13 Bühler H-630 SC real-time controlled HPDC machine, heating/cooling and vacuum units.

simulation software MAGMAsoft. The pictures in Figure 9.14 show the heat distribution inside the heavy loaded shot chamber and the die during the solidification of the suspension 15 s after the mould filling for the first and the tenth cycles of a serial production (cycle time ~500 s). The surface near regions of the die tools, and also the shot chamber, reach temperatures above 700 °C, causing a dramatic loss of hardness. Therefore, the solidification time should be reduced to a minimum to avoid massive heating through the die tools and thereby a dramatic loss of surface hardness.

In consideration of the numerical results, the new shot chamber is constructed as a multipart system with separate pressure and charging sections connected by a clamping ring (Figure 9.15). Multiple electric heating cartridges, giving a maximum temperature of about 600 °C, individually temper both parts. Due to the clamping and the selective heating, both radial and axial thermal stresses that occur can be effectively reduced. The device allows a fast and easy tool change in case of any malfunction. Moreover, the chamber can be switched from thixocasting to fully liquid casting within a few working steps.

In order to protect the chamber from massive abrasive wear, a special coating is applied before every casting operation.

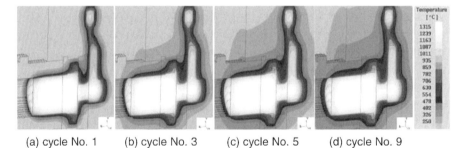

(a) cycle No. 1   (b) cycle No. 3   (c) cycle No. 5   (d) cycle No. 9

Figure 9.14 Simulated heat expansion (heat flow 1000 W m^{-2} K^{-1}) through the shot chamber and the die 15 s after the mould filling for up to nine consecutive cycles at a cycle time of about 500 s.

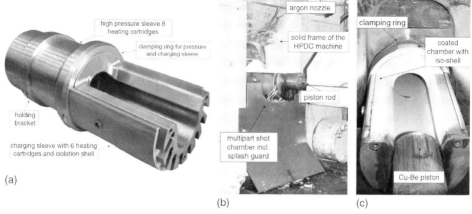

Figure 9.15 Process adapted multipart shot chamber system (a) with separate pressure and charging sleeves, complete shot chamber including the piston rod, wiring and an argon supply for the charging area (b), and coated chamber prepared for a thixocasting cycle (c).

The newly designed process-adapted shot sleeve system shows very promising results in terms of high temperature stability and wear resistance and also the influences on the properties of semi-solid steels castings. A detailed analysis of the performance is currently under investigation.

9.3.2.3 Technological Design Layout

For the form, layouts were supported by the use of MAGMAsoft. For the simulation of the thixocasting process, a specific thixo-module that permits the calculation of the behaviour of semi-solid metals was used. In addition, the database had to be enlarged with sheer rate-dependent viscosity values which serve for the description of the thixotropic material behaviour according to the Ostwald–de Waele potential law of [35]. These simulations need to be validated by using 'step-shooting' experiments.

These partial fillings of the form cavity represent a basic aid for the determination of the form filling characteristics of semi-solid thixomaterials. The shot piston is stopped at a time at which the metal has reached an interesting change in flow direction or metal velocity because of the mould geometry. In the context of several tests, the dynamics of the filling stage can be shown by 'freezing' the different filling stages. In this way it is possible to distinguish between laminar and turbulent form filling. Furthermore, important knowledge is gained for the optimization of the form construction.

The gating system and the ingate exert a decisive influence on the form filling behaviour of the slurry and at least on the mechanical quality of the parts in the thixocasting process. In principle, the ingate system must be designed in such a way that no inclusions, such as oxides, reach the component which would degrade the mechanical qualities. It is a further requirement that an even and laminar form filling

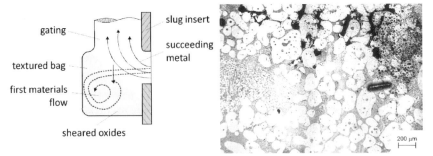

Figure 9.16 The bag serves for the restraint of the oxide face of the slugs. Thus, only oxide-free material from the middle of the billet reaches the gating, on the right: The peeled oxide outer surface found in the restraining cavity confirms the functionality of the system.

is achieved to minimize gas inclusions and other specific casting defects. Already the ingate system of the first mould sample featured an oxide restraining system (Figure 9.16). By pressing the reheated billet through a hole in the slug insert, part of the outer skin and of the oxide skin at the side is removed in a circular cavity around the biscuit. The oxide-resistant skin from the front of the billet is moved into a bag behind the slug insert bag. This preferentially 'clean' material reaches the inside of the gating channel.

The cross-sectional area of the gating channel is slightly decreased with respect to the area of the ingate of the order of 1.1:1. By pressurizing the gating system, the laminar metal flow is promoted.

The geometry of such a cavity is of special significance for the process flow. A high volume of the cavity leads to great metal losses whereas too small a system can take the outer skin only partly and oxide inclusions can arise in the component.

9.3.3
Model Tests

Different test geometries were developed and used in the characterization of the process properties of SSMs. Fork and finger samples were the first moulds, followed by a step die with variable ramps, a meandering die, a sample with a T-shaped geometry and a plate sample, which were used for model tests. All mould inserts are designed so that they can be integrated into the framework of a modular die-casting tool. Three tool frameworks in different measurements were available for the examinations. The plate dies are integrated into a multiple mould so that with this form four test parts can be produced in one shot whereas all other dies have only one cavity.

T Geometry The T-shaped sample is part of the basic flow experiments. The used sample serves for the characterization of high-melting material with the help of step-shot experiments. It is also suitable for first tool material examinations owing to its segmented inserts.

9.3.3.1 Investigations on Mould Filling Behaviour (Step Die)

The segmented step die consists of six interchangeable inserts with different ramps. The test pieces produced have wall thicknesses between 0.5 and 25 mm and represent the wall thickness range of components manufactured industrially. The step form is used to analyse the influence of changing flow crosscuts, wall thicknesses and therefore different metal velocities on the mechanical qualities and the microstructure.

The part illustrated in Figure 9.17 was produced from the cold work steel X210CrW12. At a constant piston velocity of 0.05 m s^{-1}, the metal velocity increases

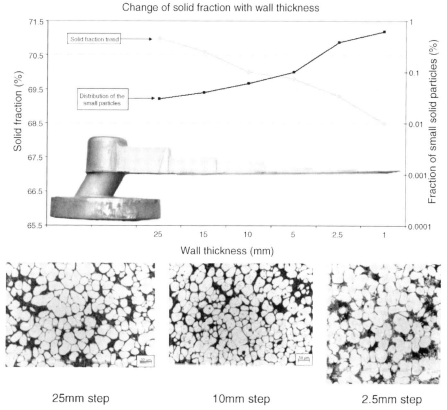

25mm step   **10mm step**   **2.5mm step**

Figure 9.17 Thixocast X210CrW12 steel part at a solid fraction of about 70% ($T_c \approx 1290\,°C$). The component shows outstanding surface quality and dimensional accuracy. The micrographs show a slight increase in liquid content towards the tip of the step specimen associated with an increasing fraction of fine dispersed solid particles.

from 0.2 to 11.5 m s^{-1} according to the wall thickness of the component. For this set of parameters (mould temperature 250 °C, piston velocity of 0.05 m s^{-1}), a 1 mm step could only be partially filled. The observed surface quality and the obtainable wall thicknesses demonstrate an excellent mould filling behaviour of the semi-solid steel suspension. Mechanical properties were evaluated by a Rockwell hardness test.

Excellent mould filling was achieved using slugs with an initial solid fraction of 60–70%, the best results. The large amount of solid phase reduces the shrinkage so that even compacted parts can be produced with a relatively small ingate cross-section and supports a more easily controllable flow tip due to its higher viscosity. Compared with aluminium alloys, a higher forming pressure is required at the beginning of the forming operation. After breaking up the meshed structure, the viscosity drops immediately, so that the semi-solid steel suspension can be processed at a forming pressure comparable to that of aluminium alloys. It has been shown that for steel alloys with their higher density and their resulting mass inertia, the processing parameters have to be adjusted in terms of reduced maximum metal velocities.

The evolution of the solid fraction with the flow length and the wall thickness is shown in Figure 9.17 for the step specimen cast of X210CrW12. The micrographs illustrated indicate a fine globular structure of the solid particles embedded in the eutectic phase.

Based on the 25 mm step, a slight increase in the liquid fraction is observed, which might have been caused by squeezing out of liquid phase from the incompletely broken structure during the initial compression of the billet. The strong supercooling of the partially solidified suspension permits a dendritic growth initiated from the globular solid particles. The grown dendrites break due to permanent shearing during the die filling and form globules. Depending on the flow length and the metal velocity, a larger amount of fine dispersed globular solid particles occurs on the flow front. According to the digital picture analysis, the microstructure of the semi-solid steel parts exhibit an average grain size of 35 µm and a shape factor of 0.7 of the primary globules.

9.3.3.2 Determination of Flow Length Capabilities (Meander Die)

The meander die was developed for the comparative characterization of the flow length in tools. It does not represent any standardized experiment for the rheological characterization of the flow properties, like the capillary experiment (cf. Chapter 3). It is used to correlate the process conditions such as velocity of flow, thermal boundary conditions and initial liquid content, with consideration of material-specific qualities such as morphology. The test geometry consists of a flow channel with a trapezoidal crosscut of area 87 mm^2 which is integrated in a meander tools in a length of 2830 mm.

Experiments with the meander die evaluate the flow length of semi-solid materials in correlation with process and material parameters obtained from prior investigations on the mould filling behaviour. The optional vacuum system allows investigations of the influence of a slight evacuation of the cavity on the flow length. If one considers the relatively low mould temperatures that cause a high temperature gradient between the mould and SSM, these investigations are necessary to evaluate

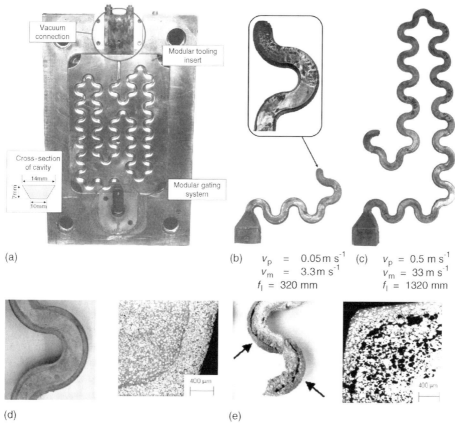

(a)

(b) v_p = 0.05 m s^{-1} (c) v_p = 0.5 m s^{-1}
v_m = 3.3m s^{-1} v_m = 33 m s^{-1}
f_l = 320 mm f_l = 1320 mm

(d)   (e)

Figure 9.18 (a) Meander tool for flow length investigations of
SSM alloys; (b, c) X210CrW12 cold work steel for two different
piston velocities at a given fraction solid of about 75%
($T_f \approx$ 1290 °C); (d) flow lines of cast metal at beginning of
meander, interface healed with eutectic melt; (e) at high piston
velocities no healing at the interface, here at the end of the
meander.

castable part geometries [31]. As illustrated by two meander parts in Figure 9.18, the
different piston velocities have a significant influence on the flow length. While at a
piston velocity of 0.05 m s^{-1} (metal velocity $\sim$3.4 m s^{-1}) a total flow length of about
320 mm can be achieved, the flow length increases to more than 1.3 m at a piston
velocity of 0.5 m s^{-1}. The flow tip shows typical laminar characteristics. Detailed
investigations on the evolution of the microstructure depending on the flow length
have shown a very similar development of the solid phase to that already observed in
the described step specimen. The formation of a distinctive boundary layer caused by
a thin fluid film at the tool surface due to the high cooling rate is observed in the
meander experiments where the skin forms a flow channel. While the boundary layer
is healed with eutectic phase in most of the experiments, at very high piston velocities
at the flow tip of the meander healing can no longer be achieved.

Standardized Form Filling Experiment   This form filling experiment was developed with the aim of providing a standardized test tool which is suitable for characterizing the qualification of a new material for the thixocasting process by measuring the form fill and flow properties.

Moreover, the geometry was designed with the support of multiphase simulation in cooperation with the Aachen Chemical Engineering Institute of RWTH Aachen University. A maximum influence of the rheological qualities of the material on the form filling characteristics can be deduced.

Fourfold Plate Sample   The plate sample is part of the preferential experiments due to the segmentation and execution as a multiple die for the characterization of different tool materials and the statistically based data for parameter variations. In turn, the block-shaped tools are fixed by holders without stress, which permits the use of full ceramic inserts.

9.3.3.3   Examinations of the Wear Behaviour of the Form Tools

Due to the high process temperatures in connection with the amplitude of the thermal shock cycles and the corrosive attack by melt, there is a good chance, particularly in the areas near the surface of the tool, for damage during the processing of steel. Propagation of cracks, strong abrasive wear, adhesive damage and deformation were frequently observed in the macroscopic area.

For this reason, the choice of the characterization and assessment of different tool materials and surface treatments, such as coating and mould release agents, were at the forefront of the design experiments. At the same time, the dispersion-hardened, powder metallurgical titanium–zirconium–molybdenum (TZM) alloy was tested as a high-temperature material. A promising and comparatively high thermal shock resistance ceramic Si3N4 was selected.

By applying suitable mould release agents, the thermal resistance can be increased and the heat transmission into the form tool can be reduced. At the same time, mould release agents reduce the stickiness and protect the tool surface against mechanical destruction. No special products have been available until now. A selected zirconium oxide mould release agent has been developed for iron and steel in sand casting or investment casting methods. The mould release agents were applied evenly by spraying on the form that was cleaned and newly prepared after every shot. The tests showed that the mould release agent is taken differently and partly sticks after forming at the component depending on the metal speed but also the distribution of the liquid phase. Damage occurred because of the high thermal load in the region of the oxide bag and also at the edges of the modular step units (Figure 9.19).

Interchangeable step modules of different wall thicknesses and coatings were placed in the tool of the T-shaped sample up to full ceramic plates. Figure 9.20 shows coated modules after a combination of approximately 50 aluminium and five steel design cycles. No damage can be observed on the flat areas but the edges show detachments. During the casting and cooling of the component, the annealing temperature of the tool steel is exceeded in the near-surface region. The hardness loss

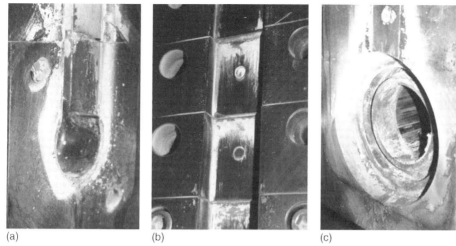

(a)   (b)   (c)

Figure 9.19 (a, b) Wear appears in the form of grooves and flattenings in the region of the oxide bag and the edges of inserts by tilting the component when ejecting; (c) the shot chamber with the oxide restrainer shows strong wear despite the careful coating with mould release agents due to the duration of the temperature influence and the movement of the shot piston.

favours deformation distortions, grooves and detachments if the component cants insignificantly under extraction.

The bulk ceramic plates made from silicon nitride were mounted largely unstressed. Unevenness of the metal surface, warmth delay of the holders and thermal shock of the ceramic led, however, to failure of the plates by break. With respect to chemical stability and resistance against abrasive wear, the silicon nitride plates show excellent results.

The *fourfold plate sample* was used for secondary proof testing in other series of experiments in which the different tool materials with coating systems were installed

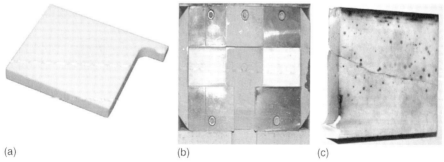

(a)   (b)   (c)

Figure 9.20 Bulk ceramic insert made of silicon nitride (a) [36]. Essential difficulties affected the unstressed tailing of the ceramic in the form in the test preparation. (b) Thermal shock of the ceramic and also warmth delay of the steel mounts led to breaking of the plate.

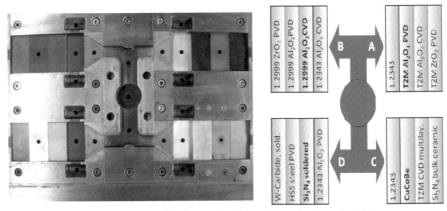

Figure 9.21 Plate sample mould with use of tool materials and schematic division.

in 16 installation spaces according to the order shown in Figure 9.21. Using this order, the different materials were tested at the same time, with four particularly promising candidates having the metal stream over one edge. As parent metals the steels 1.2343 and 1.2999 and the molybdenum-based alloy TZM were used. Steel HSS 6-3-2, the temperature-resistant copper-based alloy CuCoNiBe (2.1285) and silicon nitride ceramic, soldered to a metal socket and as a bulk insert were investigated in detail. With the steel X210CrW12 at 1290 °C, clear conclusions show the suitability already after 10 shots; TZM pieces were broken off by brittle failure and blocked the ejectors. The soldered armour plating made of silicon nitride was broken repeatedly and the uncoated steel was damaged by abrasion in the area of the overflows and redirections.

For the optimization of the coatings and comparison with the model tests performed at the collaborating institutes, the quantification of the load spectrum is of great importance. For its description, the piston pressure during form filling and pressurization and also the surface temperature of the tools can be measured directly after casting the components. Therefore, the measurement of the surface temperatures is of great importance.

The suitability of different materials for the thixocasting process in terms of mechanical strength was estimated by comparing the individual mechanical strengths at the real temperatures appearing on the tool surface. The piston force delivers the hydrostatic pressure to the semi-solid material.

A high resolution, a short-wave thermal image camera was used in the temperature measurement, since the measurements cannot be carried out with thermocouples on the surfaces. For metals, this method leads to strong deviations of the measured temperature because of the very different emissivity in the infrared area. The regulation for any single tool material or coating by calibration measurement increased the precision considerably.

Table 9.2 gives information about the emissivity coefficients found and the temperatures corrected for the tool surfaces.

Table 9.2 Temperature determination with and without correction of the emission coefficient.

Unit	Emission	Before determination of ε			After determination of ε			
		Min. (°C)	Max. (°C)	Av. (°C)	Emission	Min. (°C)	Max. (°C)	Av. (°C)
1.2343	1	84.2	160.8	114.1	0.506	127.5	243.5	173.1
TZM Al₂O₃ IOT	1	152	185.2	162.6	0.711	187.2	227.9	200.2
TZM Al₂O₃ MCh	1	95.1	147.7	132.7	0.729	115.4	179.1	161
TZM Al₂O₃ IOT	1	97	113.7	108	0.678	123	144.3	137.1
1.2343 Al₂O₃IOT	1	102.1	189.8	169.3	0.506	155	287	256.2
1.2999 Al₂O₃ IOT	1	135	171.2	153.8	0.742	162.1	205.3	184.5
1.2999 Al₂O₃ IOT	1	100.3	160	137.9	0.762	118.4	188.9	162.8
1.2999 Al₂O₃ IOT	1	104.8	134.8	123.7	0.725	127.7	164.1	150.6
1.2343	1	66.5	107.6	75	0.506	100	163.3	113.3
CuCoBe	1	40.8	64.5	48.3	0.389	67.3	113.5	82.4
TZM multilayer MCh	1	62.3	144.9	129.5	0.788	71.5	167.6	149.8
Si₃N₄ massive	1	94.3	121.7	111.2	0.691	118.2	152.6	139.5
1.2343	1	86.7	195.5	167.9	0.506	131.4	295.6	254.1
Si₃N₄ soldered	1	165.2	259.1	228.4	0.977	167.6	262.8	231.6
HSS	1	45.5	120	63.9	0.469	68.9	190.6	100.1
1.2343	1	66.3	118.8	94.6	0.506	99.6	180.3	143.5

Figure 9.22 shows a thermograph of a component approximately 5 s after the conclusion of the filling of the form. The biscuit itself is no longer very hot at 520 °C because it is only a few millimetres thick, which considerably increases the process reliability. The area of the oxide restrainer belongs, at over 730 °C, to the hottest zones. The part does not show any rips which could be recognized as fine, bright lines. It becomes visible that the different component thicknesses of 10 mm (in step 1) and 5 mm cause a difference in temperature on the surface of over 250 °C. The temperature seems to rise within 5 mm of the plate end. This can be explained by a transport of ceramic mould release agents by the flowing front and the consequent increase in the emission coefficient.

The thermography of the tool surfaces directly after the removal of the component shows clear differences for the single tool areas and materials. The centrally inrushing material leads to strong heating of the runner and particularly the exposed corners (315 °C). The redirection of the vertically flowing SSM makes a strong energy contribution to the surfaces above and below the gating.

The influence of the different tool materials can be clearly recognized. The silicon nitride ceramic shows the highest temperature, over 300 °C.

Against this, the copper insert on the other side is at only 80 °C because of its extremely high heat conductivity. The tests showed that the strength of the low-melting copper-based material is great enough for that process because the heat conductivity is improved by around a factor of 10 compared with steels of the same specific heat capacity.

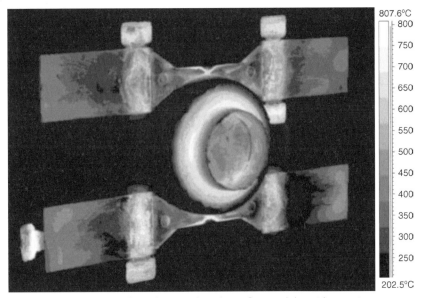

Figure 9.22 Thermograph of the plate sample with overflows and the oxide restrainer.

The first of the four inserts are manufactured from hot work tool steel. The temperatures of the inserts seem on the left and on the right differ strongly in the thermal imaging (Figure 9.23). With different emission coefficients of 0.85 (Al_2O_3 coating) and 0.506 (steel ground), the temperatures coincide with 256.2 and 254.1 $°C$

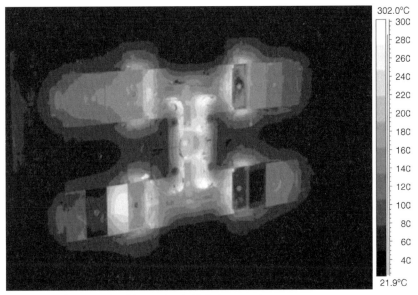

Figure 9.23 Thermograph of the moving half of plate sample mould; for materials division, see Figure 9.21

and are plausible on the left. Furthermore, it can be recognized that the coated tools allow a simpler interpretation of the temperature because of the uniform emission coefficient of more than 0.75 (see Table 9.2).

9.3.4
Demonstrators and Real Parts

9.3.4.1 Impeller
In consideration of the previously obtained good results of the fundamental experiments on the flow and mould filling behaviour of semi-solid steel casting on a high-pressure die-casting machine, a more complex part with practical application was the next step of the investigation.

Alongside the realization of a complex part in semi-solid steel casting, the experiments delivered first results on the use of rapid prototyping steel tools. The die tools were produced by direct metal laser sintering (DMLS) in cooperation with the Fraunhofer Institut für Produktionstechnik in Aachen and EOS GmbH [37, 38]. A schematic drawing of the part (diameter 51.5 mm, volume 7525 mm^3 and wall thicknesses between 1.2 and 6 mm) is shown in Figure 9.24. The completed die tool of the moving plate is built of the laser-sintered geometry layer and a welded steel back plate to reduce the production time (Figure 9.24b). The mould, displayed in Figure 9.24c, has a triple layout with a centred ingot and can be heated to 270 °C by oil tempering.

In the run up to the experiments, numerical simulation of the mould filling and solidification and the temperature distribution inside the dies were performed with the software MAGMAsoft. Figure 9.25 shows the development of the temperature of the semi-solid steel X210Cr12 during the filling of the impeller cavity. The initial temperature of the billet was 1290 °C, which approximates a solid fraction of about 70%. Due to the huge temperature difference between the cast metal and the mould material, rapid cooling occurs at the thin sections of the impeller vanes. Figure 9.25b displays the metal velocity during the mould filling process at an initial piston speed of 0.3 m s^{-1}. Due to the reduction in wall thickness at the ingate, the velocity increases to nearly 45 m s^{-1}. The validation of the simulation was done by 'step-shot'

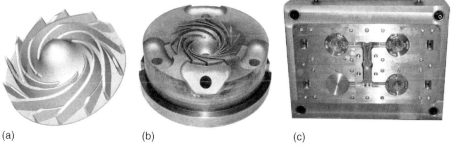

(a)   (b)   (c)

Figure 9.24 (a) Schematic drawing of the impeller geometry [38]; (b) rapid prototyping tool insert [39]; (c) complete moving plate with centred ingot and triple tool set.

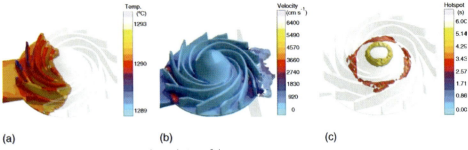

(a)　　　　　　　　　　**(b)**　　　　　　　　　　**(c)**

Figure 9.25 Numerical simulation of the temperature development during mould filling. Rapid cooling occurs at the thin impeller vans (a); distribution of the metal velocity inside the cavity at an initial piston speed of 0.3 m s^{-1} (b); simulation of 'hotspots' (c).

experiments shown in Figure 9.26. The semi-solid steel suspension shows typical laminar flow characteristics.

A comparison of the predicted porosity and the failures that occurred was performed by digital X-ray analysis (Figure 9.26). According to the simulation, slight microporosity could be observed in the casted part (see the white circle in Figure 9.26).

A more detailed investigation of the properties of the parts was carried out by metallographic analysis. Figure 9.27a shows a micrograph of the thicker cone section of the impeller with a typical thixo globular formation of the primary phase and a very fine solidified liquid phase. Figure 9.27b displays the microstructure at the top end of the thin impeller vans. As in the fundamental experiments, a slight separation of solid and fluid phases and dendritic growth of the liquid phase can be observed at the thin sections of impeller vanes.

Variation of the process parameters reduces segregation phenomena and leads to superior properties of the parts, as can be seen from Figure 9.27.

The experiments have confirmed the outstanding flow capabilities of semi-solid steel slurries processed on a high-pressure die-casting machine. The ability to fill

Figure 9.26 Corresponding step-shot experiments. The mould filling behaviour shows similar characteristics to the numerical simulation. The comparison of the simulated distribution of 'hotspots' with an X-rayed thixocast steel part shows light microporosity in the predicted sections of the part.

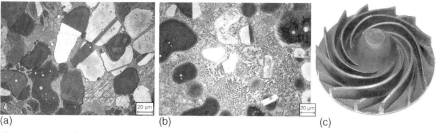

(a) (b) (c)

Figure 9.27 Development of the microstructure. (a) The microstructure at the thicker middle section of the cone. (b) Metallography displays the microstructure at the top end of the thin impeller vans; a slight separation of fluid and solid phase occurs due to the increasing metal velocity and an intense shearing of the semi-solid suspension during the mould filling. (c) Impeller wheel.

sections of less than 2 mm wall thickness combined with the excellent surface reproducibility of the steel suspensions provides exceptional potential.

9.3.4.2 Kitchen and Diving Knives

To demonstrate further the potential of the thixocasting process for steel alloys, a component which uses the characteristics of the tool steel appropriately and uses the processing potential was selected in consideration of the excellent results with the impeller geometry.

As already explained in Chapter 2, the steel alloy X210CrW12 used within this work is part of the group of the sub-eutectic or ledeburitic steels for cold work. The high carbon content of 2.1 wt% in combination with the chromium content of 12 wt% produces a content of fine distributed carbides of 15–18 vol.% at the end of the solidification, which increases the hardness and the wear resistance. The application areas are cutting tools for room temperature and high-resistance punch dies at increased temperatures. With regard to the application areas, the knife geometry shown in Figure 9.28 was investigated. The classical method for the production of high-quality full-steel knives is a multistage smithy process with recrystallization annealings between the stages depending on the degree of deformation and a punching process to remove the forging flash.

Furthermore, the knife geometry is a combination of step die and meander die already used in the pilot experiments and is therefore suitable for the validation of the previous knowledge regarding the flow and form filling behaviour of semi-solid steels. The knife has a volume of 48 751 cm^3 and a total length of 315 mm. The blade was formed over a length of ~200 mm with an even thickness of 3 mm. The manufactured multisectional tool consists of two plates with the knife geometry and a modified gating system.

For a numerical simulation of the fill and solidification process with MAGMAsoft, a maximum metal speed of $V_{metal} = 60$ m s^{-1} was predefined in the area of the knife point to avoid turbulent flow or presolidification phenomena. The calculations were therefore performed with a piston speed of $v_{piston} = 0.3$ m s^{-1}. Starting at the handle

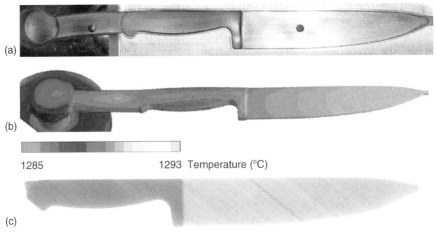

(a)

(b)

1285   1293   Temperature (°C)

(c)

Figure 9.28 (a) Tool insert of the moveable die-half: shown are the knife cavity, the oxide wiper, the gating and the boreholes of the ejectors. The knife blade is 200 mm long and 3 mm thick. (b) Temperature distribution of the cast metal (X210CrW12; $T_c = 1290\,°C$; $v_{piston} = 0.3\,m\,s^{-1}$) at 100% filled form. (c) X-ray analysis confirms the good quality of the cast parts.

of the knife, laminar filling takes place with $V_{metal} \approx 10\,m\,s^{-1}$. The metal speed increases with reduction in the cross-section to the knife shaft. The filling of the shaft area is favoured by the further diminution of the crosscut to the knife blade and the ram pressure effect resulting from it. The 3 mm thick blade then becomes filled at $V_{metal} \approx 20\,m\,s^{-1}$ durchströmt. The analysis of the temperature development of the semi-solid slurry during the form filling shows the formation of a channel flow in the area of the knife handle. The described ram pressure effect at the transition from the shaft to the thin blade leads to a strong reduction in the average metal speed in the knife handle, which leads to the formation of a solid shell in connection with the high undercooling at the form surface. Through this, a reduction in the effective flow crosscut to a middle channel arises.

The step-shot experiments carried out for the validation of the simulation correlate with the numerical calculations. In the course of the experiments the piston speed was increased for the optimization of the surface quality of the components at $v_{piston} = 0.4\,m\,s^{-1}$. The geometry showed high precision and the contour acuity filled completely and is almost free of burrs. The surface quality is influenced strongly by ceramic mould release agents sprayed on before the shot (Centricoat).

A first check of the component quality was carried out with X-ray analysis. The overview representation of the thixocast structure of the ingate area shows a distinctive hem of increased liquid phase which correlates with the shell formation observed in the results of simulation and the step-shot experiments and the channel flow resulting from it.

There is occasional formation of micro porosity at the top of the knife blade due to the high metal speed and lack of feeding. Furthermore, ceramic particles washed into some places can be observed as small non-metallic inclusions. The analytical analysis with respect to form factor, particle size and area and also the share of the globulitic phase confirms the phenomena. The distribution of the solid phase and the fine particles at the ingate area of the knife geometry shows clearly the strong segregation of the phases which is caused by the distinctive channel flow. A considerably more homogeneous structure formation can be seen at the knife tip, and a considerably higher content of fine particles which leads to a decrease of the form factor appears.

The mechanical quality of the components was checked by measuring the Rockwell hardness (HRC) of the knife in the application-relevant areas. The parts were extracted after a solidification time of 5 s and cooled in air. The matrix of the hardness distribution for a knife in the as-cast condition shows a decrease in the hardness values for the knife point from about 47 to 39 HRC as a result of the long flow distance and the steady increase in the metal speed.

By a post-connected annealing treatment of the knifes for 2 h at 550 °C, a hardness increase appears around $+20$ HRC. Furthermore, easy homogenization of the hardness distribution is achieved by heat treatment in the area of the knife point which is caused by the transformation of the metastable austenite into bainite. On average, the hardness values are 62–66 HRC, which corresponds to the quality of a forged knife. Considering the process- and material-specific boundary conditions, high-quality steel components have been produced by the thixocasting process.

Because a more complex real component was sought, a diving knife with an integral handle was selected and provided by Böker Baumwerk (Figure 9.29). This knife fitted the needs of a thixo-steel component also regarding economic issues, because of a high added value (the price of the forged knife is more than €160) and a low charge quantity of 1000 knives.

It was the aim to produce a diving knife which only needs to be freed from gating and overflows to be able to obtain a finish-worked article in the process of sharpening. As opposed to the kitchen knife, the greatest thickness of 6 mm is the same in the handle and blade areas. The handle has round openings of 22 mm diameter for weight reduction. The blade has the full height of 6 mm up to 30 mm in front of the tip because of the arched blade. To produce a knife ready for the sharpening process, the smallest wall thickness of 0.8 mm must be achieved at the cutting edges of the knife and hatchet.

Due to the component weight of ~282 g a double mould lend itself. Two blades were arranged vertically in the die cast frame of the step sample.

For the ingate the height was $0.8 H_{part} - H_{ingate}$. For the positioning of the ingate at the knife and the area of the ingate, two in principle different variants were possible.

Figure 9.29 Diving knife as-cast (overflows and gating removed).

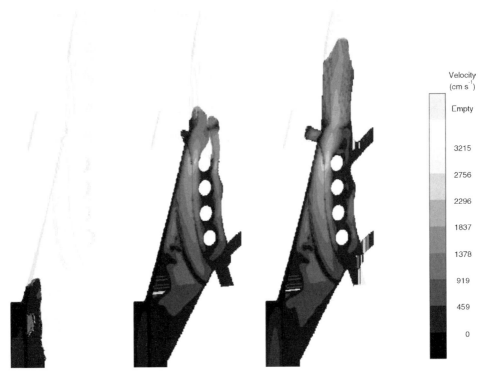

Figure 9.30 Validation of the MAGMAsoft simulation by step-shots; comparison of Figures 9.30 and 9.31 shows almost identical results in flow fronts.

One lets the metal flow in at the back end of the handle and directs it along the knife longitudinal axis. This simplifies the design of the gating and the tool construction because it does not require any strong redirection in the cavity. The other variant lets the material flow through a fan-shaped ingate with much larger crosscut, which makes the form filling faster but more complex.

The MAGMAsoft simulation program predicted for the first variant a filling with the metal speed accelerated at the first core and subsequently with a nearly constant velocity up to the end of the form filling (Figures 9.30 and 9.32). The flowing fronts split in front of the first core and unite almost without a time lag before pouring into the blade. The small cross-section of the runner has a negative effect since pre-solidification cannot be definitely prevented and because feeding problems with the risk of shrink marks have to be assumed.

This is considerably less the case with the second variant. The great ingate length generates a broad metal front contemporaneously flowing through the cores, which makes it possible to reduce the air inclusions in the region of the handle. For the experiments this variant was chosen (Figure 9.30).

At a piston speed of $1\,\mathrm{m\,s}^{-1}$, the initial ingate velocity of metal is approximately $16\,\mathrm{m\,s}^{-1}$, which increases up to almost $30\,\mathrm{m\,s}^{-1}$ when the front passes the last core.

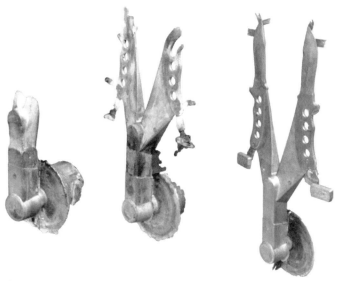

Figure 9.31 Simulation of the metal velocity in different phases of filling.

Smaller air locks can be seen directly at the surface of the parts. The step-shot experiments were validated again the simulation. The simulation results represented very well the ruling conditions both in the fan area and passing the cores (Figure 9.31).

The examination of the components produced for X210 (Figure 9.29) yielded a very good surface quality. The area of the blade could be produced throughout with a projected thickness of 0.85 mm. A single hot crack is observed repeatedly in the

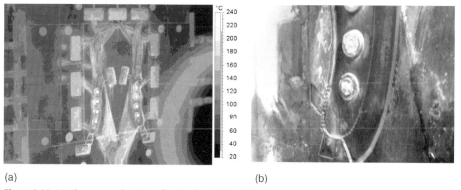

(a)   (b)

Figure 9.32 (a) Thermography 4.5 s after the form filling indicates the variation of the annealing temperature of the steel. In (b) , the gating and TZM cores (with a brittle demolition) are shown. After 10 shots, weldings in the area of the burr trains appear.

bulged part of the blade. This location already stood out as a sustained hotspot in the simulation and a simple local reduction of the blade strength could overcome this.

The metallographic examination shows only a small trend for segregation of liquid material in the tip. The hardness increase at the blade tip is related to the increased solidification rate. The steel X46Cr13 was also examined and achieved full filling only at liquid states of more than 50% because of its mushy solidification morphology. The process temperature is, at 1485 °C, again 200 K higher than for X210 and led to much more presolidification for the same tool temperature. The parts produced are only about 95% filled and show strongly segregated areas consisting exclusively of forward slipped liquid phase. For these low-carbon steels the self-heated ceramic tool could be a working solution.

9.4
Rheoroute

To overcome the disadvantage of the expensive reheating step in thixocasting process routes, the development of rheo-processes in recent years was aimed at producing semi-solid slurries directly from the melt. In manufacturing rheo-feedstock material, the dominant factors to achieve a fine and globular microstructure are the grain density and the cooling rate. Slow cooling and a high grain density result in a globular microstructure. This is caused by the longer time for diffusion of alloying elements in the melt, whereas the risk of treeing as a consequence of constitutional supercooling rises with an increase in cooling rate. In addition, a high grain density leads to a fine and spheroidal microstructure, not least because of the smaller space between the grains, whereby the formation of wide dendrite arms is limited. Another influence is the increasing latent heat per unit time due to the larger solid–liquid interface, which is caused by the higher grain density. Hence the supercooling decreases and the formation of dendrites is inhibited.

Constant Temperature Process (CTP) The development of the so-called 'Haltever-fahren' (Figure 9.33) to manufacture a globular microstructure directly from the melt is based on technical expertise that were acquired during the first period of the collaborative research centre SFB289. In this process, liquid aluminium (here A356) is cooled to the semi-solid state under controlled conditions. For this purpose, the slightly superheated melt is poured into a steel mould, located in an electric furnace to keep the melt temperature slightly above the solidus temperature. Due to the slow cooling, a globular microstructure will be obtained. At the required solid fraction of the semi-solid slurry, the billet is pushed out of the mould using an ejector (Figure 9.33) to assure continuous HPDC processing.

With respect to the process stability and the quality of the manufactured semi-solid slurry, the results could not meet expectations (the reasons for this are described later). In contrast, the developed cooling channel process provides much better semi-solid precursor material, hence the cooling channel process was mainly focused on in the following.

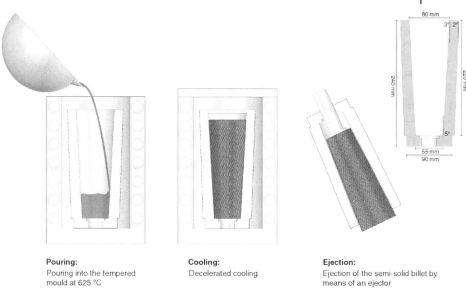

Pouring:
Pouring into the tempered
mould at 625 °C

Cooling:
Decelerated cooling

Ejection:
Ejection of the semi-solid billet by
means of an ejector

Figure 9.33 So-called Constant Temperature Process shown schematically.

9.4.1
Cooling Channel Process

The cooling channel process (Figure 9.34) was developed to deliver a fine and globular microstructure for batch processing with an HPDC machine. The significant constituent parts are the cooling channel and the ceramic mould. Accordant to this configuration (Figure 9.34), the cooling channel process is separable into two process steps with essential effects on the resulting microstructure of the precursor billets: the pouring of the melt over the inclined steel channel and the decelerated cooling of the liquid to the casting temperature.

The cooling channel process will be considered by means of the conventional alloy A356 (AlSi7Mg0.3). A strontium-refined and titanium grain-refined alloy, Anticorodal-70 dv from Rheinfelden, was used.

The process step 'pouring' includes the melt flow of the liquid metal over the inclined, 480 mm long and oil-tempered steel channel. The effect of this first process step on the microstructure becomes apparent by pouring into the mould with and without the channel. If the melt is poured directly into the ceramic mould with otherwise the same process parameters, a dendritic microstructure will be obtained. Using the channel, the microstructure forms globularly (Figure 9.35). In this respect, the mechanisms concerning the channel have an essential effect on the microstructure.

While flowing over the channel, the slightly superheated melt cools below the liquidus temperature ($T_L = 615$ °C), forming seed crystals, and in the ceramic mould. As a result of the heat extraction, a thin, solidified metal layer remains on the channel. Additional seed crystals are formed during first contact of the melt with the cold

Figure 9.34 Cooling channel process; automated pouring over the channel.

ceramic mould, but obviously insufficient for globular grain growth (Figure 9.35). Due to the flow, the seed crystals disperse homogeneously in the container. In the wake of the flow over the cooling channel, the grain density increases and a fine and globular microstructure will be obtained.

The second process step is the cooling of the semi-solid suspension to the process temperature. For slow, decelerated cooling, a ceramic mould was chosen.

As mentioned previously, the dominant factors to achieve a fine and globular microstructure are the grain density and the cooling rate. The effects of these factors are considered in the following. Table 9.3 gives an overview of the modifiable parameters in the cooling channel process.

9.4.1.1 Parameters: Seed Crystal Multiplication on the Cooling Channel

Using experimentally determined parameters, molten, slightly superheated metal was poured over the inclined (5°) and tempered (120 °C) steel channel (Figure 9.34). The tempering of the channel is caused by constant process conditions and the minimization of the remaining metal upon the channel (metal loss).

With an automatic robot, high process stability is achievable, but also with manual pouring high reproducibility is obtained. Because of the changing heat input caused by a varying pouring rate or pouring temperature, the thickness of the remaining solidified alloy on the channel also changes. Due to the isolation effect of a thicker or thinner metal layer between the melt and the channel, the resulting temperature remains approximately constant (Table 9.4).

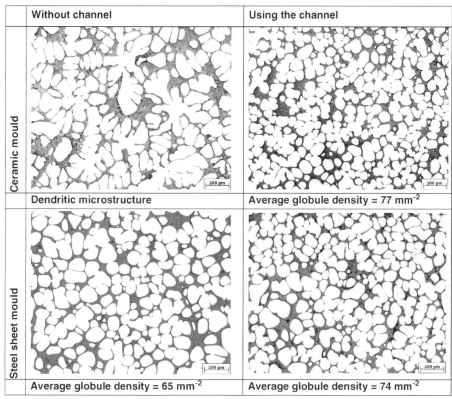

Figure 9.35 Comparison of the microstructures with and without the channel, using a ceramic and a steel sheet mould.

Table 9.3 Variable parameters in the cooling channel process.

Parameter channel	Parameter container
Cooling:	*Cooling time:*
Contact time melt channel: inclination, length	Temperature of the container
Material	Material
Temperature (T_C) of the channel	Wall thickness
Pouring temperature (T_P)	Dimensions and form
	Additional cooling
Flow rate:	*Supercooling:*
Pouring rate	Temperature of the melt
	directly after flowing into the container
Inclination	
Length (duration of acceleration)	

Table 9.4 Variables of the cooling channel process.[a]

No.	T_C (°C)	T_{Pv} (°C)	T_{Gn} (°C)	m_B (g)	m_R (g)	m_G (g)	Δm (%)	t_P (s)	Δt_{12} (min)
1	141	653	613	1623	347	1970	17,6	5,9	5:58
2	137	642	613	1759	382	2141	17,8	5,2	—
3	137	633	613	1496	425	1921	22,1	4,7	—
4	144	638	614	1575	360	1935	18,6	4,2	4:30
5	144	633	611	1561	402	1963	20,5	3,5	4:29
6	140	629	613	1465	467	1932	24,2	3,6	5:54
7	143	628	612	1517	487	2004	24,3	3,4	5:08
8	144	634	613	1646	371	2017	18,4	2,9	4:45
9	139	642	614	1682	331	2013	16,4	2,9	5:27

[a]T_C, temperature of the channel; T_{Pv}, temperature of the melt before pouring; T_{Gn}, temperature of the melt after pouring; m_B, weight of the billet; m_R, weight of the remaining metal on the channel = metal loss; m_G, poured metal = $m_B + m_V$; Δm, percentage metal loss; t_P, pouring time; Δt_{12}, time between pourings.

The increase in the grain density is affected by two mechanisms: the cooling of the melt due to the channel contact and the shearing of formed dendrites as a result of the flow.

The residence time of the melt on the channel is the decisive factor for the temperature decrease of the melt, provided that the channel temperature is not varied. This residence time is changeable on the one hand by varying the channel length. Over the channel length the melt cools continuously. With a longer flow length the melt becomes increasingly colder, but in addition the metal loss also increases. Because the pouring temperature should be as low as possible (less hydrogen absorption and lower costs for heating of the melt), pouring of 15 °C supercooled melt is useful. In this respect, the channel length is chosen appropriately. Simulations accomplished with MAGMAsoft confirmed this result.

On the other hand, the contact time between the melt and the channel can be influenced by the pouring rate and the channel inclination. Both affect the flow rate of the melt and, as a consequence, the shearing of the solid phase and the residence time. As a result of this, the flow length could be chosen to be shorter if the channel were to be more tilted. In that case, the shearing rate would be increased.

Under the influence of shear, the transformation from dendritic to globular grains occurs [43]. This effect was shown [41] while using the so-called cooling slope of the new rheocasting process.

Comparing the microstructures of the remnant metal on the cooling slope from the top of the slope with those from the bottom, a transition from dendritic to globular microstructure takes place. Due to the fluid flow, melting of the dendrite arms and deformation of the dendrites occur, resulting in crystal multiplication [39, 40]; the grain density increases.

The main difference between the cooling slope and cooling channel is the inclination (45–5°). This forces a much higher shearing on the cooling slope. To evaluate this effect in the cooling channel process, the remnant metal on the cooling channel was investigated.

If rounding of the grains is noticeable, in contrast to reports [39, 40], an increase in the grain density during channel contact is to be expected.

The remnant solidified metal from the cooling channel and the cooling slope was investigated by microscopy. Images from the top and bottom of slope and channel were compared. In contrast to the 45°-tilted cooling slope described in Ref. [41], no globularization of the grains is noticeable during flow in the cooling channel process, notwithstanding the double flow length. However, grain coarsening occurs. The decisive factor for the spheroidization seems to be the inclination of the slope of 45° compared with only a 5° tilt of the channel. Hence it can be assumed that the main effect on nucleation in the cooling channel process (only 5° inclination) is the cooling of the melt below the liquidus temperature and not the shearing.

When melt flows into the mould, additional cooling of the melt takes place and nucleation appears. To evaluate the effect of this nucleation on the microstructure, samples were taken from precursor material, with and without using the cooling channel, one cast into a ceramic mould and another cast into a steel sheet mould (Figure 9.35).

Without the channel, on pouring directly into both moulds the results differ distinctly. With the ceramic mould, a coarse and dendritic microstructure appears, whereas the microstructure in the steel sheet mould is globular. From this it follows that in the steel sheet mould an explicitly higher nucleation takes place than in the ceramic mould. The reason for this is in all probability the higher heat conductivity of the steel, which leads to stronger supercooling. In contrast, the heat conductivity of the ceramic is so low that the surface area of the ceramic mould heats up rapidly and nucleation is diminished. As a result of the small grain density, a dendritic microstructure is formed.

On pouring over the channel, the grain density is additionally raised. In consequence, the microstructures formed are fine and globular. Considering the average globule density (always determined by image analyses from five samples), the use of the channel changes the microstructures to higher values, from 65 mm^{-2} (without channel, steel sheet mould) to 74/77 mm^{-2} (steel/ceramic) [note: the standard deviation (SD) was always about 3 mm^{-2}] with the channel. In this case, it seems that it does not matter if it is cast into the ceramic mould or into the steel sheet mould. This leads one to assume that nucleation only has an effect on the grain density to a certain steady state, because the additional nucleation from the steel sheet mould seems to have no effect on the microstructure. Additional nucleation will probably be eliminated by the increased latent heat, at least with actual process and cooling parameters. Another consequence is that an additional seed crystal enhancement due to the shearing on the channel is of no use.

These and further experiments accomplished with the RCP (cf. Section 9.1.2) show that a globular microstructure is also achievable without pouring over a channel. Indeed the channel leads to a finer microstructure, but it does not differ concerning the roundness of the globules. Hence a good workability is given.

A decisive role when using the steel sheet mould is played by the wall thickness of the mould. In the Constant Temperature Process, a massive steel mould with about a 1 cm wall thickness was used. Thereby an amount of dendrites leads to a worsened workability of the semi-solid slurry. But using a steel mould with a 1 mm wall thickness, the effective heat capacity is reduced decisively so that no chilling layer or dendritic growth appears, because the supercooling is not sufficiently high to form dendrites and only nucleation occurs. There is also the advantage that no holding to spheroidize the dendrites is necessary.

During inflow of the melt, the seed crystals will be dispensed homogeneously, just as well as using the channel. Hence all necessary factors for building a globular microstructure are fulfilled.

An additional advantage of using the steel sheet mould is the heat conductivity, which is about 20 times higher than that of the ceramic used. Because of the thin-walled steel mould (1 mm), the decisive factor for cooling to the processing temperature is the heat transfer from the mould to the air. Hence additional cooling by pressurized air, for example, is completely unproblematic.

However, a disadvantage of using a channel for feedstock material production is the loss of metal due to remnants on the channel. Comparing Table 9.4, the wastage is about 20% of the cast metal. This wastage can be minimized by shortening the channel length, but this will result in changes to the inclination, the pouring rate and/or the pouring temperature, because the key task of the channel in the cooling channel process is cooling of the metal below the liquidus temperature. With the current length of 480 mm and an inclination of 5°, the pouring rate (Table 9.4) and the pouring temperature of 630 °C are set quite well.

The metal loss on the channel can also be reduced by increasing the channel temperature, but this would influence the cooling of the melt. Indeed, investigations showed that a variation of the channel temperature from 20 to 230 °C does not affect the resulting microstructure significantly. A temperature of at least 120 °C was chosen to produce back-cooling of the channel to a steady state in less than 150 s.

To investigate if the tempering works well, a Flir ThermaCam P640 thermographic camera was used. The surface temperature was found to be very homogeneous with a median deviation of less than 1 °C over the channel length.

Summarizing, Table 9.5 shows the found and set parameters of the cooling channel process.

In the following, the cooling to the process temperature is examined.

Table 9.5 Parameters of the channel in the cooling channel process.

Parameter	Advice	Set parameter
Length	Preferably short	48 cm
Inclination	Balanced (effect subsidiary)	5°
Pouring rate	Balanced	$\sim$250 mL s^{-1}
Temperature	Preferably low	100–150 °C
Cycle time	Preferably short	>150 s

9.4.1.2 Parameters: Cooling to the Process Temperature

The second important process step of the cooling channel process is the decelerated cooling to the process temperature. Regarding sufficiently slow cooling, the mould material ceramic was chosen for homogeneous cooling of the precursor billets. The mould was designed in such a way as to ensure easy handling and transfer to the shot sleeve of the HPDC machine. To evaluate the influence of the mould temperature on the resulting microstructure, the mould temperature was varied between 20 and 400 °C. Heating to 400 °C resulted in doubling of the cooling time from about 500 to 1000 s. Furthermore, the increased temperature unfortunately led to a coarser microstructure. It was found that the mould should be cooled to below 80 °C before pouring to prevent disadvantageous grain coarsening.

In contrast to the decelerated cooling, the cycle time should be as short as possible for useful process times. As a result, it has to be evaluated how fast cooling is possible to achieve nevertheless a globular microstructure. Sufficiently spherical and fine globules (diameters less than 100 μm) are necessary in order to fill homogeneously thin die cavities during casting [42]. Otherwise, if dendrites are formed, segregation in the wake of the sponge effect occurs, which results in inhomogeneous material properties [43].

Concerning this issue, simulations using the software package MICRESS [44] coupled with ThermoCalc were carried out [55]. To perform the simulations, some starting parameters had to be determined: the real cooling rate, the seed density and the alloy composition. Using the ceramic mould without additional cooling, the solidification lasts 6.5–8 min, depending on the aimed for solid fraction. To determine the parameters of the cooling channel process for the simulations, five experiments were carried out. The casting temperature was always 630 °C. From the measured cooling curves, an average cooling rate of $0.07 \, \text{K s}^{-1}$ was determined, and considering the latent heat a heat flux of $0.94 \, \text{J s}^{-1}$ was approximated with the ThermoCalc software.

Due to the effect of alloying elements on the microstructure evolution, the exact alloy composition was needed. Hence the precursor material was analysed at 60 positions, using a spectrometer (Table 9.6). All alloying elements with concentrations less than 0.1 wt% were excluded from the simulations. The thermodynamic database TTAL5 from ThermoTech was used. The diffusion coefficients for the solid and liquid phases were taken from [45] (Table 9.7) and modelled by means of an Arrhenius plot. Cross-diffusion effects were neglected during the simulations. The initial temperature was chosen as 615 °C (a cooling of 15 °C during channel contact was assumed, based on previous investigations) and the release of latent heat was

Table 9.6 Results of the spectroscopic analysis of the precursor material (wt%).

	Si	Fe	Cu	Mn	Mg	Ti	Ag	Sr	V	Ga
Average	7.11	0.0855	0.0023	0.0017	0.340	0.1041	0.0013	0.0294	0.0119	0.0147
SD	0.28	0.0106	0.0004	0.0003	0.024	0.1177	0.0001	0.0020	0.0002	0.0007

Table 9.7 Diffusion coefficients, [46].

Element	Solid aluminium		Liquid aluminium		Content (%)
	D_0 (m² s⁻¹)	Q (kJ mol⁻¹)	D_0 (m² s⁻¹)	Q (kJ mol⁻¹)	
Mg	1.49×10^{-5}	120.5	1.49×10^{-5}	71.6	0.34
Si	1.38×10^{-5}	117.6	1.34×10^{-7}	30.0	7.11
Ti	1.12×10^{-1}	260.0	9.90×10^{-5}	36.3	0.10

taken into account during the simulations, averaged over the simulation domain, as described recently [48].

The grain density was evaluated from micrographs by image analysis. Samples were taken from the edge and the middle of each billet, etched (Barker method) and analysed metallographically. A total of 50 pictures (five per sample) were analysed in respect of the number of globules, and an average grain density of 77 mm⁻² was found (SD = 2.61 mm⁻²). The average spacing of the globules was 114 µm. This globule density does not represent the seed density directly after pouring, because the dissolution of small grains due to Oswald ripening was not considered. As a starting parameter for the simulation, this was accurate enough, especially as in the simulation only stable grains were considered.

The MICRESS simulations were carried out on a 500×500 µm calculation domain. Twenty seeds are needed to achieve an average globulitic spacing as determined from the experimental results.

In order to determine the boundaries of the globular–equiaxed transition from the simulations, a series of simulations with varying heat flux rate and seed density were carried out. The heat flux rate was varied from the experimentally determined 0.94–10.0 J s⁻¹ and the grain density from 5 to 40 grains per calculation domain. The seeds were set at random locations in the calculation domain.

To analyse the results, a shape factor (F) for the simulated grains was calculated by the following equation [47]:

$$F = \frac{U}{4\pi A} \tag{9.1}$$

where U is the circumference and A the area of a grain.

For $F = 1$ all grains are circular and for $F > 1$ the grains exhibit an increasingly complex shape. Due to the influence on the viscosity, the shape factor should be <2 for sufficient castability [47]. On balance, the criteria for good castability are assumed to be $F < 2$ and $D < 100$ µm.

9.4.1.3 Simulation Results

The simulation results are shown in Figure 9.36, where t is the solidification time, f_s is solid fraction, F is the median shape factor and D is the grain diameter of a circle with an area equal to that of the grain. The calculations were always stopped at similar temperatures (∼575 °C) and solid fractions (∼44%). Using the experimentally determined parameters (0.94 J s⁻¹, 80 grains mm⁻²) a close to reality globular and

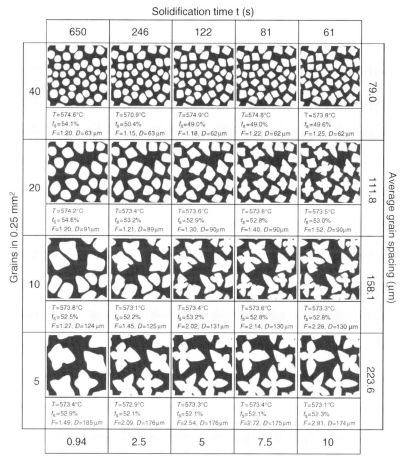

Figure 9.36 Results of the 2D-MICRESS simulation.

fine microstructure ($F = 1.21$, $D = 97\,\mu m$) was obtained (Figure 9.35). With a decrease in the grain density and an increase in the cooling rate, the microstructure changes from globular to dendritic. Increasing the cooling rate with the grain density from the experiments, the morphology changes to a more dendritic form, but the postulated criteria were nevertheless maintained. Comparing Figure 9.37, in reality a higher coagulation of the grains is expected and hence the castability diminishes. The reason is that the simulation does not take fluid flow and thermal convection into account, but coagulation is forced by grain impact. Apart from the observed coagulation, minimization of the cooling time to less than 3 min seems to be possible.

The increase in the grain density has a particularly positive effect on the microstructure, because very fine and spherical globules were formed. Moreover, they remain relatively round at high heat fluxes ($Q = 10\,J\,s^{-1}$, $F = 1.25$). At low heat fluxes and high grain densities, coagulation appears; hence the real grain diameters and the

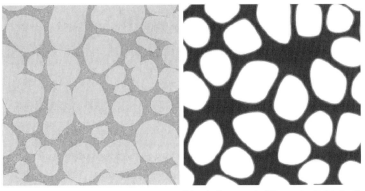

Example of real microstructure   Simulation result (20 seeds, 0.94 J s⁻¹)

Figure 9.37 Comparison between the simulation and reality; the displayed detail edge length is always 500 µm.

shape factors rise. With fewer grains the formation of large dendrites is advantaged, but for casting such semi-solid slurries are unsuitable.

On balance, an increase in the heat flux to $5\,\mathrm{J\,s^{-1}}$ seems to be useful. In that case, the aimed for solid fraction of about 50% will be achieved in less than 3 min.

To check the results, experiments were carried out using a pressurized air cooling device coupled to the already described steel sheet moulds, shown in Figure 9.38. To decrease the temperature loss over the front wall of the mould, Al_2O_3 insulating mats were used. Before this insulation was inserted, the temperature differed by $8\,^{\circ}\mathrm{C}$ from the centre to the billet edge, and afterwards it was only $3\,^{\circ}\mathrm{C}$. This leads to a

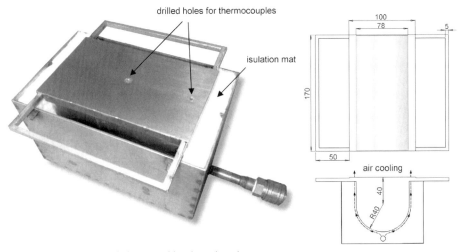

Figure 9.38 Steel sheet mould with cooling device.

homogeneous microstructure and no difference concerning the microstructure was observed between samples from the edge and centre.

Using a simple modulating valve, the cooling air flow could be altered. With different lever positions, varying cooling times were achieved. Retrospectively the heat transfer was determined from measured cooling curves. Eight experiments were performed. In experiment number 8 the highest cooling rate was achieved by water cooling. From the middle between the edge and centre of the cast billets, samples were taken and image analysed, as for the preliminary experiments with MICRESS simulation.

The results of the image analyses (with the software package IBAS) are given in Table 9.8. The change in the microstructure with different cooling rates is illustrated in Figure 9.40. Furthermore, this figure shows the measured frequency distributions to illustrate the effect of the cooling rates on the microstructures. A transition from a globular to a dendritic grain shape becomes apparent. Moreover, the microstructure formed is much finer with higher cooling rates (see Figure 9.40, where the measured grain amount is plotted against the cooling rate). Because a mixed microstructure forms during the transition a from globular to a dendritic shape, no clear borderline could be set; the change from globular to dendritic microstructure is fluent.

To estimate nevertheless the highest possible cooling rate to achieve a useful globular microstructure, the measured and simulated shape factors were plotted against the cooling rate (Figure 9.40). For an easier depiction, the reciprocal of the shape factor from Equation 9.1 was used. As expected, the shape factor decreases with an increase in cooling rate. The reading points could be approximated by a linear

Table 9.8 Results of the validation experiments for the MICRESS simulations.

Parameter	Experiment No.							
	1	2	3	4	5	6	7	8
Average starting temperature[a] (°C)	607.0	603.5	606.0	608.5	608.0	607.3	607.4	606.7
End temperature[a] (°C)	585.0	585.0	585.0	585.0	585.0	585.0	585.0	585.0
Cooling duration (s)	335.5	245.5	143.0	147.5	138.5	100.0	83.0	48.0
Cooling duration extrapolated (613–585 °C) (s)	427	372	191	176	169	126	104	62
Cooling rate $(K s^{-1})$	0.066	0.075	0.147	0.159	0.166	0.223	0.270	0.452
Grain density[d] (mm^{-1})	72	88	99	104	90	120	120	170
Grain density[b,c] (mm^{-1})	57	67	76	73	59	82	85	86
Omitted portion[c]	0.09	0.12	0.12	0.22	0.22	0.25	0.24	0.43
Solid fraction[d] (%)	54.79	53.94	53.45	54.98	54.62	52.62	50.89	49.63
Averaged shape factor[d]	0.79	0.76	0.78	0.71	0.69	0.72	0.72	0.63
Average grain diameter[d] (µm)	81.37	75.52	69.45	68.03	69.57	62.89	61.87	50.93

[a]Linear cooling condition.
[b]Without edge-truncated grains.
[c]Only grains with area $\geq 1500 \mu m^2$.
[d]Determined using image analysis software; because of ripening effects during quenching, the measurements may differ from the real microstructure at the quenching temperature, but this effect is negligible.

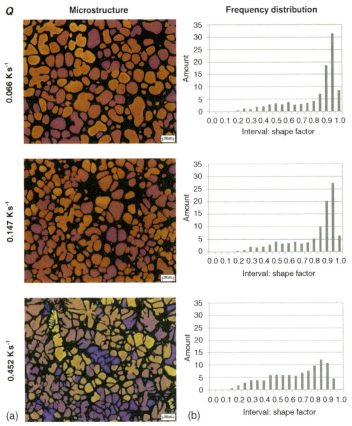

Figure 9.39 At different cooling rates: (a) microstructures; (b) frequency distribution of the shape factor normalized on 100%. The distribution flattens to higher heat fluxes. The profiles at 0.066 and 0.147 $K s^{-1}$ are very similar to each other, in contrast to the water-cooled sample (0.452 $K s^{-1}$). This indicates the probable castability: sample 1, very good; sample 2, good; sample 3, bad.

graph. Comparing Figure 9.39, the water-cooled sample forms a dendritic microstructure containing globules.still

The comparatively large change in the shape factor from 0.78 at a cooling rate of 0.147 $K s^{-1}$ to 0.69 at 0.166 $K s^{-1}$ leads to the supposition that cooling should not be faster than 0.150 $K s^{-1}$. In that case, the process time will be reduced to about 180 s. This matches the simulation results fairly well.

The difficulty in appraising an exact borderline is the two-dimensional view. The more significant 3D structure is in contrast not observable. To verify the process window found, forming experiments regarding segregation and mechanical properties have to be performed. In addition, the effect of the cooling on the mechanical properties, regarding the finer microstructure, should be demonstrated.

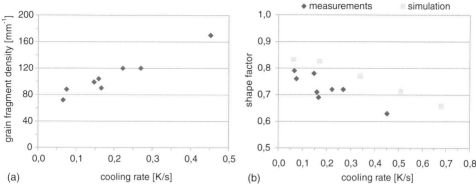

Figure 9.40 Grain fragment density (a) and shape factor (b) versus cooling rate.

9.4.1.4 Oxidation

To assure good mechanical properties of manufactured components, prevention of deleterious oxide layers embedded in the cast parts is essential. In that case, the high affinity of aluminium to oxygen is also advantageous, because instantaneously during pouring over the channel a tube-like oxide layer, where the melt flows through, forms and shelters from the atmosphere. Hence no additional encapsulation or similar is necessary.

9.4.1.5 Summary

The cooling channel process is an easy to use and stable process to provide excellent semi-solid precursor material for subsequent HPDC processing. The process window was definitely determined. The cooling channel enhances nucleation and hence benefits the globular microstructure. An additional air cooling can minimize the cooling time to less than 3 min, with enhancing effects on the grain sizes and maybe also the mechanical properties, which has still to be proved.

In addition, in forming experiments very good mechanical properties were achieved. Investigations have shown that a simple and cheap implementation to industrial manufacturing is possible and particularly with regard to the expensive thixoroute, the cooling channel process is an advantageous option.

9.4.2
Rheo Container Process (RCP)

The RCP, shown in Figure 9.41, was developed on the basis of the Constant Temperature Process and the cooling channel process. The special feature is the encapsulation and consequently the suitability for processing of highly reactive alloys such as magnesium or aluminium–lithium alloys.

To diminish the contact with the atmosphere of a high oxygen-sensitive alloy, the slightly superheated molten metal is poured directly into a 1 mm wall thickness

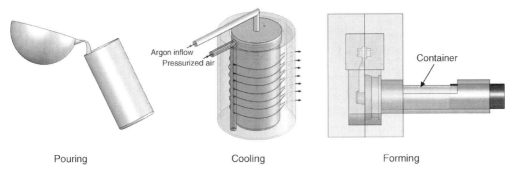

Pouring   Cooling   Forming

Figure 9.41 The RCP (schematic).

aluminium can (container), ensuring wall contact during melt inflow. At first contact with the room-temperature container wall, seed crystals are formed and subsequently, because of the flow, are finely dispersed in the container. The heat capacity of the container suffices to cool the melt below the liquidus temperature. Because of this, and as a result of a high grain density, a globular microstructure forms (cf. cooling channel process). Therefore, the container is placed directly after pouring into a cooling device and covered immediately afterwards and filled with an argon atmosphere to prevent combustion. To minimize the process time, the container is cooled with pressurized air. At the aimed for process temperature, and at the aimed for solid fraction, the container is transferred, together with the semi-solid billet, to the shot chamber of the die-casting machine (Figure 9.42) and pressed into the die. The folded container remains completely in the biscuit [48] (Figure 9.42).

The use of a non-returnable container, consisting of 99.5-aluminium, avoids recycling [49]. Apart from this, the use of steel cans and so on is also possible and

Figure 9.42 RCP: (a) container with semi-solid slurry placed in the shot sleeve; (b) folded container in the press biscuit.

can be considered for rheocasting of high-melting metals, and in particular for steel. Conventional 1 mm wall thickness aerosol can blanks, of inner diameter 80 mm and with a total length of 280 mm, were chosen, because of their exact suitability for the thixo shot sleeve of the high-pressure die-casting machine. The thixo shot sleeve was designed for 170 mm long and 78 mm wide circular billets of conventional thixo feedstock material. Because the aerosol cans could be easily cut to the necessary length, no new shot sleeve was needed and excellent comparability to the thixo process was observed.

9.4.3
Demonstrator: Tie Rod

Preliminary forming investigations by means of the meander die and the step die showed the suitability of the both of the developed processes: the cooling channel process and the rheo container process. To evaluate the usability also for real industrially manufactured components, comprehensive casting tests were performed.

As a demonstrator component, a tie rod was chosen (Figure 9.43), with a minimum thickness of 2.5 mm at a total length of 400 mm and a weight of ~350 g (A356), to validate the thixo-process window. Due to the challenging castability, it was also used to investigate the difficult to cast aluminium–lithium wrought alloy AA1420 in a semi-solid processes. Moreover, the suitability of the cooling channel process and the RCP could be observed using the tie rod mould.

The mould was adjusted regarding the rheological behaviour of semi-solid suspensions. The ingate was modified to produce a laminar flow of the SSM. Additionally, the already described oxide wiper rings were used. The bearings excite

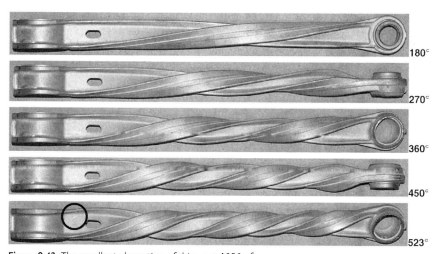

180°

270°

360°

450°

523°

Figure 9.43 The excellent elongation of thixocast A356 of more than 22% is demonstrated by means of the twist test. At a twist of 523° a first crack appears at the large bearing, marked by the circle.

Table 9.9 Process window of A356 for thixocasting and rheocasting [6].

Parameter	Symbol	Interval	Optimal value	Reason
Processing temperature of the billet (°C)	T_C	580–585	580	<580 °C: increasing risk of silicon precipitation; >585 °C: decreasing fraction solid results in increasing shrinkage porosity
Solid fraction	f_S	0.40–0.44	0.44	
Mould temperature (°C)	T_F		250	For an optimal heat balance
Piston velocity (m s^{-1})	v_P	0.05–3	0.5	Depends on metal velocity
Metal velocity (m s^{-1})	v_M	7–30		Depends on mould-cavity cross-section; warranty of laminar form filling with closed flow front
Closing pressure (bar)	P_C		900	To reduce porosity

a critical two-channel flow at both ends of the component, which result in cold runs or oxide layers.

9.4.3.1 Thixocasting: A356

With the modified mould, a small thixocasting batch was produced using the optimal process parameters shown in Table 9.9. The cast components were torsion tested. For that purpose, the tie rods were twisted until the first crack appeared (Figure 9.43). All tested tie rods could be twisted about 525°, which relates to a tested elongation of 22%. Furthermore, the tie rods were investigated metallographically. In spite of thin and long flow paths, no segregation, oxide layers or porosities could be determined.

This shows the high quality of the cast components. The industrially required properties could even be outperformed.

9.4.3.2 Aluminium–Lithium Alloy

Wrought alloys in general are hardly liquidly castable. This is due to the high solidification ranges, which lead to a high volume shrinkage. Furthermore, insuffi-cient feeding in hot tearing results. Oxygen-sensitive elements such as magnesium and lithium impede conventional casting processes, and casting under particular conditions is necessary to prevent combustion. To demonstrate the usability of the RCP, the highly reactive wrought alloy AA1420 (AlLi2.1Mg5.5 + Sr + Zr) was chosen. Porosity, volume shrinkage and hot tearing susceptibility should be signifi-cantly improved due to the lower liquid phase content of 40–60% during the thixoforming process.

9.4.3.3 RCP: AlLi2.1Mg5.5

Using the tie rod mould, the suitability of the RCP [6, 50] for the processing of highly reactive Al–Li alloys was investigated [53].

The molten alloy (modified AA142, cf. Chapter 4) was poured at 630 °C into the container and processed at a temperature of 600 °C and a solid fraction of 0.5

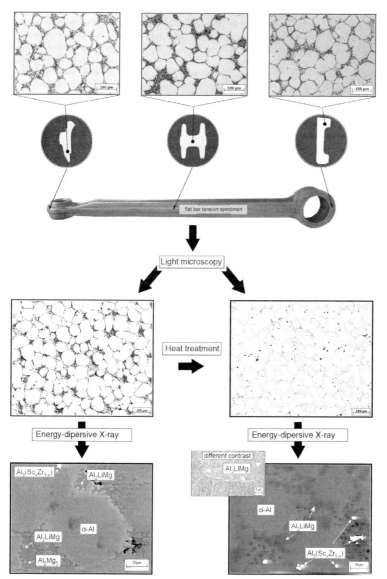

Figure 9.44 Microstructure of Zr/Sc-microalloyed AA1420:
homogeneity (top) and phase detection (bottom): as-cast (left)
and heat treated (right); sampling position of the flat bar tension
specimen.

(calculated by ThermoCalc) in the HPDC machine. Manufactured tie rods were
analysed microscopically, and heat treated and tensile tested. During HPDC processing, rapid solidification of the remaining melt took place in non-equilibrium by the
formation of an 'unexpected' eutectic, including the phases Al_2LiMg, $Al_{12}Mg_{17}$ and
Al_8Mg_5 [51] (Figure 9.44). During heat treatment for 24 h at 460 °C (the solidus

temperature was determined by differential thermal analysis and found to be 480 °C), these phases (Al_2LiMg, Al_8Mg_5, $Al_{12}Mg_{17}$) were to a large extent dissolved in the alpha solid solution, leading to a more or less single-phase structure (Figure 9.44). The added elements Sc and Zr formed $Al_3(Sc_xZr_{1-x})$ dispersoids, which inhibited undesired grain growth during the solution treatment [52, 54]. After water quenching from the solution temperature, the components were aged for 17 h at 160 °C under an argon atmosphere. According to the literature, the primary strengthening phase Al_3Li was formed [52].

From the cast tie rods, flat bar tension specimens were sectioned and tested. Figure 9.44 shows the sampling locations and the good homogeneity of the microstructure over the total length of the component. Via heat treatment, the mechanical properties of A1420 (AlLi2.1Mg5.5 + Sc and Zr) were improved significantly. In this way, the tensile strength was increased to over 400 MPa, the yield strength up to 250 MPa and the elongation to fracture to an average of 6.4%, almost three times higher than in the thixocast condition [53]. Chemical analyses showed that the tie rods were manufactured homogeneously [53].

To investigate the improvement potential of the heat treatment, sectioned samples of the connecting rod were solution annealed (24 h at 460 °C), quenched in water and subsequently aged for various times in an oil bath at different temperatures (140, 160, 180 °C) at the Swiss Federal Institute of Technology (ETH) Zürich. The pyramid diamond hardness [Vickers hardness (HV)] (HV5) was measured and plotted against ageing time. The hardness before heat treatment was 84 HV and after solution annealing 94 HV. On increasing the ageing temperature, the maximum hardness (130 HV in each experiment) requires a shorter ageing time. In order to achieve maximum hardness, the following annealing parameters are advisable: 180 °C/10 h, 160 °C/25 h or 140 °C/>35 h [53]. This revised heat treatment was tested during thixocasting experiments with the modified AA1420 (see the next section).

Despite casting in the semi-solid state, hot cracks appear at the bearing points. To solve this problem, simulations regarding the hotspots at the bearings were carried out using MAGMAsoft and verified during thixocasting of modified AA1420.

9.4.3.4 Thixocasting of Tailored AlLi2.1Mg5.5

The as-cast feedstock billets (Sc- and Zr-modified AA12; see Chapter 2) were produced to a diameter of 78 mm and a length of 160 mm for the thixoforming process. By means of MAGMAsoft simulations, the hot-tear problem was solved. Because of the absence of thermomechanical data for AlLi2.1Mg5.5, A356 was used as a cast alloy for the simulations. A removable steel core was used for building up the large bearing. If the steel core was hot, a hotspot directly behind the core was the result. Otherwise, with a cold steel core the hotspot is displaced into the over flow. The simulation results using a 60 °C cold steel core compared with a 300 °C hot steel core show this effect (Figure 9.45). Regarding this, the steel core was cooled before each shot, and the effect of this is visualized in Figure 9.46. On decreasing the core temperature, the hot tearing tendency also decreases. Using a cold core that was inserted only directly before forming, no further hot tear appears.

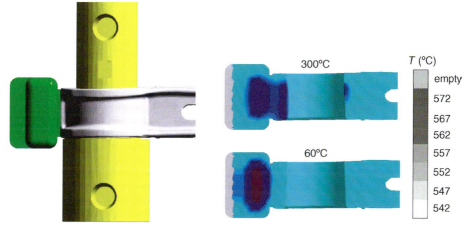

Figure 9.45 MAGMAsoft simulation: comparison between steel core temperature: 60 and 300 °C. The simulation shows the displacement of the hotspot.

Twelve tie rods were cast and consecutively numbered T1–T12. The samples T1–T3 were not further treated. All other samples were heat treated at the IME (Department for Process Metallurgy and Metal Recycling of RWTH Aachen University) under an argon atmosphere. First, all nine samples were solution annealed for 24 h. With reference to the improvement tests at the ETH Zürich, the samples were subsequently aged: the samples T4–T6 were aged for 10 h at 180 °C, T7–T9 for 25 h at 160 °C and T10–T12 for 35 h at 140 °C. As shown in Figure 9.44, flat bar tension specimens were taken from both sides of the double-T beam of the tie rods (six samples per treatment condition) and tensile tested. The results are illustrated by means of a bar diagram in Figure 9.47.

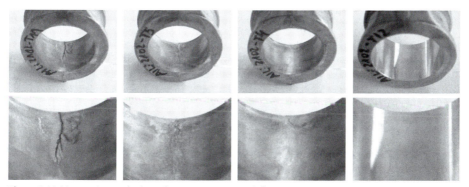

Figure 9.46 Hot tearing at the large bearing point. From left to right: a decreasing steel core temperature results in a decreasing hot tear tendency.

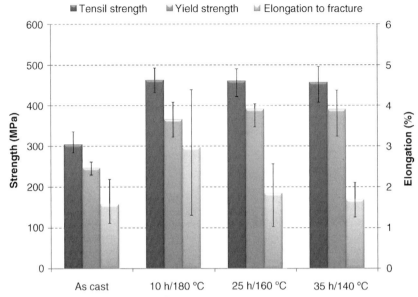

Figure 9.47 Mechanical properties of thixo-cast
AlLi2.1Mg5.5ScZr; all heat-treated samples were solution
annealed at 460 °C for 24 hours.

Due to the heat treatments, a large increase in the mechanical properties could be achieved. As a result of the longer ageing time, concomitantly with the reduced ageing temperatures, the elongation to fracture decreased; notwithstanding the strength remained at a similar level.

The mechanical properties were improved during the heat treatment, the tensile strength to almost 500 MPa and the yield strength to a maximum of 437 MPa. The elongation to fracture could be doubled during ageing at 180 °C from an average of 1.57 to 2.98%, but these results fall short of the expectations (cf. the results of the RCP experiments).

During the examination of the fracture surfaces, in almost at all samples dark microstructure areas were observed. Because these are probably the reason for the substandard elongations, the fracture surfaces were investigated by means of scanning electron microscopy, which showed porosity to be the cause of early fracture. This porosity is the result of the particularly high volume shrinkage of this special alloy, which likewise shows the importance of an accurate mould layout.

9.4.3.5 Cooling Channel Process: A356

To demonstrate the serviceability of the cooling channel process for industrial application, a small batch of a real component was produced. For this purpose, the tie rod mould was also used to ensure comparability to the other investigations. In this case, batch production means a continuous steady-state casting of components without any longer break. No additional pressurized air cooling was implemented.

The cooling to the process temperature of 585 °C lasts 6.5 min. Manually unfastening the oxide-wiper ring leads to an overall cycle time of about 15 min. This cycle time could be easily reduced by using more containers simultaneously, assuming that the batch time for high-pressure die-casting is fast enough.

The casting velocity during form filling was always $0.4 \, \mathrm{m \, s^{-1}}$. A scavenging gas treatment to reduce hydrogen was applied before casting. The hydrogen index was 2.02 at the beginning of the experiments and 2.63 at the end. To ensure a steady state, the first five cast parts were discarded. During subsequent experiments, 20 tie rods were produced and partitioned into three sets for subsequent heat treating and tensile testing. One-third (six samples: 1st, 4th, ...) were tested as-cast, the remaining samples were T6 heat treated with the object of high strengths [heat treatment (HT) A: seven samples: 2nd, 5th, ...] and high elongations (HT B: seven samples: 3rd, 6th, ...). The solution annealing was applied at 535 °C for 17 h and the ageing temperature in all experiments was 160 °C. The difference between HT A and HT B was the ageing time. To achieve high strengths, the ageing time was 2.5 h for high elongations in contrast to 7 h. The increase in the ductility is caused by the rounding of the silicon precipitations during ageing. The success of flushing and heat treatment and the suitability of rheocasting becomes apparent on examining the surface of the cast parts, where almost no blistering is visible.

Similarly to the sampling in thixocasting of microalloyed AA1420, flat bar tensile specimens were taken from all cast parts (40 samples in total) and tested. The results are illustrated in Figure 9.48. As the result of heat treatment, the mechanical

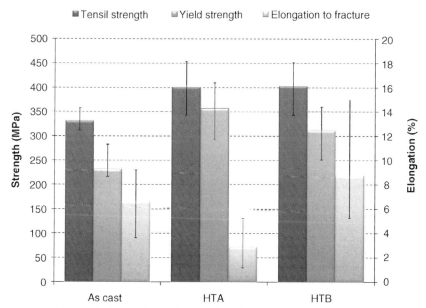

Figure 9.48 Mechanical properties of rheo-cast tie rods (A356); HT A, aged 2.5 h; HT B, aged 7 h (HT, heat treatment).

properties could be significantly improved. Using HT A, yield strengths of 410 MPa could be obtained but the measured elongation to fractures were insufficient. Treatment HT B, in contrast, results in very good ductility (up to 15%) in combination with excellent mechanical properties at the same level as with HT A. In comparison with sand casting and permanent mould casting, an increase in the mechanical properties is achievable due to the fine microstructure caused by high-pressure die casting and the suitability for heat treating. This demonstrates the high potential of the cooling channel process.

9.5
Assessment and Outlook

9.5.1
Rheo Processes

The developed cooling channel process and rheo container process provide an opportunity of industrial implementation. In addition to simple use, the processing of highly reactive alloys such as aluminium–lithium and magnesium is feasible, due to an optional enclosure. Furthermore, with the determined detailed process window, the reproducible mechanical properties demonstrate the high potential of the processes for industrial applications.

9.5.2
Thixocasting of Steels

Despite existing challenges regarding the tool concept, thixocasting of steels provides a promising outlook for industrial application in the intermediate and long term, due to the significantly lower temperatures and the high solid fractions. For a reproducible feedstock material quality, the foundations were laid with the cartridge concept and the energy-controlled reheating concept. The challenges that remain will be approached in the coming years.

List of Symbols and Abbreviations

A	area (of a grain)
A356	AlSi7Mg0.3
A357	AlSi7Mg0.6
AA1420	AlLi2.1Mg5.5
AA1420*	AlLi2.1Mg5.5ScZr
AA2024	AlCuMg2
AA5086	AlMg4.5Mn
AA6012	AlMgSiPb
AA6082	AlSi1MgMn
AA7075	AlZn5.5MgCu

D	grain diameter of a circle
DGM	Deutsche Gesellschaft für Materialforschung
DMLS	direct metal laser sintering
ETH	Swiss Federal Institute of Technology
F	shape factor (e.g. for the simulated grains)
f_l	liquid fraction
f_s	solid fraction
H_{ingate}	ingate height
H_{part}	part height
HPDC	high-pressure die casting
HRC	Rockwell hardness
HT	heat treatment
HV	Vickers hardness
IBAS	interactive image analysis system
MMC	metal matrix composite
ϕ	diameter
Q	heat flux
RCP	rheo-container process
RWTH	Rheinisch Westfälische Technische Hochschule
SD	standard deviation
SIMA	stress-induced and melt-activated
SSM	semi-solid metal
STC	shot-through cartridge
t	solidification time
J_c	casting temperature
T_L	liquidus temperature
TZM	titanium–zirconium–molybdenum
U	circumference
v_{metal}	metal speed
v_{piston}	piston speed
X210	X210Cr12

References

1 Hirt, G., Nohn, B., Morjan, U. and Witulski, T. Thixogiessen – Verfahrenstechnik und Werkstoffaspekte, Glusserei Praxis Nr. 2 (1999), p. 58

2 Gabathuler, J.-P., Erling, J. (1994) Ergebnisse der Werkstoff-Forschung, Band 6: *Aluminium als Leichtbauwerkstoff in Transport und Verkehr*, ed. M.O. Speidel, P.J. Uggowitzer, Verlag, Thubal Kain, p. 63.

3 Bernhard, D. (1999) Zur Prozessoptimierung des Thixogiessens, Shaker Verlag, Aachen.

4 Witulski, T. (1999) Ein Beitrag zur Beschreibung der Prozessparameter beim Thixoforming von Aluminium-Legierungen, Shaker Verlag, Aachen.

5 Müller-Späth, H. (1999) Legierungsentwicklung unter Einsatz des SSP-Verfahrens und Umsetzung intelligenter Materialkonzepte beim Thixogiessen, Shaker Verlag, Aachen.

6 Fehlbier, M. (2003) Herstellung, Charakterisierung und Verarbeitung teilflüssiger metallischer Werkstoffe am

Beispiel verschiedener Aluminiumund Magnesiumlegierungen, Shaker Verlag, Aachen.

7 Kaufmann, H. *et al.* (2000) Proceedings of the 6th International Conference on Semi-solid Processing of Alloys and Composites, Turin, 27–29 September 2000, pp. 457–463.

8 Giordano, P. (2003) Neue Formgebungstechnologien für Leichtmetalle, Neuere Entwicklungen in der Massivumformung, K. Siegert (ed.), Fellbach, 2–4 June 2003, pp. 155–173.

9 Aguilar, J. *et al.* (2003) Proceedings of the 6th International Conference on Mg Alloys and Their Applications, Wolfsburg.

10 Aguilar, J., *et al.* (2002) Processing of semi-soild Mg alloys, P6, pp. 789–794.

11 Dworog, A. *et al.* (1999) Thixomoulding: Neue Märkte für Spritzgiesser. *KU Kunststoffe, Jahrg,* **89** (3), 34–38.

12 Garat, M. and Maenner, L. (2000) State of the art of thixocasting. Proceedings of the 6th International Conference on Semi-solid Processing of Alloys and Composites, Turin, 27–29 September 2000, pp. 187–194.

13 Achten, M. (1995) Herstellung und Verarbeitung thixotroper Aluminium-Basiswerkstoffe, VDI-Verlag-Düsseldorf, Reihe 5, Grund- und Werkstoffe Nr. 394.

14 SAG (Salzburger Aluminium AG)– Lüchinger, H. and Wendinger, B. (2002) Thixoforming – der rationelle Weg zur Herstellung von Premium-Bauteilen aus Aluminiumwerkstoffen. Giesserei-Rundschau, 49, Nr 314, 34–37.

15 Mercer Management Consulting (2002) Automobile Technology 2010, Technological Changes to the Automobile and Their Consequences for Manufacturers, Component Suppliers and Equipment Manufacturers.

16 Winterbottom, W.L. (2000) Semi-solid forming applications. High volume automotive products. Proceedings of the 6th International Conference on Semi-solid Processing of Alloys and Composites, Turin, 27–29 September 2000, pp. 73–78.

17 Hirt, G. *et al.* (1995) Thixoforming – Ein Verfahren mit Zukunft, Neuere Entwicklungen in der Massivumformung, Fellbach b. Stuttgart.

18 Engler, S., Hartmann, D. and Niedick, I. (2000) Alloy development for automotive applications. Proceedings of the 6th International Conference on Semi-solid Processing of Alloys and Composites, Turin, 27–29 September 2000.

19 Gabathuler, J.P. (1998) Thixoforming: Ein Neues Verfahren für die Produktion von Near-Net-Shapeformteilen, Innenhochdruck- und Thixoformen von Stahl und Al, Automobil-Arbeitskreise, Tagung, Praxis-Forum, Automobil-Arbeitskreise, Band 2, Technik + Kommunikation, Berlin, pp. 153, 155–168.

20 Neue Aluminium Giessverfahren, Ergänzung oder Revolution, in DVM-Tag (1999) Berlin.

21 Wendinger, B. (2004) Thixoforming von Aluminium, Semi-solid Technology Seminar, RWTH Aachen, 17–18 February 2004.

22 Giordano, P. (2002) Thixo and Rheo Casting: Comparison of a High Production Volume Component, Proceedings of the 7th International Conference on the Processing of Semi-solid Alloys and Composites S2P congress, Tsukuba.

23 Garat, M. and Blais, S. (2001) Die thixotrope Version der übereutektischen Legierung A390. *Giesserei,* **88** (7), 187*ff.*

24 Hirt, G., (1998) Untersuchungen zum Thixoforming als Alternative zur Herstellung komplex geformter Stahlschmiedeteile. Schmeide-Journal, Industrieverband Massivumformung e.V.

25 Flemings, M.C. and Young, K.P. (1977) Thixocasting of steel, 9th SDCE International Die Casting Exposition and Congress, Paper No. G-T77-092, pp. 1–8.

26 Murty, Y.V., Backman, D.G. and Mehrabian, R. (1977) Structure, heat treatment and properties of rheocast alloys, Rheocasting, Proceedings of the MCIC Work Shop, pp. 95–107.

27 Doppelbauer, M. (2003) Wirtschaftliche Motoren dank Kupferrotoren, in: Technikca Proceedings; Band 52, Heft 13/14, pp. 58–61.

28 Baur, J. (1997) Thixoforging einer Messing-Knetlegierung, Proceedings of the International Conference: Neuere Entwicklungen in der Massivumformung, DGM Fellbach b. Stuttgart, 3–4 May 1997.

29 Baur, J., Wolf, A. and Fritz, W. (1999) Thixoforging von Aluminium und Messing – Produkte, Werkzeuge und Maschinen, Proceedings of the International Conference: Neuere Entwicklungen in der Massivumformung, Fellbach b. Stuttgart, 19–20 Mai.

30 Siegert, K., Wolf, A. and Baur, J. (2000) Thixoforging of aluminium and brass, Production Engineering. *Annals of the German Academic Society for Production Engineering* Vol. II/1, pp. 21–24.

31 Forschungsbericht P322 , Untersuchungen zum Thixoforming als Alternative zur Herstellung Komplex Geformter Stahlschmiedeteile, Studiengesellschaft Stahlanwendung (1999).

32 Young, K.P., Riek, R.G. and Fleming, M.C. (1979) Structure and properties of thixocast steels. *Metals Technology*, **130**, 130–137.

33 Kapranos, P., Kirkwood, D.H. and Sellars, C.M. (1993) Semi-solid processing of aluminium and high melting point alloys. *Journal of Engineering Manufacturing*, **207**, 1–8.

34 Cremer, R. (1996) Dr.-Ing. Thesis, RWTH Aachen University, VDI-Verlag, Reihe 5, Nr. 216, Düsseldorf, p. 50.

35 Kirkwood, D.H. (1996) Semi-solid processing of high melting point alloys, Proceedings of 4th International Conference on Semi-solid Processing of Alloys and Composites, Sheffield, pp. 320–325.

36 Beyer, C. and Füller, K.-H. (2004) Werkzeugtechnologien für das Thixoschmieden von Stahl, Semi-solid

Technology Seminar 2004, RWTH Aachen, 17–18 February 2004.

37 Meuser, H. and Bleck, W. (2000) Untersuchungen zur Gefügeentwicklung des Stahles X210CrW12 im teilflüssigen Bereich, Steel Research 71.

38 Shimahara, H. and Kopp, R. (2004) Investigations of basic data for the semi-solid forging of steels, Proceedings of the 8th S2P International Conference on Semi-solid Processing of Alloys and Composites, Limassol, Cyprus.

39 Bower, T.F. and Flemings, M.C. (1967) *Transactions of the Metallurgical Society of AIME*, **239**, 216.

40 Motegi, T., Tanabe, F. and Sugiura, E. (2002) *Materials Science Forum*, **396–402**, 203–208.

41 Legoretta, E.C., Atkinson, H.V. and Jones, H. (2007) Cooling slope casting to obtain thixotropic feedstock, Proceedings of the 5th International Conference on Semi-solid Processing SP07, pp. 582–586.

42 Apelian, D. (2002) Semi-solid processing routes and microstructure evolution, Proceedings of the 7th S2P International Conference on Semi-solid Processing of Alloys and Composites, Tsukuba, pp. 25–30.

43 Gullo, G.-C. (2001) Thixotrope Formgebung von Leichtmetallen – Neue Legierungen und Konzepte, Dissertation, ETH Zürich, ETH Nr. 14154.

44 MICRESS®. The MICRostructure Evolution Simulation Software, http://www.micress.de (2007).

45 Du, Y., Chang, Y.-A., Huang, B.-Y., Gong, W.-P., Jin, Z.-P., Xu, H.-H., Yuan, Z.-H., Liu, Y. and He, Y.-H. (2003) Diffusion coefficients of some solutes in fcc and liquid Al: critical evaluation and correlation. *Materials Science and Engineering*, **A363**, 140–151.

46 Böttger, B., Eiken, J. and Steinbach, I. (2006) Phase field simulation of equiaxed solidification in technical alloys. *Acta Materialia*, **54**, 2697–2704.

47 Uggowitzer, P.J., Gullo, G.-C. and Wahlen, A. (2000) Metallkundliche Aspekte der semi-solid Formgebung von

Leichtmetallen. Proceedings of the First Ranshofener Leichtmetalltage, pp. 101–107.

48 Aguilar, J., Grimmig, T. and Bührig-Polaczek, A. (2003) Rheo-container-process (RCP): new semi-solid forming method for light metal alloys. Proceedings of the 6th International Conference on Magnesium Alloys and Their Applications, DGM, pp. 767–773.

49 Grimmig, T., Aguilar, J. and Bührig-Polaczek, A. (2004) Optimization of the rheocasting process under consideration of the main influence parameters on the microstructure. Proceedings of the 8th S2P International Conference on Semi-solid Processing of Alloys and Composites, Limassol, Cyprus, 21–23 September 2004.

50 Aguilar, J., Fehlbier, M., Grimmig, T. and Bührig-Polaczek, A. (2004) Processing of semi-solid Mg alloys. Proceedings of the 8th S2P International Conference on Semi-solid Processing of Alloys and Composites, Limassol, Cyprus, 21–23 September 2004.

51 Belov, N.A., Eskin, D.G. and Aksenov, A.A. (2005) *Multicomponent Phase Diagrams: Applications for Commercial Aluminium Alloys*, Elsevier, Amsterdam, p. 261.

52 Eswara Prasad, N., Gokhale, A.A. and Rama Rao, P. (2003) *Mechanical behaviour of aluminium–lithium alloys*, Sādhanā, Vol. 28, Parts 1 and 2, pp. 209–246.

53 Sauermann, R., Friedrich, B., Bünck, M., Bührig-Polaczek, A. and Uggowitzer, J. (2007) Semi-solid processing of tailored aluminium–lithium alloys for automotive applications. *Advanced Engineering Materials*, **9** (4), 253–258.

54 Raade, G. (2003) Scandium. *Chemical and Engineering News*, **81**, 66–68.

55 Bünck, M., Warnken, N. and Bührig-Polaczek, A. (2007) Defining the boundaries of globular to dendritic transition of A356 in the cooling channel process, Proceedings of the 5th International Conference on Semi-solid Processing SP07, pp. 615–619.

10
Thixoforging and Rheoforging of Steel and Aluminium Alloys*

Gerhard Hirt, René Baadjou, Frederik Knauf, Ingold Seidl, Hideki Shimahara, Dirk Abel, Reiner Kopp, Rainer Gasper, and Alexander Schönbohm

10.1
Introduction

Thixoforging is closely related to the drop forging process. In contrast to the thixocasting process, the semi-solid material is inserted into the lower half of horizontally sectioned dies. On closing the dies, the material is deformed until the die cavity is completely filled in one step. The force transmission for the forming and densification step is applied across the whole surface of the die so that the component is compressed during the complete solidification of the material [1–3].

The thixo lateral extrusion process involves charging the semi-solid material through an aperture into already closed dies (Figure 10.1). As opposed to thixocasting, the plunger reaches into the die cavity and regulates the densification pressure directly in the dies. Furthermore, the injected material equals the weight of the finished component as in thixoforging, because of the abolition of an ingate as used in thixocasting processes [3].

The modification of the tools and dies for thixoforging and thixo lateral extrusion by inserting additional components makes it possible to produce composite parts. The flowing material surrounds inlays, which act like additional tools. In contrast to the compound casting process, there is the possibility that alloys with a comparable melting point are applicable without destroying the geometry by melting [4].

Figure 10.1 shows a schematic view of thixoforging and thixo lateral extrusion, which enable similar components to be obtained by different tool concepts.

Specific advantages of this process compared with thixocasting are shorter flow lengths for the complete filling of the dies and a higher compression during the solidification that minimizes pores, which are caused by shrinking. For example, in the case of thin-walled components made of aluminium, the gas inclusion is very low so that a weldable microstructure can be produced even for filigree parts as used in the automotive industry. The process also has disadvantages compared with

---

* A List of Symbols and Abbreviations can be found at the end of this chapter.

Thixoforming: Semi-solid Metal Processing. Edited by G. Hirt and R. Kopp
Copyright © 2009 WILEY-VCH Verlag GmbH & Co. KGaA, Weinheim
ISBN: 978-3-527-32204-6

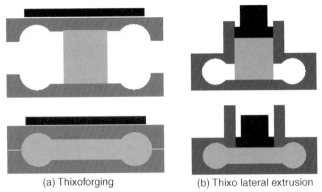

(a) Thixoforging   (b) Thixo lateral extrusion

Figure 10.1 Schematic view of semi-solid forging processes.

thixocasting. Oxides on the surface of the material can be introduced into the bulk material, the geometry complexity is limited and lower liquid-phase fractions are used.

Compared with drop forging, the resulting mechanical properties are similar to those of forging parts whereas geometries adapted to requirements of lightweight components can be achieved more easily because of existing component potentials. Additionally to the increase in the possible component geometries and material savings, aggregates with lower capacities can be used owing to the reduced capacities that are needed for the forming operation. Concerning the mechanical loads of the tools, the low forces that are required for the forming operation reduce the mechanical wear and permit a longer tool life in comparison with the drop forging process. However, there are also disadvantages regarding higher cycle times, higher thermal load of the tools and limited suitability of alloys for the forming operation.

One goal of thixoforging is to replace forged components. In comparison with industrial forging of complex geometries, advantages of thixoforging could be the realization of more complex geometries with slight spline and long flow lengths. Even the shortening of several processing stages (roll forging, upsetting, forging, deburring, piercing and calibration) that are not required for semi-solid forming operations (Figure 10.2) are of interest. The process is shortened right at the beginning because pre-forming of the precursor material is not necessary. There are significant differences in the forging operation itself. Whereas conventional forging requires few forming steps until the end geometry is reached, semi-solid forming is carried out in just a single step. Draft angles can be designed very flat and core pullers enable near-net-shaped geometries to be produced without piercing the component after forming. Also, deburring of the component can be minimized or even avoided. In particular, this leads to advantages regarding process costs, mainly based on material savings [5, 6].

Open challenges are the development of economic working tools (as shown in Chapter 8), even though first investigations with promising results were started in the

Figure 10.2 Required forming steps for forging a wheel trunk. Material: C70S6. [5].

1970s [7]. Also, the quality of the components and the reproducibility of the whole process have to be investigated thoroughly due to the narrow temperature interval in which the material should be worked up [8].

The goal of this chapter is to give an overview of the technology and specific topics related to the family of thixo- and rheoforging processes.

10.2
Forging in the Semi-solid State

Based on first investigations on the thixotropic behaviour of semi-solid metal alloys that were carried out by Spencer's group in the mid-1970s [9], different approaches to forming processes were developed and investigated to benefit from the advantages of the specific material properties. Investigations in research centres all around the world concentrated on the forming of both light metal and high-melting alloys in the semi-solid state and investigated elementary fundamentals for the production of complex parts such as the rheology and material characterization, tooling and forming strategies [10–18]. Whereas thixocasting and thixomoulding of aluminium and magnesium alloys are applied on an industrial scale by several producers [19] (see Chapter 9), thixoforging of aluminium is applied only for the production of rims by SSR-Wheels as an industrial producer. The development of semi-solid forging of high-melting alloys such as steel is still under way.

In the case of high-melting steel alloys, technological problems slow the development of the process for implementation as an industrial application. This is mainly caused by higher processing temperatures, which increase the thermal loads of the tools, and the requirements on process control. Therefore, one basic part of the investigations has concentrated on the development of adapted tool concepts by using coatings and ceramic tools (see Chapter 8) and the automation of the process [20–25]. Nevertheless, different components have already been produced by the thixoforging process and show the feasibility of thin-walled and complex parts (Figure 10.3) using steel alloys.

Demonstrators   Wheel hub   Demonstrators   Pump housing
IFUM-   BMBF/Daimler–Chrysler -   IBF - Aachen
Hannover   Stuttgart

Figure 10.3 Selected thixoformed components made of different steel alloys [3].

Analogous to the thixocasting process, different routes for preparation of the feedstock can be applied for forging in the semi-solid state (Figure 10.4).

The conventional route begins with the casting of the precursor material, which is cooled to below the solidus temperature and indirectly reheated into the semi-solid interval (thixoforming). The rheo-process is characterized by cooling of the melt into the semi-solid interval so that it is directly utilized for the forming operation (rheoforming) (Section 10.3.7). In both cases, the forming operation in the semi-solid state is dependent on precise heating and reproducible handling of the preheated precursor material [25]. In comparison with the processing of aluminium, small variations of the process duration lead to critical differences in the temperature in the steel billets due to the higher billet temperature and increased radiation losses [6, 25].

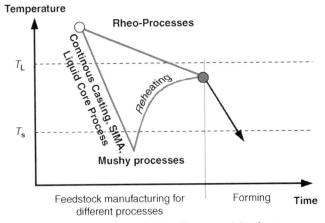

Figure 10.4 Temperature control of thixoforming and rheoforming route.

10.3
Heating and Forming Operations

The following sections present selected results and examples concerning semi-solid forging using an inductive heating unit and a hydraulic forging press. As already described in Chapter 1, an essential part of any thixoforming unit is the heating system providing the semi-solid material. General questions concerning the requirements are discussed in Chapter 1 and specific solutions for inductive heating of steel billets for thixocasting are described in Chapter 9, where the billet is heated in a sleeve and ejected by the shot piston. This solution cannot be transferred to thixoforging. The billet has to be placed in the lower half of the die without the sleeve. One possible solution is the heating of freestanding billets. For this purpose, specific process control features for steel billets have been developed and tested on experimental pilot equipment.

The further processing of the semi-solid material is investigated with regard to different materials and process alternatives such as thixoforging, thixo lateral extrusion and thixojoining. After first investigations, it became apparent that especially the processing of steel requires high precision regarding the reproducibility so that an automated thixoforming unit was developed.

Amongst already mentioned advantages of the thixo-route, further saving potential regarding process time and energy consumption can be achieved by rheoforging. Two different steel alloys were investigated concerning the applicability in the semi-solid state by forming experiments that use the same tools as in thixoforging.

10.3.1
Induction Process

In the following, inductive heating, which is applied for the thixoforging operation, and important fundamentals are described. As already shown, the induction furnace is an essential part of a production unit for thixoforging parts (Figure 10.5), which mainly consists of a resonant circuit and a converter. The resonant circuit consists of a capacity connected to an induction coil via a transformer. Energy is fed into the circuit by the converter, the frequency of which is automatically adapted in order to match the characteristic frequency of the resonant circuit. The heating power is brought into the billet by eddy currents, which are mainly induced near the billet surface by the induction coil's alternating magnetic field. The manipulated variable of the process is the electrical power of the converter, which means that the coil current peak value is varied since the converter voltage is constant. In the experimental facility used, the billets are heated in the upright position to ensure simplified handling and less heat transfer into the billet carrier.

Disturbances of the inductive heating process are caused by the temperature of the environment, the cooling water temperature and the changing surface properties of the billet. The losses during heating consist of convective and radiation losses. These losses lead to a large difference between the core and the edge temperature if a conventionally designed coil is used. Also, losses caused by heat conduction at the bottom surface occur.

Disturbances:

- temperature of the environment

- cooling water temperature

- changing surface properties
  of the billet

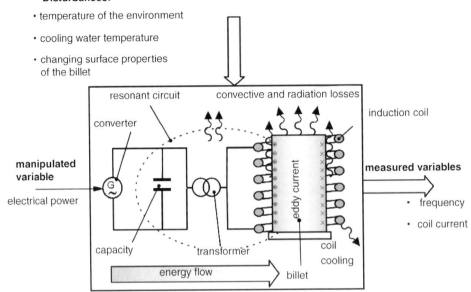

Figure 10.5 The induction furnace [26].

An optimized coil geometry is used to increase the energy entry at the edges. Thus, the losses at the edges are compensated and a homogeneous temperature distribution is produced. The coil consists of two separate parts positioned above each other, which are independently adjustable in height. The distance between these two parts can be changed and thus the energy entry to the face and the bottom surface can be influenced [20].

The eddy currents cause a power distribution in the billet, the so-called volume power density. The volume power density can be determined through Maxwell equations. Due to the optimal geometry of the coil, the variation of the magnetic field strength height (z-direction) can be neglected. Therefore, only deviations in the radial direction have to be considered. This assumption and the cylindrical form of the billet yield the following Maxwell equations in cylindrical coordinates:

$$0 = \frac{\partial^2 \hat{H}_z}{\partial r^2} + \frac{1}{r}\frac{\partial \hat{H}_z}{\partial r} - \kappa\mu j\omega \hat{H}_z$$

and

$$\hat{E}_\varphi(r) = -\frac{1}{\kappa}\frac{\partial \hat{H}_z}{\partial r} \tag{10.1}$$

where $\hat{H}_z$ is the magnetic field strength, $\hat{E}_\varphi$ the electric field, ω the circular frequency of the resonant circuit, κ the electric conductivity and μ the magnetic permeability.

With the solution of the Equations 10.1, the volume power density can be calculated:

$$\dot{\Phi}(r, t) = \frac{1}{\kappa} \frac{H_0^2}{\delta^2} \left| \frac{J_1\left(\frac{\sqrt{j}\sqrt{-2}r}{\delta}\right)}{J_0\frac{\sqrt{j}\sqrt{-2}R_B}{\delta}} \right|^2 = g(t)f(r) \tag{10.2}$$

with

$$\delta = \sqrt{\frac{2}{\mu \kappa \omega}}$$

and

$$g(t) = \frac{N_c^2}{\kappa \delta^2 L_c^2} \hat{I}_c^2(t), \quad f(r) = \left| \frac{J_1\left(\frac{\sqrt{j}\sqrt{-2}r}{\delta}\right)}{J_0\frac{\sqrt{j}\sqrt{-2}R_B}{\delta}} \right|^2$$

where J_0 and J_1 are modified Bessel functions of first kind, zero- and first-order respectively, δ is the penetration depth, which depends on the frequency and the magnetic material parameters; the penetration depth specifies how deep the magnetic field penetrates the billet; N_c is the number of turns, L_c is the coil length and $\hat{I}_c$ is the coil current peak value and is the control variable of the converter. The magnetic material parameters (magnetic permeability and electrical conductivity) are temperature dependent. Therefore, the magnetic behaviour changes during heating. A more detailed description of the induction process can be found in [26–28].

The solution of the Maxwell equations is not correct for the heating of steel because ferromagnetic material has a nonlinear relationship between flux density and magnetic field intensity. To consider this circumstance, the volume power density in Equation 10.2 has to be multiplied with 1.47 [28].

Figure 10.6 shows the dependence of the volume power density on the ratio of the penetration depth to the billet radius. If the penetration depth is small compared with the billet radius, the induced power is concentrated at a small surface area only. This behaviour is called the skin effect [27]. Thus, a higher frequency leads to a higher energy entry at the surface areas. In order to compensate for the radiation losses at higher temperature, it is necessary to choose the converter frequency to be as high as possible [26].

10.3.1.1 Heat Transfer Equations

In the following, we discuss the partial differential equation that describes the heating process of the billet. For more details on the heat transfer, see [29]. We assume that the heat transfer occurs mainly in the radial direction because the coil geometry ensures a homogeneous temperature distribution along the height. The heat conduction through the billet can be described by the following partial differential equation in cylindrical coordinates:

$$\rho c_p \frac{\partial \vartheta(r, t)}{\partial t} = \frac{1}{r} \frac{\partial}{\partial r} \left[\lambda r \frac{\partial \vartheta(r, t)}{\partial r} \right] + \dot{\Phi}(r, t) \tag{10.3}$$

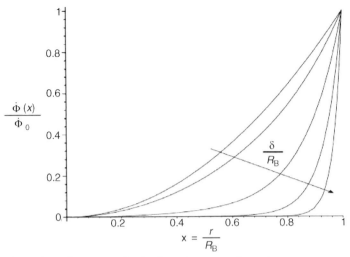

$$x = \frac{r}{R_B}$$

Figure 10.6 Volume power density dependence on radius and ratio of penetration depth to billet radius [26].

where $\dot{\Phi}(r, t)$ is the power volume density as shown above, c_p is the specific heat capacity, λ is the heat conduction coefficient and ρ is the density of the alloy. The specific heat capacity and the heat conduction coefficient are temperature dependent; see Section 10.4.

At the beginning of a heating cycle, the billet temperature is supposed to be uniformly equal to the constant environment temperature ϑ_e, which leads to the following initial condition:

$$\vartheta(r, 0) = \vartheta_e \quad 0 \le r \le R_B \tag{10.4}$$

Because of the rotational symmetry with respect to the point $r = 0$, no heat transfer through the axis can occur. This leads to the first boundary condition:

$$\frac{\partial \vartheta(r = 0, t)}{\partial r} = 0 \tag{10.5}$$

The second boundary condition is determined by the heat transfer from the jacket to the environment. This heat transfer is a combination of radiation and convection. The second boundary condition is described by

$$\lambda \frac{\partial \vartheta(R_B, t)}{\partial r} = -\alpha[\vartheta(R_B, t) - \vartheta_e] - \varepsilon\sigma\left[\vartheta^4(R_B, t) - \vartheta_e^{\ 4}\right] \tag{10.6}$$

where α is the heat transfer coefficient, ε is the emissivity factor, ϑ_e is the temperature of the environment and σ is the Stefan–Boltzmann constant. In the case of the heating of steel, the losses caused by radiation are much larger than through convection. The emissivity factor and the heat transfer coefficient are temperature dependent; see Section 10.4. Therefore, the heat transfer is a highly nonlinear problem.

10.3.1.2 Modelling of Inductive Heating with the Finite Difference Method (FDM)

For simulating the heat transfer, we have to discretize the heat transfer (Equation 10.3) in time and over the radius. The billet is split into different scales via the radius. This yields the following equation: for more details, see [29]:

$$
(\rho c_p)_i^k \frac{\vartheta_i^{k+1} - \vartheta_i^k}{\Delta t} = \left(\frac{\lambda_{i+1}^k - \lambda_{i-1}^k}{2\Delta r} + \frac{\lambda_{i-1}^k}{r_i} \right) \frac{\vartheta_{i+1}^k - \vartheta_{i-1}^k}{2\Delta r}
$$
$$
+ \lambda_i^k \frac{\vartheta_{i+1}^k - 2\vartheta_i^k + \vartheta_{i-1}^k}{\Delta r^2} + P_i^k
$$

(10.7)

where P_i^k is the induced power in the scale i at the time-step k. The induced power is the integral of the volume power density over the scale i:

$$
P_i^k = \int_{r_i - \Delta r/2}^{r_i + \Delta r/2} \dot{\Phi}(r, t_i) \, dr = \frac{N_{S_p}^2}{\kappa \delta^2 L_{S_p}^2} \hat{I}_{S_p}(t_i) \int_{r_i - \Delta r/2}^{r_i + \Delta r/2} \left| \frac{J_1\left(\frac{\sqrt{-2jr}}{\delta}\right)}{J_0\left(\frac{\sqrt{-2jR_B}}{\delta}\right)} \right|^2 dr
$$

(10.8)

A similar approach to model the heat flow can be found in [30].

For the inductive heating of aluminium, the temperature dependences of the material properties can be neglected and the calculation of the heat transfer is considerably easier.

10.3.1.3 Control of Inductive Heating

The reproducible heating of the billet into the semi-solid state is a very important part of the production of thixoforging parts. The results of the subsequent forming step depend heavily on the quality of the heating. The billet should have a uniform temperature distribution in order to obtain good forming results. At the same time, the billet should be heated to the target temperature as fast as possible. However, it must be guaranteed that the outer area of the billet does not begin to melt prematurely.

With the existing control system for the heating process [20], it is possible to follow a predefined power over time trajectory. This trajectory has to be found in time-consuming and expensive test series. This method, like all open-loop control strategies, is not robust against any changes in the conditions of the production. Hence reproducibility cannot be guaranteed. Another problem is that in an industrial production environment it is not possible to equip each billet with thermocouples to measure the temperature. Therefore, closed-loop control is not feasible. For this reason, two approaches were developed to avoid the need for thermocouples. The first approach is to measure the temperature at the surface of the billet with a pyrometer and to detect the entry into the melting phase with the characteristic slope change of the temperature. The second approach is to calculate the current trajectory needed for the induction unit from a predefined temperature trajectory at the middle of the billet. This method is the so-called flatness-based control. In the following, these two approaches are described.

Introduction to Flatness-based Control

Flatness is a property of a control system which allows us to trivialize the trajectory planning tasks for the open-loop control, without solving differential equations. The curved coordinates of a nonlinear flat system can be converted into flat coordinates, similar to the coordinates of a linear system [31]. If a system is flat, we can directly express the states and inputs, without integrating any differential equation, in terms of the so-called flat output and a finite number of its derivatives [32]. More precisely, if the system has states x and inputs u then the system is flat if we can find outputs y of the form [33]

$$y = h\left(x, u, \dot{u}, \ldots, \overset{r}{u}\right)$$

such that

$$x = \varphi\left(y, \dot{y}, \ldots, \overset{q}{y}\right)$$

and

$$u = \alpha\left(y, \dot{y}, \ldots, \overset{q+1}{y}\right) \tag{10.9}$$

The number of inputs has to be equal to the number of outputs (dim y = dim u). If the system is flat, we can calculate the inputs with respect to the desired outputs which guarantee that the outputs follow the desired behaviour. The desired trajectories for the outputs y have to be functions which are $(q + 1)$-times differentiable. The trajectories can be expressed by polynomials which are sufficiently often differentiable [31]. If a system is flat, it is an indication that the nonlinear structure of the system is well characterized and one can exploit that structure in designing control algorithms for motion planning, trajectory generation and stabilization [33]. There are numerous applications of flatness to systems with concentrated parameters [32] and lately to distributed systems also. Flatness-based control is therefore chosen to control the inductive heating and the forming process.

Examples of flatness-based control for heat transfer problems can be found in [34–37]. These systems are boundary controlled and the material properties are constant, that is, the system is influenced by a controlling element at the systems boundary and the material properties are not temperature dependent. The problem with inductive heating is that the system has a distributed input, the volume power density $\dot{\Phi}(r, t)$ and, especially for the inductive heating of steel, the material properties are highly temperature dependent. Hence the open-loop control schemes mentioned previously cannot be applied to the given problem. Therefore, a new approach for feed-forward control with the finite difference method is explained in the following.

Flatness-based Control of Inductive Heating

Depending on the model of the heat transfer, there are two methods to determine a flatness-based control. One method was applied to the inductive heating of aluminium. This approach can be found in [38] and will not be described here because the material properties have to be constant and not, as in this problem, temperature dependent. In the following, the approach for the feed-forward control, which uses the inverted numerical model, is described.

The temperature in the middle of the billet is assumed to be the output of the system and can be predefined. This temperature can be used to calculate the temperatures next to the middle and so on until the last scale at the boundary of the billet.

Due to the skin effect, we can assume that the power is only induced in the outer scale.

If we discretize Equation 10.3 via the radius and convert the equation to the temperature in the $i + 1$ scale, we obtain the equation

$$\vartheta_{i+1} = \frac{(\Delta r - 2r_i)}{(\Delta r + 2r_i)}\vartheta_{i-1} + \frac{4r_i}{\Delta r + 2r_i}\vartheta_i + \frac{2\chi_i r_i \Delta r^2}{(\Delta r + 2r_i)}\frac{\partial \vartheta_i}{\partial t}$$

with

$$\chi_i = \frac{\rho_i c_{pi}}{\lambda_i} \qquad (10.10)$$

To calculate the temperature ϑ_2 next to the middle of the billet, we have to consider the first boundary condition. With Equation 10.10, this yields the following equation:

$$\vartheta_2 = \vartheta_1 + \frac{\chi_1 \Delta r^2}{2} \cdot \frac{\partial \vartheta_1}{\partial t} \qquad (10.11)$$

The temperatures in the third to the nth scale can be calculated with Equation 10.10. At the outer scale n we have to consider the second boundary condition and the induced power, which yields the following equation for the induced power:

$$P(t) = A_{\text{Billet}}\left\{\frac{\left(RB + \frac{1}{2}\Delta r\right)\Delta r}{2R_B}\left[\rho c_{pn}\frac{\partial \vartheta_n}{\partial t} + 2\frac{\lambda_n}{\Delta r^2}(\vartheta_n - \vartheta_{n-1})\right] + q_{\text{loss}}\right\}$$

with

$$q_{\text{loss}} = \alpha(\vartheta_n - \vartheta_e) + \varepsilon\sigma\left(\vartheta_n^4 - \vartheta_e^4\right) \qquad (10.12)$$

If we consider the equation above, we can conclude that we need the temperature in the outer scale and the first time derivative to calculate the induced power. If we go backwards to the temperature in the middle of the billet, we can conclude that the desired temperature trajectory has to be n times differentiable.

The manipulated variable of the induction furnace is the electric power of the converter. Therefore, we have to integrate Equation 10.2 from 0 to R_B. This yields the equation

$$P(t) = \sqrt{2}\pi\frac{N_{S_p}^2}{\kappa\delta L_{S_p}^2}\hat{I}_{S_p}^2(t)R_B\frac{ber\left(\frac{\sqrt{2}R_B}{\delta}\right)ber'\left(\frac{\sqrt{2}R_B}{\delta}\right)bei\left(\frac{\sqrt{2}R_B}{\delta}\right)bei'\left(\frac{\sqrt{2}R_B}{\delta}\right)}{ber^2\left(\frac{\sqrt{2}R_B}{\delta}\right) + bei^2\left(\frac{\sqrt{2}R_B}{\delta}\right)}$$

$$(10.13)$$

where $ber()$ and $bei()$ are Kelvin functions. To ensure that the needed power is induced into the billet, the coil current has to be controlled. We can convert Equation 10.13 to achieve the coil current peak value:

$$\hat{I}_{S_p}(t) = \frac{L_{S_p}}{N_{S_p}} \sqrt{P(t) \frac{\kappa \delta}{\sqrt{2}\pi R_B} \frac{ber^2\left(\frac{\sqrt{2}R_B}{\delta}\right) + bei^2\left(\frac{\sqrt{2}R_B}{\delta}\right)}{ber\left(\frac{\sqrt{2}R_B}{\delta}\right) ber'\left(\frac{\sqrt{2}R_B}{\delta}\right) + bei\left(\frac{\sqrt{2}R_B}{\delta}\right) bei'\left(\frac{\sqrt{2}R_B}{\delta}\right)}}$$

(10.14)

The converter is controlled with a simple proportional–integral–derivative (PID) controller to ensure that the actual current match the desired coil current trajectory and thus the needed power is induced into the billet.

Control with a Pyrometer

In the following, we will discuss another approach to heating steel into the semi-solid state. For a more detailed description of the control scheme, see [39]. Because direct measurement of the temperature via thermocouples is not feasible in a production environment, a radiation pyrometer was used as a contactless measurement device. The accuracy of the pyrometer depends heavily on the exact knowledge of the radiation coefficient, which can vary from billet to billet due to different surface properties and which is subject to change during the heating process. These uncertainties prohibit the implementation of a closed-loop control scheme since the exact temperature cannot be measured with the required accuracy. In order to be independent of the measurement errors, the proposed control scheme relies only on the slope of the temperature. By detecting the distinct change of slope that occurs when the solidus temperature is crossed, the beginning of the melting process can be determined. The energy fed into the billet from this point onward determines the resulting liquid fraction. By feeding the same amount of energy to each billet, it is guaranteed that the billets reach the desired liquid fraction despite the uncertain absolute value of the temperature and small variations of the alloy composition.

During the heating of X210CrW12, there are two distinct changes in the slope of the heating curve, one at about 800 °C where the billet loses its magnetic properties and therefore the efficiency falls considerably. The second appears around 1230 °C and this is the beginning of melting of the alloy. By detecting the slope change and controlling the energy fed to the billet after this point, the liquid fraction can be controlled. After the required melting energy has been induced into the billet, the billet temperature needs to be homogenized since the core temperature lags behind the surface temperature. This is achieved by reducing the converter power to a level, which ensures that the surface temperature remains nearly constant and only the radiation losses are compensated.

The proposed heating control scheme consists of four steps [39]:

1. heating of the billet with constant converter power until the crossing of the solidus temperature;
2. automatic detection of the entry into the melting region by evaluation of the temperature slope;

3. supplying the required melting energy to reach the needed liquid fraction;
4. homogenization step: achieving a constant surface temperature and allowing the billet temperature to homogenize by compensation of the radiation losses only.

The power level for the first step can be chosen arbitrarily since the beginning of the melting process is automatically detected. The power level and duration for the melting phase can be derived from the required melting energy, which can be roughly estimated using chemical calculations of the enthalpy or can be determined by experiments. The power level and duration of the homogenization step can also be determined by experiments. For the detection of the beginning of the melting process, a threshold for the slope of the temperature has to be chosen, which triggers a stopwatch to obtain the melting time when the slope drops below this threshold. The threshold has to be adapted to the induction furnace setup and the chosen alloy. A limiting factor for the achievable liquid fraction is the vertical heating setup, because the billet develops an elephant foot for higher liquid fractions and tends to tilt with increasing softness caused by gravity.

Experimental Results

For the experiments, commercially available rod material of X210CrW12 alloy was used; the billets had dimensions of 34 mm diameter $\times$ 50 mm and a weight of 355 g. The induction furnace has a variable converter power of 50 kW with a maximum frequency of 2 kHz. To reduce the heat losses, the coil was encapsulated, surrounded by isolating material and flushed with inert gas to prevent surface oxidation. The inert gas has also a positive effect on the heat losses, because a blank surface exhibits a lower radiation coefficient than an oxidized surface.

Pyrometer Control

The temperature can be measured by a quotient pyrometer through a small hole within the coil jacket. Figure 10.7 shows the results of two different experiments. Also, the four control steps are shown for one experiment. Despite the different temperature behaviours, all billets have the same liquid fraction at the end of the heating. It can be concluded from the results that it is possible to heat the billets reproducibly to the same liquid fraction for fixed parameters of the control scheme. The proposed control scheme has been integrated successfully into the sequential control of the automated thixoforming plant (Section 10.3.6).

Flatness-based Control

To ensure that the needed power was fed into the billet, the converter is controlled by a PID controller. The converter is a nonlinear system, hence the PID controller has to be tuned carefully because the coil current tends to oscillate.

Figure 10.8 shows the calculated and measured coil currents. The coil current does not follow the calculated trajectory exactly because the PID controller is not tuned perfectly. Figure 10.9 shows the measured temperatures in the middle of the billet, at the surface and the desired surface temperature. There are deviations between the desired and the measured temperatures because the needed power is not exactly

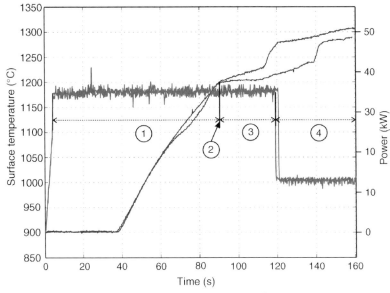

Figure 10.7 Experimental results with pyrometer control, surface temperature and power of the converter over time.

induced into the billet. The steady-state temperature shows a difference between the surface and the axis temperature. The assumption that the energy is only fed into the outer scale yields this deviation. Furthermore, the material data were not adjusted exactly. However, taking into consideration that open-loop control will always have some deviations, the results achieved are very good. Further improvements of the results could be obtained by using a model-based controller for the converter. Additionally, optimization could be used to adjust the material properties [30]. In order to reduce the deviation between the inner and outer temperatures of the billet, it could be considered in the open-loop control that the energy is not only fed into the

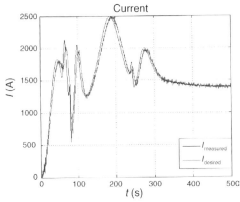

Figure 10.8 Experimental results with flatness-based control, desired and measured coil current.

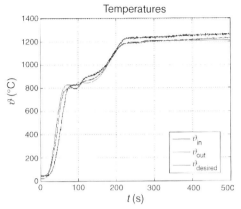

Figure 10.9 Experimental results, desired surface temperature ($\vartheta_{desired}$), measured temperature at the surface (ϑ_{out}) and in the middle (ϑ_{in}) of the billet.

outer scale. However, this would lead to an extensive calculation of the coil current trajectory.

10.3.2
Thixoforging and Thixo Lateral Extrusion of Aluminium

This section takes a closer look at investigations in the field of thixoforging and thixo lateral extrusion of aluminium alloys and shows possible geometries and typical defects based on rheological causes. Several components with a large range of different geometries were produced to analyse the feasibility of thixoforging and thixo lateral extrusion by using the low-melting alloy AlSi7Mg (A356) (Figure 10.10). The tools were made of hot forming tool steel X38CrMoV5-1 and are preheated to 300 °C before starting the forming. The increase in the geometric complexity in comparison with forging is clearly demonstrated with the selected components shown due to specifications of their geometry. Typical constructions such as under-cuts and bore diameters, which are realized by using core pullings (Figure 10.10a)

(a)   (b)

Figure 10.10 Selection of produced components (AlSi7Mg) by thixoforging.

and are also used in casting processes, verify the possibility of producing complex components. Additionally, the process chain is shortened by reducing the forming steps while comparable material properties are obtained, as exist for forging components. The main advantage of the thixoforging process over thixo lateral extrusion and the related thixocasting is the larger effective surface area during compression after the complete die filling has been reached [40]. Subsequently, shrinkage holes can be avoided due to the constant effective pressure during the solidification. Depending on the pressure distribution during the forming operation, for both processes segregations of the two phases can be observed. Especially for long flow lengths, an increasing pressure during the forming operation results in increasing segregation. Because of the better flexibility of the liquid phase, it is pressed out and separated from the semi-solid bulk material. This leads to an inhomogeneous phase distribution and local variations of the mechanical properties [40].

Material Flow and Segregation

To clarify the influence of the geometry and forming parameters on the development of segregation, a demonstrator part was investigated. As shown in Figures 10.11 and 10.12, the geometries of the flange and the upper cup of the component shown in Figure 10.10b were varied.

Investigations regarding the design of the flange show different flow behaviour of the semi-solid material in macroscopic view. Apart from different filling sequences of the cups and the flange depending on the tool movement, the main difference appears in the formation of the flow front. In the case of the flat flange, the flow front breaks because of the growing surface of the flow front with increasing diameter (Figure 10.11b). Semi-solid material leaks out of the bulk material and can lead to inclusions in the flange due to its inhomogeneous filling. The angle of the modified

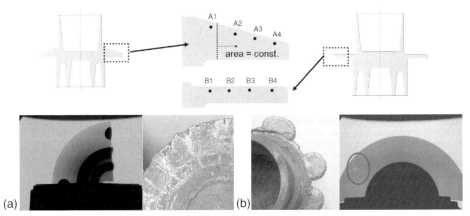

(a) (b)

Figure 10.11 Filling behaviour of different shapes for thixoforging of AlSi7Mg and X-ray analysis of resulting components [41].

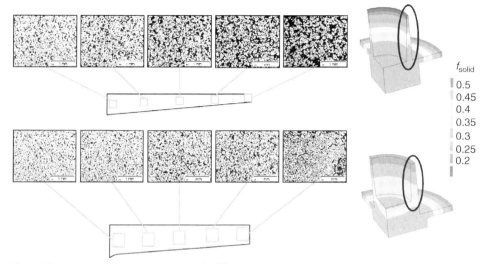

Figure 10.12 Segregation in components with different geometries.

flange ensures that the flow front surface area remains nearly constant with increasing diameter so that the material flow and the die filling are more homogeneous, as shown in Figure 10.11a. According to this behaviour, complete filling is possible without any inclusions.

The comparison of different relations between wall thickness and flow length in the upper cup shows basically tendencies to higher segregation in the case of longer flow lengths (Figure 10.12). The metallography shows a homogeneous contribution of the solid and liquid phases at the bottom of the cup. In the course of material flow for the thinner cup, the liquid-phase fraction increases until the top of the cup is reached. In the case of a greater wall thickness, the gradient between the distribution of solid and liquid phases is much lower. Two-phase simulations (see Chapter 6), which started with a contribution of 50% solid-phase fraction, confirm these results. The analyses of different velocities show generally the tendency for higher segregation with increasing strain rates. This segregation mechanism is based on the better fluidity of the liquid-phase fraction whereas the solid-phase particles could interfere with each other and slow down in their movement. In conclusion, an adequate forming strategy for aluminium alloys has to consider a short forming procedure regarding the temperature development and also adapted geometries to ensure good form filling of the dies without causing high chemical gradients by segregation of the liquid and solid material during the forming operation.

10.3.3
Thixo Lateral Extrusion of Steel

Within the SFB289, first investigations of steel were realized with the thixo lateral extrusion process. This section shows the general feasibility of producing first

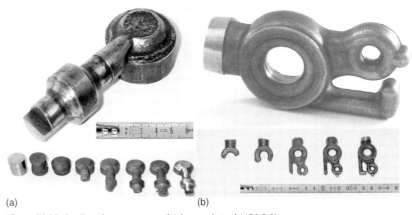

(a)   (b)

Figure 10.13 Produced components (high-speed steel HS6-5-3) by thixo lateral extrusion with their different stages of die filling.

components by using steel. Based on early experience with aluminium alloys, different geometries were chosen to study the applicability of steels for forming techniques in the semi-solid state. In addition to simple geometries shaped like a disk with an arm to clarify the fundamental applicability of different steel alloys, further components were realized (Figure 10.13). In case of the component in (a), the speciality of the geometry is the filling of an undercut and the realization of sharp contours. Investigations with different steel alloys (HS6-5-3, X210CrW12, 100Cr6, C60, 41Cr4) demonstrate sufficient form filling in so-called step-shot trials except for 41Cr4 [42]. Even when the diameter of the channel increases, the flow front is homogeneous for the testing procedure up to the chosen maximum punch speed (80 mm s^{-1}). Against first estimations, a steady filling from the beginning of the gravity up to the end enables defects that would be caused by a transient and a turbulent material flow, respectively, to be avoided or reduced. However, the properties of the different materials show different values of the microstructure depending on the location in the component due to segregations or pores that appear in the component. Especially for the materials HS6-5-3 and 100Cr6 in locations near the surface the fast cooling leads to higher strength, which was proven by hardness measurements.

Further investigations to widen the range of possible geometries for the thixoforming process were carried out with the concept of the component shown in Figure 10.13b. Whereas the earlier components made of steel conformed to a simple geometry of a drop forged part, with this part the flowability of semi-solid material could be shown in its full potential. Illustrated by core pullings, which cannot be realized in traditional forging operations, the excellent form filling and the welding in areas of two flow fronts is demonstrated. To avoid problems of shrinking of the component on to the tools, relatively large draft angles of about 5° were used.

10.3.4
Thixoforging of Steel

In this section, thixoforging parts show the possibility of producing near-net shaped components, which are not realizable in the same way in drop forging processes. They were carried out by using standard dies of hot forming tool steels X38CrMoV5-1 and X45MoCrV5-3-1. Two different geometries were chosen: a demonstrator in the form of a wheel hub and a pump housing. As thixoforming material, X210CrW12 and 100Cr6 billets were preheated into the semi-solid interval with about 35% liquid-phase fraction and positioned upright in the lower die. The upper tool was lowered to close the die and, subsequently, a flexible cylinder below the lower tool frame carried out the rest of the filling operation and maintained the pressure in the cavity until the solidification was completed. The part was ejected immediately and allowed to cool in the appropriate atmosphere or medium.

The demonstrator (Figure 10.14a) was chosen for basic experiments to clarify the filling behaviour of the die depending on the different diameters of the flange and different wall thicknesses of the bush [11]. It was also investigated with the rheoforming process (Section 10.3.7) and was the main part for simulations (Section 10.4.2) because of it axisymmetric geometry and the thereby reduced calculation time. Generally, the combination of a sufficient tool velocity and compression force ensures good form filling. Especially for thin-walled components there is a risk of early solidification with too slow forming operations. In the region of the bush, the material can be cooled below the solidus temperature so that the flow front breaks open and the semi-solid material of the inner part surrounds this frozen area. This can lead to cold shuts, which are characterized by strongly weakened material properties.

The investigation of the pump housings produced regarding the microstructure and local inside hardness showed that no shrinkage holes and porosity can be identified on the examined cross-sections. However, some oxidized outer shell of the billet is found inside the part. Several microcracks exist around the oxide inclusion

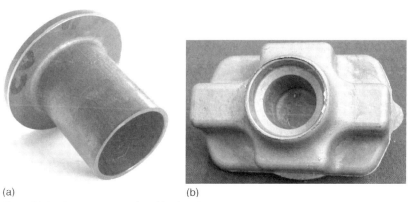

(a)   (b)

Figure 10.14 Components produced by thixoforging.
Demonstrator, 100Cr6; pump housing, X210CrW12.

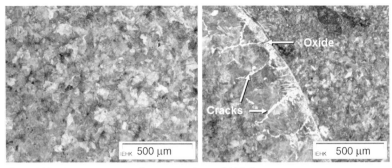

Figure 10.15 Microstructures of a processed part (100Cr6).

(Figure 10.15). The use of inert gas shielding is effective in avoiding or reducing the oxidation depending on the chemical composition of the material.

Figure 10.16 shows the results of local hardness investigations along the displayed lines inside the forged part. The hardness increases from the centre to the outside in a range of 250–350 HV10. This is related to the locally different cooling rates and can additionally be influenced by the occurrence of segregation. The results of the trials demonstrate the potential of semi-solid forging as a forming operation for steels but also reveal the necessity for adequate measures against scale inclusions. Counter-measures were taken by applying a fully automated semi-solid forging plant including a thorough protective atmosphere. Morphological studies of the material thus produced verify high suitability for semi-solid forging (Section 10.3.6).

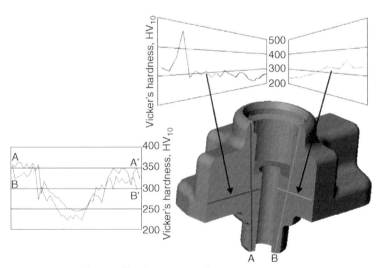

Figure 10.16 Distribution of hardness in pump housing (100Cr6) in vertical and horizontal directions.

10.3.5
Thixojoining of Steel

In the following, the process of thixojoining shows the increase in the functionality and complexity of components by including additional inserts in the semi-solid matrix during forming. In first investigations, the thixo lateral extrusion process as described in Section 10.3.3 produced several components (Figure 10.17) with integrated inserts. Metallographic analyses showed a good connection between the bulk material and the inserts without any pores at the contact area. Only at the border area did a change in the grain size indicate a thermal interaction in the form of a heat-affected zone. Further investigations were made to clarify the possibility of joining lower melting materials such as copper-based alloys.

One of the tool sets utilized for producing thixojoining components with different insert materials is shown in Figure 10.18a. It consists of an upper and a lower die and a plunger. These components are made of hot forming tool steel X38CrMoV5-1. The upper die and the plunger can be moved in the vertical direction by hydraulic cylinders. The plunger is guided with a sliding bush to reduce the friction in the narrow gap to the lower die. It also protects the lower die surface against possible damage. The functional elements (bearing bush and shaft) are just mounted on/in the plunger.

The materials of the functional elements were varied in the trials. In the case of the shaft, stainless-steel X5CrNi18-10 and free cutting steel 9SMn28 were used. For the bearing bush the sinter bronze GBZ12 and copper–beryllium alloy CuCoBe were tested. The feedstock of cold forming tool steel X210CrW12 (34 mm diameter, 50 mm height is heated to approximately 1300 °C.

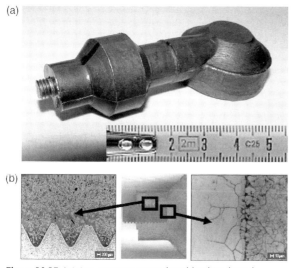

(a)

(b)

Figure 10.17 Joining component produced by thixo lateral
extrusion (a) and metallographic analyses of the contact area
between semi-solid matrix and insert (b).

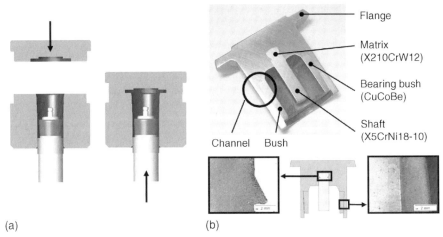

(a) **(b)**

Figure 10.18 Sequence of the tool movement (a) and component produced by thixojoining (b) with metallography of the contact surfaces between semi-solid matrix and inserts.

Concerning the forming parameters, the following procedure gave the best results. After the insertion of the semi-solid billet into the lower die, the forming operation began by closing the upper die. This was followed by an upward movement of the plunger with a velocity of $75\,\text{mm}\,\text{s}^{-1}$ to perform the filling of the cavity. In the last stage of the forming operation, the semi-solid material was solidified under a compression force of 300 kN (which approximately equals 260 MPa) to avoid typical material defects due to the solidification. A compaction time of 5 s was applied in the trials. Finally, the part was slowly cooled to room temperature in air for the subsequent metallographic inspections. The vertical cross-section of a produced part (Figure 10.18b) shows good form filling in the area of the flange and the bush. Also, the slot at the top of the shaft is completely filled with material. Complete filling is achieved in all parts independent of the materials utilized for the functional elements. Also, the metallographic analyses show that the functional elements are completely embedded in a semi-solid matrix. At the contact surface the metallographic analyses confirm a good form closure between the thixoformed basic material and the inserts (Figure 10.18b). The two kinds of steel that were utilized for the shaft do not show any significant influences on the joining quality with the chosen parameters. The geometry of the bearing bush made of the copper–beryllium alloy was kept in its original state, whereas the bearing bush made of sinter bronze showed minimal fusion at the upper edge where the first contact with the semi-solid X210CrW12 occurred.

A tight mechanical connection between the joining elements and the matrix was achieved for most of the studied materials and parameter combinations. This shows that various kinds of joining elements of different materials (higher melting steel, nonferrous metals, etc.) can be integrated for different purposes such as a screw thread or a nut as a junction for assembling, a gliding functional element or a reinforcing element. The possibilities of new part designs are extended by this thixojoining technique with an additional advantage of process shortening at the same time.

10.3.6
Automated Thixoforming – Thixo Cell

The high demands on the thixoforming process regarding the reproducibility of the whole process chain require a very repeatable procedure, especially in the case of applying steel alloys. Even very short interruptions or changes of the experimental procedure have significant influences to the results. To cope with these requirements, automation of the process and its containing aggregates is highly recommended. The automated thixoforming cell at the IBF consists of a heating unit (1), an industrial robot (2) and the forming aggregate (3) (Figure 10.19), which are detailed in the following sections concerning their construction and integration.

10.3.6.1 Feedstock Magazine and Unloading Unit
The billets are stored in an upright feedstock magazine. A pneumatic cylinder moves the billet to a position where the robot can pick it up and transport it to the inductive heating unit. At the other end of the production cycle, an unloading unit was integrated to remove the finished part from the tools. A hydraulic cylinder of the press ejects the formed part [20].

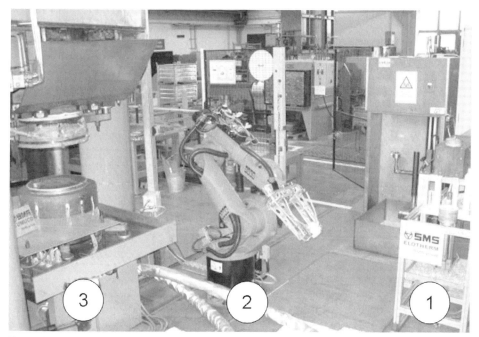

Figure 10.19 Automated thixo cell at the Institute of Metal Forming.

Table 10.1 Hydraulic press main parameters.

Parameter	Value
Nominal press load	6.3 MN
Maximum press ram velocity	120 mm s^{-1}
Maximum press ram stroke	600 mm
Stroke accuracy	±0.1 mm
Auxiliary cylinder load	1 MN
Auxiliary cylinder maximum piston velocity	100 mm s^{-1}
Auxiliary cylinder maximum piston stroke	300 mm

10.3.6.2 Robot with Gripper

To increase the reproducibility of the thixoforming process, a industrial six-axis robot (KUKA KR 15/2) is used to transport the heated billet from the induction furnace to the hydraulic press. The robot has to place the billet in the cavity. Therefore, the gripper tongs were elongated to avoid damage to the drives or cables by radiation or contact. Furthermore, the gripper tongs are equipped with ceramic inlays, which reduce the radiation losses, and with an inert gas rinsing that prevents oxidation of the skin.

10.3.6.3 Forming Aggregates

To perform the thixoforging experiments, a hydraulic press with a nominal load of 6.3 MN was used (see Table 10.1). The lower plunger movement is controlled by an auxiliary cylinder, which is positioned under the lower tool frame. The press ram and auxiliary cylinder piston velocities, working load, press ram and auxiliary cylinder piston strokes and also time delay during the forming process are controlled by the press CNC programme. Electrically heated tool frames that can be heated to 500 °C are equipped with tool inserts, which can be easily changed.

10.3.6.4 Sequence Control of the Plant

The production of semi-solid parts is very sensitive to disturbances or changes in the production conditions. To assure constant boundary conditions and to minimize disturbances, the automated production plant connects the following components with a master process controller [20] (Figure 10.20):

1. feedstock magazine
2. robot with gripper
3. induction furnace
4. hydraulic forging press with tools
5. unloading unit for forged parts.

10.3.6.5 Master Process Controller

The production process is controlled by a master controller using the programmable logic controller (PLC) of the robot controller. The automation components are connected with a remote I/O, which itself is connected with the robot controller

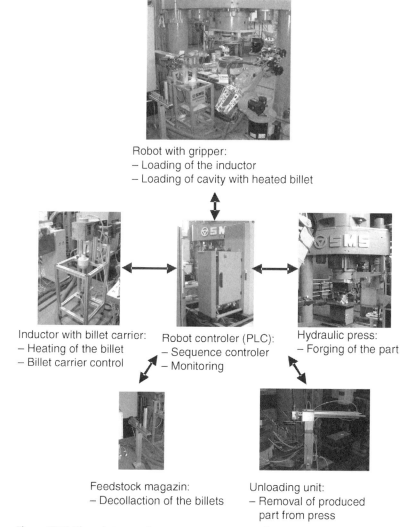

Figure 10.20 The robot controller acts as a master controller which triggers the other components to start their respective operations [20].

via a CAN BUS. Many sensors detect if operations were successfully done or an error occurred.

An operator has to start the production chain by starting the programme of the robot. The programme sends a signal to the feedstock magazine that separates a billet to provide it to the robot. When this operation has been done, the feedstock magazine sends a signal for clearance. The robot engages the billet and puts it into the induction furnace. Then the robot controller gives the start signal for the heating of the billet. A separate PC controls the heating with the control scheme using a pyrometer explained in Section 10.3.1. When the billet has reached the desired temperature, the PC sends a signal to the robot controller. The robot can now engage the heated billet and transport it into the forming press. The forging process will be started as soon as the billet has been placed in the cavity and the robot has left the area of the forging press. After the forming process, the finished part can be removed from the forming press by the unloading unit.

10.3.6.6 Control of the Forming Process

The requirements for handling material in the semi-solid state are very demanding. On the one hand, the forming process has to be done as fast as possible. On the other hand, the flow of the material must not stall because this would lead to imperfections in the work piece. Therefore, the velocity of the piston has to be limited. Commonly, the velocity is constant during the forming process and the flow velocity of the material is limited by the minimum width of the cavity. Hence the forming process is not time optimal. If we could follow an optimal trajectory from form filling simulations, we would obtain a time-optimal process without imperfections in the parts. Therefore, feed-forward control for the trajectories is needed. For these reasons, the flatness-based approach is used (see Section 10.3.1). However, the velocity of the piston is not a flat output for the system of a differential cylinder [43]. To calculate the servo valve voltage from a given velocity trajectory of the piston, one has to measure the pressures in the two chambers of the differential cylinder. Therefore, offline computing of the servo valve voltage is not possible.

In order to achieve a system of the differential cylinder, which has the velocity as a flat output, the compressibility of the hydraulic fluid is neglected. This leads to a simplified system without the pressure dynamics. With the simplified system, we achieve the following equation to calculate the servo valve voltage:

$$u(t) = \frac{1}{K_V B_V} \sqrt{\frac{A_K^3 (1 + \alpha^3)}{A_K p_V - m\dot{v}(t) - F_R - F_L}}$$

$$\times \left\{ T_V \dot{v}(t) + v(t) \left[1 + \frac{T_V}{2} \frac{m\ddot{v}(t) + \dot{F}_R + \dot{F}_L}{A_K p_V - m\dot{v}(t) - F_R - F_L} \right] \right\} \qquad (10.15)$$

The voltage $u(t)$ depends on the velocity $v(t)$ of the piston, the friction force F_R and the load F_L. In a first approximation, the load is considered to be proportional to the velocity. Also, the friction force is a function of the velocity. Therefore, we can conclude that the piston velocity is a flat output of the simplified model and we can calculate the servo valve voltage from a smooth velocity trajectory. For more details, see [44].

We have to consider that the feed-forward control does not contain the pressure dynamics. Therefore, a controller is needed to make sure that the piston follows the desired velocity trajectory. A simple PID controller is sufficient.

The experiments were carried out with a differential cylinder where the behaviour of the thixotropic material was simulated by a damper. Figure 10.21 shows the desired

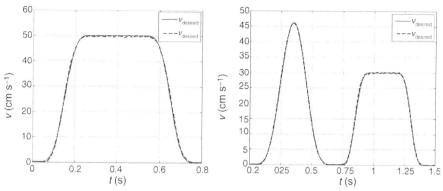

Figure 10.21 Experimental results without and with load.

and measured velocities of the piston over time for the unloaded and loaded cases. One can see that the results are very good. The piston follows the desired velocity with only a small error. This applies to both the unloaded and the loaded cases.

10.3.7
Rheoforging of Steel

As an alternative semi-solid processing route for components, rheoforging of steel alloys is in the focus of different research studies [45, 46]. The semi-solid rheoforming processes include the production of the semi-solid billet starting with a liquid state and the subsequent forming operation without prior cooling and reheating, for example using the cooling channel process (Figure 10.22) [47]. Microstructure

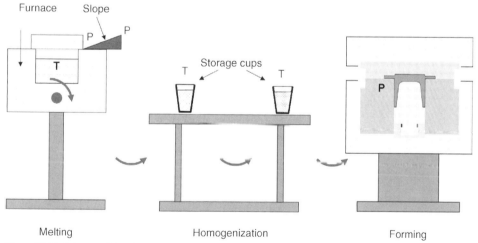

Figure 10.22 Scheme of the rheocasting process with its three steps and positions for temperature measurement with thermocouples (T) and pyrometers (P).

evolution, being dominated by nucleation, forced convection and decelerated cooling, is characteristic for all rheo processes. The rheoforming process has already been applied successfully to serial production using light metal alloys. However, the rheo process in principle has greater potential concerning energy consumption and cost reduction compared with the thixo route due to the shorter process chain. To exploit this potential, several new variations of the process are also under development for light metal alloys [48–52]. For the application of steel alloys, the requirements for the process chain and the metallurgical development of the utilized materials were basically analysed. Analogous to the thixoforming process, the duration of the process steps has to be kept as short as possible due to temperature losses and variations in the chemical composition because of melting losses and slagging.

The study considered two different steel grades, X210CrW12 and 100Cr6. The steel melt is produced in a tilting induction furnace with a maximum power of 45 kW and a frequency of 2 kHz. To avoid oxidation, the furnace and the slope are rinsed with argon. The melt is just slightly overheated so that the temperature drops below the liquidus temperature during flowing on the slope. This induces homogeneously distributed nucleation in the storage cup. Figure 10.22 gives an overview of the schematic sequence and the positions of temperature measurement by thermocouples and a pyrometer during the rheocasting operation for steels.

The supercooled melt is collected in a manually manipulated and argon pre-rinsed cup at the end of the slope. It is insulated and preheated (400 °C) to avoid rapid cooling of the semi-solid billet. The volume of the cup suits the size of a single part. To guarantee a constant volume content and to remove slag or other impurities on the surface, the excess slurry is cut off with a special mechanism directly after casting (Figures 10.23 and 10.24). Then, the cup is immediately closed with an insulation plate to avoid oxidation during a given holding time.

By inverting the holding cup, the billet is ejected and placed in the open lower die for subsequent forging (Figure 10.25). The dies made of working steel X38CrMoV5-1 are preheated to 300 °C and rinsed with argon shortly before the forging operation to avoid fast cooling and oxidation of the billet. After the insertion of the billet (step 1), the upper die closes and the lower plunger (step 2) carries out the main forming operation. During the complete forming operation, the upper die is locked with a holding force of 480 kN. The peak force of the plunger reaches 190 kN (equal to 160 MPa) at the final position and it is kept during the solidification

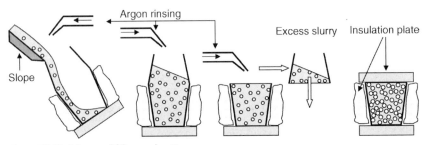

Figure 10.23 Scheme of filling and cutting sequence.

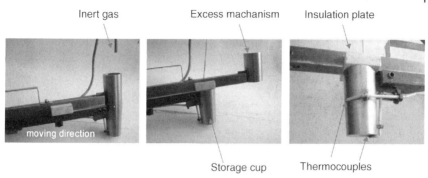

Figure 10.24 Design and working principle of the holding cup with cutting system of excess slurry.

of the part in the dies (step 3). Finally, the upper die opens and the part is ejected by raising the plunger.

The given part geometry is characterized by a long flow length in the bush and a large decrease in cross-section (ratio of area A to B) in Figure 10.26. To investigate the microstructure state at four chosen points, all parts are divided along the axis of symmetry.

The investigation on X210CrW12 shows a similar microstructure in the flange and the bush (positions 1, 3 and 4) in Figure 10.27. It consists of a mixture of austenitic grains and former liquid phase, which is solidified as a eutectic structure or partially as ledeburite for fractions with higher carbon content. The mean grain size of this region is around 50 µm. In contrast to the flange and the bush, austenitic phase dominates in the centre of the part's body (position 2). The different development of

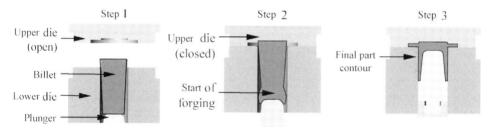

Figure 10.25 Schematic view of forming steps.

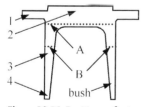

Figure 10.26 Positions of micrographs in vertical cross-section (1–4) and different sizes of the cross-sections (A and B).

Position 1   Position 2   Position 3   Position 4

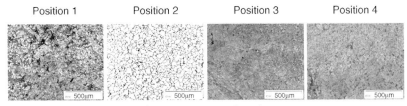

Figure 10.27 Micrographs at different positions in a forged part (X210CrW12).

the microstructure is assumed to be a result of solid–liquid separation during the forming. The average grain size is 70 μm and therefore larger than in other parts of the workpiece, which can be traced back to the slower cooling in the middle of the component.

For 100Cr6, homogeneous microstructures are observed all over the part (Figure 10.28). Between the dark coloured martensite, an area of remaining austenite (white part) occurs, which originates from the former liquid phase enriched with chromium and carbon. Due to the transformation, the grain size of 100Cr6 cannot be determined by conventional microstructure analysis. However, the final structure at room temperature (~50 μm) is much finer than that obtained via the thixo-route (~500 μm).

In detailed observations of local positions of the parts, several defects, for example incomplete filling and cold welded areas, were found for both steels, which mostly appeared in the upper head, the flange and the bush (Figure 10.29). These defects are assumed to be caused by freezing of semi-solid material during the complete filling. The remaining gaps between the frozen material and die surface can be filled by mostly liquid material at the end of the process. Both defects are found more frequently in the parts from X210CrW12 than from 100Cr6.

Position 1   Position 2   Position 3   Position 4

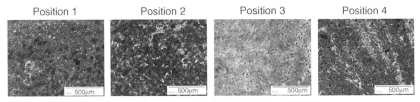

Figure 10.28 Micrographs at different positions in a forged part (100Cr6).

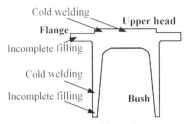

Figure 10.29 Observed local defects.

The investigations show that the rheocasting process is also possible for higher melting steel alloys. A semi-solid slurry with a fine, globular microstructure can be successfully created by controlling the casting temperature and holding time in a cup. In a subsequent inline-forming step, the semi-solid billet produced is forged to a demonstration part. This final part shows a globulitic homogeneous microstructure in a large part of the volume but still includes several local failures, which have to be taken into account for the design of the forming strategy and advanced component geometries.

10.4
Simulation of the Thixoforming Process

To clarify the boundary conditions and the temperature distributions during the process, two types of simulation software were applied to analyse the thixoforging process by numerical calculation. Due to high deformation during the forming operation, it is necessary to have the possibility of remeshing to retain the functionality of the mesh. With LARSTRAN/SHAPE, which is based on viscoplastic, modelling a remeshing is applicable. It is also possible to take the material properties into account, which consider the high dependence of the flow stress in the solidus interval (Section 6.1.6.3). The commercial standard software ABAQUS has its main advantages in the calculation of complex thermal problems. The used solver ABAQUS-Standard is a general-purpose finite element program that is suitable for static, thermal, electrical and transient problems. In this work, in the thermal coupling of component and tools during the cooling operation is of particular interest. In addition to the advantage of consideration of the tools as elastic or plastic dies that support detailed information about temperature gradients in the tools, it is also possible to define the latent heat individually for a given temperature in contrast to LARSTRAN/SHAPE (Section 6.2.3).

10.4.1
Material Properties and Boundary Conditions

As shown in Chapter 6, several values are necessary as input data for the simulation. Most of the properties can be found in the literature, but in case of the flow stress, radiation and heat transfer additional analyses have to be made. The determination of the flow stress for thixoforging processes with lower liquid-phase fraction has already been described with the compression test (Section 6.1.6.3). For the determination of boundary conditions such as radiation and heat transfer, a combination of experimental and simulative analyses is necessary to ensure precise reproduction of the temperature development.

10.4.1.1 **Radiation**
Radiation losses, which greatly influence the process steps during heating and while closing the dies, are determined by using cylinders (20 mm diameter × 20 mm), that

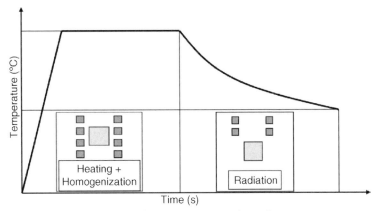

Figure 10.30 Scheme of the radiation experiment and typical temperature development of the sample.

are equipped with thermocouples in the centre of the sample. To consider different effects such as temperature, surface properties and atmosphere, different conditions have to be tested. After heating the sample into the semi-solid interval (Figure 10.30), it is cooled in the defined atmosphere and the temperature is measured. The resulting experimental temperature curve is adjusted with simulated curves so that a value for the radiation can be found. The analyses for X210CrW12 and 100Cr6 show in both cases a high dependence of the cooling behaviour on the atmosphere. Whereas samples cooled in argon as an inert gas have radiation values of about 0.38 (X210CrW12) and 0.28 (100Cr6), respectively, the values for oxidized samples lie above 0.52. Therefore, inert gases ensure the avoidance of oxide layers, which can lead to failures in the forming operation, and also lead to a significant reduction of heat loss during and after heating.

10.4.1.2 Heat Transfer Coefficient Between Workpiece and Die

The experimental setup (Figure 10.31) for determination of the heat transfer is realized by a modified compression test. Flat cylinders (100 mm diameter × 30 mm) are heated into the semi-solid interval. After sufficient homogenization, the induction coil is removed and the sample is pressed against an upper tool. Thermocouples that are fixed in the sample and the tool measure the temperature development. Analogous to the radiation trials, the resulting temperature curves are compared with simulated curves to define the heat transfer between forming material and tool. In these trials, the tool materials X38CrMoV5-1 and the molybdenum-based alloy TZM were investigated with a sample of X210CrW12.

Depending on the pressure p_N that has to be below the flow stress k_{f0} of the semi-solid material to avoid deformation of the sample, the heat transfer was determined as between 2000 and 12 000 kW m^{-2} K^{-1}) ($p_N/k_{f0} = 0.04$ and 0.43) for both tool materials. A significant increase in the heat transfer coefficient starts with a relative pressure $p_N/k_{f0} > 0.2$ due to local forming on the surface of the sample. The surface is flattened and therewith the contact area increases.

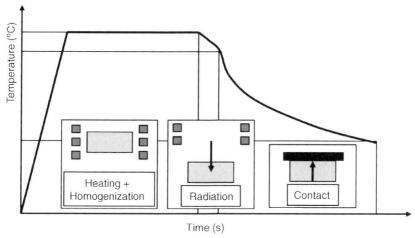

Figure 10.31 Scheme of the heat transfer experiment and typical temperature development of the sample.

Regarding the specific heat conductivity, in the case of the steel tool there will be less heat transfer. This advantage for the cooling behaviour of the sample causes on the other hand high temperatures in the surface of the tool because of retaining heat in the contact area of tool and semi-solid material. In the case of the TZM tool, transferred heat can be transmitted faster and critical temperature peaks can be avoided at and below the surface.

10.4.2
Simulation of a Thixoforged Component

The axisymmetric component in the form of a wheel hub is chosen for numerical analyses of the thixoforging process. Thermomechanical, coupled simulations are used in the case of the forming operation and thermal simulations for the cooling operation. In both simulations, thermophysical and thermodynamic data are implemented that are investigated in several experimental and numerical analyses of the utilized materials within the SFB289. Additionally, flow curve data are provided for a temperature range of 965–1280 °C, a strain of 0–0.7 and a strain rate of $0.1\ 10\ s^{-1}$. The viscoplastic approach in LARSTRAN/SHAPE allows the description of nonlinear functions of flow stress as a function of temperature, strain and strain rate. Especially for the temperature range above the solidus temperature and the resulting low flow stress, the consideration of the temperature dependence is of particular importance.

The initial temperature of the billet is 1300 °C and that of the dies is 300 °C. Due to the short process duration of about 0.5 s, LARSTRAN/SHAPE is applied for the forming operation even though the definition of rigid dies has an influence on the heat flow compared with the experimental values. It is assumed that the results after

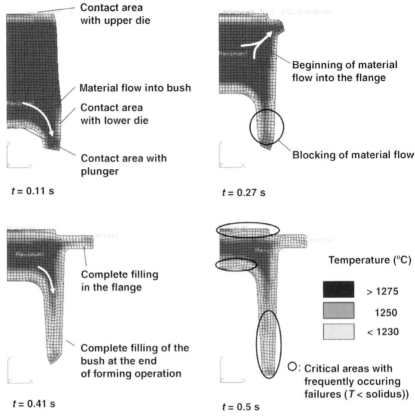

Figure 10.32 Simulated results of temperature distribution during forming operation (LARSTRAN).

the forming step coincide approximately if a few adjustments regarding the heat transfer coefficient are considered.

As shown in Figure 10.32, the temperature increases very rapidly in contact areas of the billet and the die. Initially the material flows into the bush until a narrow channel of semi-solid material occurs due to the fast solidification at the contact zones. The required flow stress increases for this area so that the material flow reverses to the flange. When the flange is completely filled, the existing force of the plunger subsequently suffices to fill the bush with semi-solid material. As shown for the experimental part, existing failures in the area of the bush such as pores and separation of solid and liquid phases can be caused by the sequence of filling and the accompanying cooling of the material near the solidus temperature.

The main heat transfer occurs after completing the forming operation. For these calculations, the geometry of the component and its resulting temperature distribution are imported to ABAQUS. The thermal interactions during the cooling procedure, which was varied in its process duration analogous to the experiment, were calculated with the algorithm 'heat transfer' and following boundary conditions

Table 10.2 Definition of boundary conditions of the cooling process (ABAQUS).

Temperature of environment	$T_{surrounding} = 25\,°C$
Temperature of tools	$T_{tool} = 300\,°C$
Convective heat transmission	$H_{surrounding} = 10\,W\,m^{-2}\,K^{-1}$
Heat transfer	$\alpha_{tool\ component} = 5000\,W\,m^{-2}\,K^{-1}$
Radiation	$\varepsilon_{X210CrW12} = 0.82$

(Table 10.2) for the component and the tools. The materials considered for the tools were hot forming tool steel X38CrMoV5-1 and the molybdenum-based alloy TZM to compare the influence of different heat conductivities on the cooling process.

Figure 10.33 and 10.34 show the results for the tool combination of TZM as die material and 1.2343 as plunger material. After complete filling of the die and a compression for 3 s, the main volume of the component is cooled very quickly below the solidus temperature. Only the centre contains a small region of semi-solid material. In the case of the tools, the surface temperatures reach values of about 600 °C (TZM) and 760 °C (1.2343). Because of the higher heat conductivity of TZM, the heat flow is enhanced so that the heat-influenced zone reaches deeper into the dies. Therefore, critical temperatures at the boundary areas can be minimized.

Figure 10.34 shows the temperature distribution of the component in different stages after finishing the forming operation. When the die is completely filled, rapid cooling starts in the flange and the bush while the centre is still in the semi-solid interval at about 1270 °C after 3 s of cooling under compression. After pulling out the plunger and cooling for another period of 10 s, the component is completely

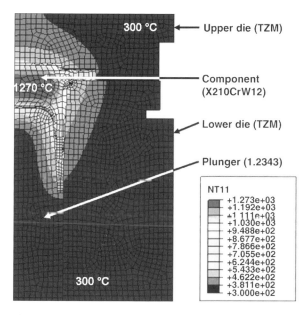

Figure 10.33 Temperature distribution in the component and the tools after forming operation and a cooling time of 3 s (ABAQUS).

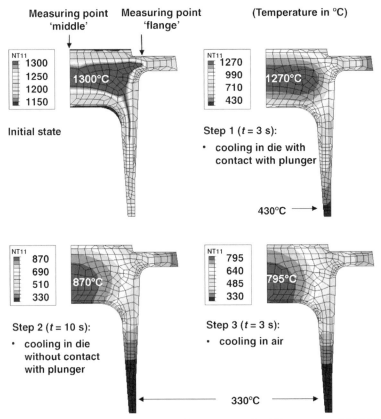

Figure 10.34 Temperature distribution in the component for different stages (ABAQUS).

solidified and more homogeneous regarding the temperature distribution. The subsequent cooling in air ($t = 3$ s) that occurs during to the ejection of the component leads to further homogenization of the temperature field in the whole component because of the reduced heat transfer into the tools.

To clarify the influence of different materials and dwell durations, the temperatures of two positions of the component are measured with pyrometers after opening the cavity. The comparison considers simulated and experimental results for the defined measuring points (Figure 10.35). Generally, the tendencies and the level of the simulated temperatures of TZM fit the measured values. The simulated results for the hot forming tool steel dies are about 100 °C lower than the results for TZM. In any case, the flange shows a lower temperature than the middle of the component because less heat has to be transmitted into the die. Depending on the measuring point and the contact time, the measured temperature can increase with time because the heat flow from the inner part of the component is greater than the heat losses due to radiation and the convective heat transmission to the surroundings.

The comparison of the simulated results between the considered tool materials shows a significant difference in the cooling behaviour of the component. Concern-

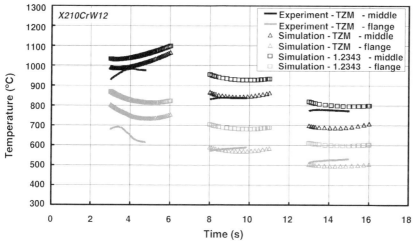

Figure 10.35 Experimental and simulated temperatures at middle and flange of the component after three cooling steps (ABAQUS).

ing the narrow temperature interval of the thixoforming process, a preheated tool material with low heat conductivity would be preferred for the component regarding the material flow during the forming operation. Accordingly, heat losses could be minimized. Concerning the tool and its concept, a material with contrasting properties would be preferred to ensure lower temperature gradients of the dies for extending the tool life [23, 53, 54].

10.5
Conclusion

Compared with the conventional route of drop forging, the process technology of forging metals in the semi-solid state shows potential with regard to the geometric complexity of the components, material savings and lower energy consumption of the required aggregates. Although the process requires only one forming step, it has an increased cycle time compared with the drop forging process because of solidification of the liquid-phase fraction. Efficient casting technologies are another strong competitor for the semi-solid forging technique in the area of aluminium components. Consequently, a possible niche for thixoforging production is not to be found in mass production but in special applications such as the described joining operations. However, analogous to thixocasting, adapted tool concepts have to be developed that ensure a higher durability, especially for economic thixoforging of steel components.

The application of the rheo-route shows further savings potential. The expensive induction heating unit can be economized and the process can be shortened. The experiments show similar results to the thixoforging route. Numerical process modelling will be a helpful development tool by providing a deeper insight into metal flow and temperature field development.

List of Symbols and Abbreviations

ω	circular frequency of the resonant circuit
σ	Stefan–Boltzmann constant
ρ	density
μ	magnetic permeability
λ	heat conduction coefficient
κ	electric conductivity
ε	emissivity factor
δ	penetration depth
α	ratio of ring side area to piston side area
ϑ_{out}	measured temperature at the surface
ϑ_{in}	temperature at the middle of the billet
ϑ_e	temperature of the environment
$\vartheta_{desired}$	desired surface temperature
$\dot{\Phi}(r,t)$	volume power density
A_k	piston area
B_v	value flow gain
$ber()$, $bei()$	Kelvin functions
CNC	Computerized numerical control
c_p	specific heat capacity
$\hat{E}_\phi$	electric field
FDM	finite difference method
F_L	load
F_R	friction force
$\hat{H}_z$	magnetic field strength
$H_{surrounding}$	convective heat transmission
$\hat{I}_c$	coil current peak value
J_0, J_1	modified Bessel functions of first kind, zero and first order
k_{f0}	flow stress
k_v	value dynamic gain
L_c	coil length
N_c	number of turns
PID	proportional–integral–derivative
P_i^k	induced power in the scale i at the time step k.
PLC	programmable logic controller
p_N	pressure
P_v	supply pressure
R	radius of coil
$T_{surrounding}$	temperature of environment
T_{tool}	temperature of tools
T_v	time constant of the value
TZM	molybdenum-based alloy
$u(t)$	voltage
$v(t)$	velocity of the piston

References

1 Koesling, D., Tinius, H.C., Cremer, R., Hirt, G., Morjan, U., Nohn, B., Wittstamm, Th. and Witulski, T. (1999) Final Technical Report P 322/08/96/S 24/23/95, Verlag und Vertriebsgesellschaft, Düsseldorf.

2 Hartmann, D., Morjan, U., Nohn, B., Zimmer, M., Hoffmann, A. and Braun, R. (2002) Final Technical Report P 483/11/2000/S 24/28/9, Verlag und Vertriebsgesellschaft, Düsseldorf.

3 Neudenberger, D. (2001) Rheologische Untersuchungen und Einflüsse auf das Prozessfenster zum Thixoschmieden und Thixoquerfliesspressen, Dissertation, RWTH-Aachen, Institut für Bildsame Formgebung.

4 Baadjou, R., Shimahara, H. and Hirt, G. (2006) Automated semi-solid forming of steel components by means of thixojoining, Proceedings of the 9th International Conference on Semi-solid Alloys and Composites (S2P). *Solid State Phenomena*, **116–117**, 383–386.

5 Hirt, G., Nohn, B. and Witulski, T. (1999) Herstellung komplexer Stahlbauteile durch Thixoforming, VDI-Tagung (Rapid-Prototyping – Verkürzte Fertigungsketten durch INSTANT-Produktion), Düsseldorf, 25. März.

6 Brooks, D., Michael, M. and Wienströer, M. (1999) Process Modeling for Forging Using Rapid Simulation, 16. Umformtechnisches Kolloquium Hannover (Umformtechnik an der Schwelle zum nächsten Jahrtausend), 25. und 26. Februar.

7 Flemings, M.C., Boylan, J.F. and Bye, R.L. (1978) Thixocasting Steel Parts, Final Technical Report, Massachusetts Institute of Technology, Cambridge, MA.

8 Breuer, H.-W. and Peddinghaus, J. (1999) Entwicklung und Fertigung komplexer Stahlschmiedeteile, 16. Umformtechnisches Kolloquium Hannover (Umformtechnik an der Schwelle zum nächsten Jahrtausend), 25. und 26. Februar, pp. 159–167.

9 Young, K.P., Riek, R.G. and Flemings, M.C. (1979) Structure and properties of thixocast steels. *Metals Technology*, **6** (44),130137.

10 Flemings, M.C. (1991) Behavior of metal alloys in the semisolid state. *Metallurgical Transactions A*, **22** (5), 957.

11 Hirt, G. *et al.* (2005) Semi-solid forging of 100Cr6 and X210CrW12 steel. *CIRP*, **54/1**, 257–260.

12 Seidl, I., Kallweit, J. and Kopp, R. (2002) New potentials for steel forming: application for semisolid forming and joining at the RWTH Aachen, Proceedings of the 7th International Conference on Semi-solid Processing of Alloys and Composites (S2P), pp. 341–342.

13 Püttgen, W. *et al.* (2005) Thixoforged damper brackets made of the steel grades HS6-5-3 and 100Cr6. *Advanced Engineering Materials*, **7** (8), 726–735.

14 Omar, M.Z. *et al.* (2005) Thixoforming of a high performance HP9/4/30 steel. *Materials Science and Engineering*, **A395**, 53–61.

15 Kiuchi, M., Sugiyama, S. and Arai, M. (1996) Mushy-state forging of cast-iron. *Journal of the JSTP*, **37** (430), 1219–1224.

16 Rassili, A., Robelet, M. and Fischer, D. (2006) Steel grades adapted to the thixoforging process: metallurgical structures and mechanical properties, Proceedings of the 9th ESAFORM Conference, Glasgow, pp. 819–822.

17 Tsuchiya, M., Ueno, H. and Takagi, I. (2003) Research of semi-solid casting of iron. *JSAE Review*, **24**, 205–214.

18 Omar, M.Z., Atkinson, H.V., Palmiere, E.J., Howe, A.A. and Kapranos, P. (2006) Thixoforming two different steels. in 9th International Conference on Metal Forming, ESAFORM 2006, pp. 847–850.

19 LeBeau, S. and Walukas, D.M. (1999) Thixomoulding of magnesium

components for electronic applications, presented at International Conference on Plastics in Portable and Wireless Electronic, Mesa, AZ, 25 January 1999.

20 Küthe, F., Schönbohm, A., Abel, D. and Kopp, R. (2004) An automated thixo-forging plant for steel parts. *Steel Research International*, **75** (8/9), 593–600.

21 Siegert, K. and Messmer, G. (2004) Thixoforging of steel alloys – facility requirements for industrial application. *Steel Research International*, **75** (8/9), 601–606.

22 Münstermann, S. and Telle, R. (2004) Model tests and analyses of corrosion resistance of ceramic tool dies for steel thixoforming. *Steel Research International*, **75** (8/9), 581–587.

23 Kopp, R., Shimahara, H., Schneider, J.M., Kurapov, D., Telle, R., Münstermann, S., Lugscheider, E., Bobzin, K. and Maes, M. (2004) Characterization of steel thixoforming tool materials by high temperature compression tests. *Steel Research International*, **75** (8/9), 569–576.

24 Kurapov, D. and Schneider, J.M. (2004) Adhesion and thermal shock resistance of Al_2O_3 thin films deposited by PACVD for die protection in semi-solid processing of steel. *Steel Research International*, **75** (8/9), 577–580.

25 Behrens, B.-A., Fischer, D., Haller, B. and Rassili, A. (2004) Introduction of a full automated process for the production of automotive steel parts, Proceedings of the 8th International Conference on Semi-solid Processing of Alloys and Composites, Limassol, Cyprus, 21–23 September 2004.

26 Fleck, Ch. and Schönbohm, A. (2004) Flatness based trajectory generation for the inductive heating of aluminium billets, Proceedings of the 8th International Conference on Semi-solid Processing of Alloys and Composites, Limassol, Cyprus, 21–23 September 2004.

27 Siegert, H. (1961) Induktive Erwärmung. Technische Rundschau. 13, 17–21.

28 Davis, J. and Simpson, P. (1979) *Induction Heating Handbook*, McGraw-Hill, London.

29 Baehr, H.D. and Stephan, K. (2006) Wärme- und Stoffübertragung, 5. Auflage, Springer, Berlin.

30 Ritt, H.M. (2000) Regelung der induktiven Erwärmung von Metallen in den teilflüssigen Zustand. Dissertation RWTH Aachen, VDI-Verlag.

31 Paulus, T. (2007) Integration flachheitsbasierter Regelungs- und Steuerungsverfahren in der Prozessleittechnik. Dissertation RWTH Aachen, VDI-Verlag.

32 Fliess, M., Levine, J., Martin, Ph. and Rouchon, P. (1995) Flatness and defect of nonlinear systems: Introductory theory and examples. *International Journal of Control*, **61**, 1327–1361.

33 Martin, Ph., Murray, R.M., Rouchon, P., Bastin, G. (1997) Flat systems, in Plenary Lectures and Mini-Courses, 4th European Control Conference ECC'97, Gevers, M.(ed.), Brussels, pp. 211–264.

34 Lynch, A.F. and Rudoph, J. (2000) Flachheitsbasierte Randsteuerung parabolischer Systeme mit verteilten Parametern. at-Automatisierungstechnik. 48, 478–486.

35 Laroche, B., Martin, P. and Rouchon, P. (2000) Motion planning for the heat equation. *International Journal of Robust and Nonlinear Control*, **10**, 629–643.

36 Fleck, Ch. and Abel, D. (2003) Eine Methode zur Bestimmung eines flachen Ausgangs für Systeme mit örtlich verteilten Parametern, GMA-Kongress, Baden-Baden. VDI-Verlag, pp. 545–556.

37 Fleck, Ch., Paulus, Th. and Abel, D. (2003) Eine flachheitsbasierte Randsteuerung für das Stefan-Problem, GMA-Kongress, Baden-Baden. VDI-Verlag, pp. 311–319.

38 Fleck, Ch. and Schönbohm, A. (2004) Flatness based trajectory generation for the inductive heating of aluminium billets, Proceedings of the 8th International Conference on Semi-solid Processing of Alloys and Composites, Limassol, Cyprus, 21–23 September 2004.

39 Schoenbohm, A., Gasper, R. and Abel, D. (11–13 09 2006) Inductive reheating of

steel billets into the semi-solid state based on pyrometer measurements, Proceedings of the 9th International Conference on Semi-solid Processing of Alloys and Composites, Busan, Korea; Solid State Phenomena, Vols 116–117, Trans Tech Publications.

40 Kopp, R., Neudenberger, D. and Winning, G. (2000) Optimisation of the forming variants forging and transverse impact extrusion with alloys in the semi-solid state, Proceedings of the 6th International Conference on Alloys and Composites, 6th SSM Conference, Turin, pp. 295–300.

41 Wang, K., Hirt, G. and Kopp, R. (2006) Investigation on forming defects during thixoforging of aluminum alloy AlSi7Mg. *Advanced Engineering Materials*, **8** (8), 724–729.

42 Kallweit, J. (2003) Entwicklung von Formgebungsstrategien zum Thixoforming von Stahl und zum Fügen im thixotropen Zustand, Dissertation RWTH Aachen.

43 Lemmen, M. and Wey, T. (1998) Flachheitsbasierte Regelung eines hydraulischen Differentialzylinders, VDI-Berichte 1397, VDI-Verlag, pp. 869–876.

44 Gasper, R., Schönbohm, A. and Abel, D. (2007) Flachheitsbasierte Vorsteuerung eines hydraulischen Differentialzylinders beim Thixoforming, GMA-Kongress, Baden-Baden, 12–13 June 2007, VDI-Berichte 1980, VDI-Verlag, pp. 487–496.

45 Seidl, I. and Kopp, R. (2004) Semi-solid rheoforging of steel. *Steel Research International*, **75** (8/9), 545.

46 Ramadan, M., Takita, M. and Nomura, H. (2006) Effect of semi-solid processing on solidification microstructure and mechanical properties of gray cast iron.

Materials Science and Engineering A, **417**, 166–173.

47 Aguilar, J., Fehlbier, M., Grimmig, T., Bramann, H., Afrath, C. and Bührig-Polaczek, A. (2004) Semi-solid processing of metal alloys. *Steel Research International*, **75** (8/9), 492–505.

48 UBE Industries (1996) European Patent EP 0 745 694 A1.

49 Yurko, J.A., Martinez, R.A. and Flemings, M.C. (2002) Development of the semi-solid rheocasting (SSR) process, Proceedings of the 7th International Conference on Semi-solid Processing of Alloys and Composites (S2P), p. 659.

50 Pan, Q.Y., Findon, M. and Apelian, D. (2004) The continuous rheoconversion process (CRP): a novel SSM approach, Proceedings of the 8th International Conference on Semi-solid Processing of Alloys and Composites (S2P), session 2.

51 Kaufmann, H., Potzinger, R. and Uggowitzer, P.J. (2001) The relationship between processing and properties of new rheocast AZ91 and AZ71 magnesium parts. *Light Metal Age*, 56.

52 Jorstad, J., Thiemann, M. and Rogner, R. (2005) SLC™- Subliquidus Casting: an economical route to rheocasting Proceedings of the 8th Esaform Conference on Material Forming, p. 1116.

53 Kopp, R., Hirt, G., Shimahara, H., Seidl, I., Küthe, F., Abel, D. and Schönbohm, A. (2005) Semi-solid forging of 100Cr6 and X210CrW12 steel, 55th CIRP General Assembly, Antalya, Turkey, 21–27 August 2005, STC F, 54/1/2005, p. 257.

54 Behrens, B.A., Haller, B. and Fischer, D. (2004) Thixoforming of steel using ceramic tool materials. *Steel Research International*, **75** (8/9), 561

11
Thixoextrusion

Frederik Knauf, René Baadjou, Gerhard Hirt, Reiner Kopp,
Simon Münstermann, and Rainer Telle

11.1
Introduction

Extrusion is a forming process where a billet is inserted in a container first and thereupon extruded through a forming die by a punch. The conventional extrusion process is industrially mainly applied for light metal- and copper-based alloys. Relatively low flow stresses compared with the tool material used, moderate temperatures and the possibility of solid-state welding during the process open up the possibility of producing light metal profiles with complex cross-section geometry by bridge dies. These profiles require little or no further surface treatment and achieve excellent mechanical properties by appropriate heat treatment.

Compared with light metal alloys, the conventional steel extrusion process is characterized by high thermal and mechanical stresses. This results in limitation of cross-section complexity and alloy selection, very short tool life and limited industrial use. Steel profiles are therefore typically produced by profile rolling, which, however, requires many rolling passes even though the cross-sections achievable are restricted to relatively simple geometries.

Hence semi-solid extrusion of steels might open up new possibilities for the production of steel profiles. The reduced flow stresses of semi-solid processes lead to lower mechanical tool stress and accordingly the extrusion of conventionally non-extrudable steel alloys or an increasing profile complexity might become possible. Also, the specific globular microstructure (see Chapter 3) which is obtained in thixoforming could be of interest for some materials and applications.

However, identifying decisive operating parameters and maintaining the process window defined by those parameters in a continuous extrusion process is a challenging task for process development. Although experience and results obtained in sequenced thixoforming methods such as thixoforging and thixocasting may be transferred to the realization of steel thixoextrusion, continuous thixoextrusion differs considerably in terms of material handling, tool design and process control.

The following sections consider first the state of the art for semi-solid material extruding separated into light and heavy metals. The fundamental requirements for tool design for the semi-solid extrusion process are pointed out and also the design of the tools realized. For these tool concepts, all variations of process parameters ensuring extrusion in the semi-solid state followed by the results of the experiments and the numerical simulation are presented.

11.2
State of the Art

As already stated in the Introduction, complex profiles of light metals are produced industrially by conventional solid-state extrusion so that there is no economic need to develop a more complicated semi-solid extrusion process for aluminium alloys. However, semi-solid extrusion of lower melting alloys is a very interesting alternative to evaluate the process characteristics without having to deal with the additional problems created by the high temperature levels of semi-solid steel.

Regarding a typical process window diagram for conventional solid bar extrusion processes (Figure 11.1a), it becomes clear that increasing the billet temperature to semi-solid condition using conventional extrusion setups will lead to product defects. If the force required to achieve a high extrusion ratio exceeds the press force limit, then the billet temperature may only be increased until the maximum extrusion ratio (A_0/A_1) is reached. According to general experience, a further increase in the billet temperature results in local defects such as heat cracks (Figure 11.1b), because the dissipation energy causes local melting.

The success of the future development of semi-solid extrusion will therefore depend strongly on the question of whether process parameters and tool designs can be found that avoid these defects. Even though this question is mostly not

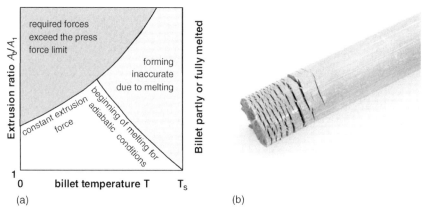

(a)

(b)

Figure 11.1 Process window diagram for conventional bar extrusion processes (a) [1] and heat crack (b).

directly addressed in the available literature on semi-solid extrusion, the reported data allow some conclusions to be drawn concerning process parameters and part quality.

Semi-solid Bar Extrusion of Aluminium and Lead   The first investigations on semi-solid extrusion were done by Kiuchi *et al.* [2] in 1979. On a horizontal forming device, they studied the influence of the required extrusion force and the tool temperature for Pb and Al alloys with a billet 40 mm in diameter and 25, 30 or 40 mm in height under different test conditions. These conditions were the extrusion channel diameter, varied between 2 and 10 mm, and the extrusion channel length, varied from 4 to 100 mm. For the investigated alloys, it was found that the required initial press force increases with decreasing liquid fraction. The extrusion ratio was varied between 16 and 400. The press velocities were chosen between 37 and 47 mm s^{-1}. The experiments were carried out using a 'simultaneous heating process', where the tool and the billet were heated to the required temperature simultaneously, and a so called 'no-preheating process', where the tool was not actively heated. A further modification of the 'simultaneous heating process' was an additional cooling of the material with compressed air after the die. The authors pointed out that the optimum liquid fraction for the semi-solid extrusion of the investigated alloys was in the range 5–10% and the required press force, which was only one-quarter to one-fifth of the required press force in conventional extrusion, increased with decreasing liquid fraction. When applying compressed air cooling directly after the forming die, the temperature of the die decreased due to cooling by 20 °C. Concerning the process variations, the bars extruded with the no-preheating tool setup showed greater strain hardening and less elongation than the bars in the simultaneous extrusion process. Concerning tensile strength, both process variants showed the same results, but the tensile strength decreased with increasing liquid fraction. The microstructure of the extruded bars was generally fine for all process variants. When the liquid fraction was higher than 20%, friction between the billet and container was negligible.

Much later, the same group [3] investigated, in addition to conventional aluminium alloys, also the particle-reinforced aluminium alloy A-5056 + 20 vol.% Al$_2$O$_3$ using an extrusion channel diameter varied from 2 to 10 mm in 2 mm steps and a liquid fraction of about 20–30%. To prevent cooling of the billet in the container, the tool was preheated up to the billet temperature. The authors also mentioned that the extrudable wall thickness decreased with increasing profile complexity. Generally, they underlined the conclusion that the alloys can be extruded easily with low strain. Heat treatment can improve the yield stress of the extruded products. Concerning the particle-reinforced materials, the extruded products showed a smooth surface and no extrusion failures. The dominant factor in semi-solid extrusion of this material is the liquid fraction, which is responsible for the embedding of the particles.

Möller [4] investigated billets of the aluminium alloy A356 + 20 vol.% SiC with a height of 100 mm and a diameter of 76 mm. With the chosen press velocity of 10 mm s^{-1} and a transverse extrusion tool setup, a maximum press force of 600 kN was required irrespective of the tool temperature, which was varied between 200 and 400 °C. The billets were extruded to a total height of about 600 mm using an extrusion channel with a rectangular cross-section of 10×10 mm. The edges were filled out well. In the microstructure analysis, the author detected the phenomenon of shell formation. The extruding material which entered the forming die first cooled on coming in contact with the extrusion channel wall. This cold material solidified and stuck to the extrusion channel wall. The following still semi-solid material flowed through the reduced extrusion channel. The phenomenon of shell formation was not solved by increasing the tool temperature. A plug to increase the pressure in the extrusion channel did not increase the heat transfer and did not ensure rapid solidification of the material before leaving the extrusion channel.

Semi-solid Impact Extrusion of Steel   Some authors have described experiments in which they 'extruded' billets into a closed die of limited length. These processes, which are similar to impact extrusion, were mainly used to study the influence of the process parameters on solid–liquid segregation and part properties.

Miwa and Kawamura [5] carried out experiments similar to semi-solid impact extrusion, denoting them thixoextrusion into a mould. They extruded billets (diameter 30 mm, height 20 mm) of the stainless-steel alloy UNS: s30400 (AISI 304), held in a ceramic container, into a metallic mould to investigate the influence of the press velocity on the tendency for segregation. Depending on the applied forming velocities, they observed effortless form filling, but phase separation of the liquid and solid fractions at 100–1800 mm s^{-1} press velocities, whereas at a low press velocity of 10 mm s^{-1} a homogeneous phase distribution was observed, but poor surface quality. An unintended increase in the required forming pressure occurred. The authors attributed the latter effects to undesired rapid solidification of the semi-solid slurry in the comparatively cold extrusion dies. Unfortunately, they did not mention either any dimensions of the forming die diameter or extrusion channel length or give an illustration of the tool setup.

Rouff *et al.* [6], in impact extrusion experiments, investigated the influence of different shear rates on the steel grade C80 (AISI 1080). The extruded billets had a diameter of 30 mm and a height of 45 mm. The desired extrusion temperature was $1430\,^\circ$C, which corresponds to a liquid fraction of about 40%. In order to reach a large range of strain rates, the authors investigated two possibilities. First, the press velocity was kept constant and the extrusion diameter was changed from 8 to 15 mm. Second, the extrusion tool geometry was kept constant and the press velocity was changed. For the first possibility, where the calculated shear rate was 243 or 215 s^{-1}, the extruded bars showed a homogeneous material flow and no kind of segregation. In the second case, when the extrusion channel diameter was kept constant at 12 mm, inhomogeneous material flow could be detected for shear rates between 10 and 95 s^{-1}, whereas a homogeneous material flow was observed for higher press velocities, which resulted in shear rates between 95 and 475 s^{-1}. An influence of different shear rates could not be detected if the shear rate was lower than 10 s^{-1}. The results of the extrusion experiments were compared with simulated values. With inverse modelling, using a viscoplastic material law, the authors concluded that the form filling behaviour of this steel grade is dependent on the shear rate. No data were provided on the applied tool materials and die temperatures.

In similar experiments, Abdelfatah *et al.* [7] reported a strong increase in the required extrusion pressure with increasing extrusion ratio. The billets used had a diameter of 30 mm and a height of 45 mm. The chosen liquid fraction of the investigated steel grade C80 was between 20 and 30%. The container diameter was 40 mm. In two test series, the diameter of the extrusion channel was reduced from 12 to 8 mm, which correspond to extrusion ratios of 11 and 25. On increasing the extrusion ratio, the required press force had to be increased from 15 to 20 tons. The ultimate tensile strength and yield strength of the extruded bars were comparable to values from conventional forged parts. Other mechanical properties such as percentage reduction of area after fracture were lower than for conventional forged parts due to the appearance of microporosity in the thixoforged parts.

Semi-solid Bar Extrusion of Steel   Sugiyama *et al.* [8] investigated the semi-solid bar extrusion process of steel. In their experiments, the steel alloy C22 (AISI 1020) (billet of diameter 18 mm and height 35 mm) was extruded through a forming die channel 7 mm in diameter, corresponding to an extrusion ratio of 10. The billet was inserted in a graphite case and the gap between the billet and case was filled with alumina powder. Then the graphite case was inserted in a graphite block which was located directly in the inductive heater, where the billet was heated to the desired temperature. With a press velocity of $8.8\,mm\,s^{-1}$, the billets were extruded at hot forming and semi-solid temperature. After the forming die, active cooling with compressed air and water dust was applied. A comparison of the resulting press force showed that the press force for semi-solid extrusion is half of the force required for conventional extrusion. Furthermore, the authors observed shell formation in the extrusion channel when the semi-solid material touched the extrusion channel wall. Due to segregation, poor quality at the top of the extruded bar was observed. The other parts of the bar showed a good microstructure – independent of the process parameters chosen. The hardness in the surface area was similar to that for conventional extruded products but the centre was about two times harder.

Möller [4] extruded billets (diameter 34 mm, height 40 mm) of the steel grade X210CrW12 with variation of the press velocity from 20 to $100\,mm\,s^{-1}$. After the extrusion channel exit, the tool was equipped with an active water dust cooling device. With variation of press velocity, the author observed defects such as porosity in the extruded bars. The phenomenon of shell formation occurred during the extrusion process. Due to the billet size and the tool temperature of 300 °C, the billet solidified quickly. In the experiments, only a small bar could be extruded. In contrast to the conventional extrusion process, no stationary press force was realizable.

Owing to the aforementioned difficulties, the number of publications dealing with steel semi-solid extrusion is low, compared with thixocasting and thixoforging of steels. The above-mentioned studies [2–8] have demonstrated the feasibility of semi-solid extrusion. Experiments were performed on the laboratory scale and the experimental results such as extrusion load development and microstructure analysis of the extruded parts were presented. Mechanical properties such as yield stress and hardness were also presented. However, in all these studies, no tool and process parameter concepts for reproducible long bar extrusion were presented. The first investigation of this aspect is represented by recent work on an appropriate determination of the solidification stage by applying numerous different die concepts and cooling devices [9, 10].

11.3
Tool Concepts

As described above, the low strength of semi-solid metals may cause defects if the extruded bar leaves the die in a semi-solid condition. On the other hand, the advantage of low forces can only be achieved if the material passes the deformation zone in a semi-solid condition. Two opposite concepts, which, however, can be combined to some extent, attempt to solve this contradiction.

11.3.1
Isothermal Tool Concept

Since the extrusion process takes some time, this favours the use of a billet container which is heated to the semi-solid temperature so that the billet condition remains the same during the process (isothermal tool concept for steel). This can only be achieved using ceramic containers, which will not be able to withstand high tensile stresses caused by high extrusion pressures, which might be desired, for example, to avoid solidification porosity, or could be caused by any problem within the process. Due to the forming process in the semi-solid state and the temperature of the tool, the solidification has to take place after the isothermal part of the forming die. This could either be a non-isothermal elongation of the forming die or strong active cooling directly after leaving the die, such as performed in continuous casting of aluminium, where the mould length is very short. In the latter case, a mechanism to carry the weight of the extruded bar is required. One possible solution is guide rollers as shown in Figure 11.2.

The first challenge of this concept is the design of the isothermal die setup, including the heating device. Further, the cooling of the bar and the guiding system, which must be carefully coordinated with the extrusion punch, has to be implemented.

11.3.2
Non-isothermal Tool Concept

The other alternative is to accept that the billet in the container forms a solid shell during extrusion and to extrude deliberately only the inner portion of the billet, thus ensuring that any oxides which might have formed on the surface will not be extruded (non-isothermal tool concept for steel). In this case, a strong metallic container can be used to apply high pressure and allow complete solidification within the extrusion die. This requires that the extrusion channel length is chosen long enough to provide

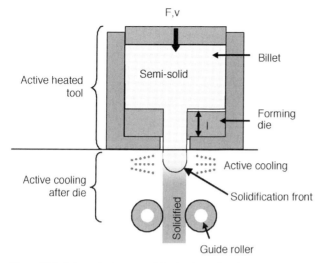

Figure 11.2 Schematic diagram of the isothermal tool concept.

sufficient time for solidification and to integrate the required cooling system to ensure stationary conditions (Figure 11.3).

For this tool concept, the challenges are the process design, especially the extrusion channel length and the press velocity, the dimensions of the cooling system and the prevention of initial solidification and shell formation of the material in the extrusion channel.

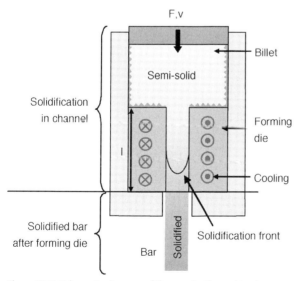

Figure 11.3 Schematic diagram of the non-isothermal tool concept.

Both concepts, which could also be combined to a certain extent, have advantages and disadvantages and will therefore be evaluated in more detail in the following sections.

11.4
Isothermal Thixoextrusion

11.4.1
Experimental Strategy, Tools and Process Parameters

The tool design and experimental setup used for isothermal thixoextrusion of steel are based on the ceramic tool concept introduced in Chapter 8. In order to determine the principal operating parameters of this novel tool setup, small-scale extrusion tests were performed on an INSTRON universal testing machine, allowing precise control of process loads and extrusion velocity. Subsequently, the tool design was adapted and scaled up to extrusion tests on the SMS Meer open die forging press.

The laboratory tool setup is shown in Figure 11.4. For these first laboratory experiments, the steel billets were manufactured from rods of 30 mm diameter, while the punch diameter was set to 28 mm. This was due to the billets being preheated in air, inducing the formation of an oxide layer on the billet surface to facilitate handling of the semi-solid billets. The semi-solid slurry was squeezed out of this oxide shell by a smaller punch, leaving the scale in the container.

The specific characteristic of this tool is the die geometry, consisting of an axial-symmetric shamrock-like cross-section (Figure 11.5), exhibiting a high ratio of circumference to cross-sectional area (Table 11.1). The tool preheating strategy was adapted to the work alloy. As discussed in Section 11.3, two temperature zones may be distinguished in isothermal extrusion: (i) the forming zone, in which the material

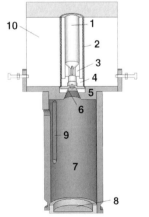

1	alumina container
2	Kanthal A1 heating coil
3	alumina die
4	alumina die socket
5	hot working steel plate
6	ceramic isolation plate
7	slotted hot working steel tube
8	sand bed container
9	window for temperature control
10	ceramic isolation body

Figure 11.4 Tool geometry as applied in laboratory-scale isothermal thixoextrusion tests.

Figure 11.5 Shamrock-like thixoextrusion die.

should be in a semi-solid state to take advantage of thixotropic behaviour, and (ii) the solidification zone, in which homogeneous and rapid cooling of the extruded bars is desired.

The respective boundary temperatures are defined by the envisaged liquid-phase content and the solidus temperature of the work alloy. In the case of X210CrW12 (AISI D6), the target value for the container was set to 1300 °C, corresponding to a liquid-phase content of approximately 55% (see Chapter 3). The steel grade X210CrW12 was selected as work material owing to its suitability for semi-solid processing (cf. Chapter 3). At the onset of forming, that is, at the die land inlet, the material should still be in a semi-solid state; hence at this point the steel temperature

Table 11.1 Tool dimensions as applied isothermal laboratory thixoextrusion tests.

Container		
Material		Al_2O_3, 99.7%
Inner diameter	[mm]	34
Outer diameter	[mm]	50
Cross-sectional area	[mm²]	907.9
Wall thickness	[mm]	8
Height	[mm]	150
Die		
Material		Al_2O_3, 99.7%
Cross-sectional area	[mm²]	108.4
Extrusion channel length	[mm]	13.3
Feeding angle	[°]	52
Extrusion ratio		8.4

should be just above the solidus temperature, for example 1250 °C for X210CrW12. The heating core constructed for the first extrusion experiments provided a peak temperature of approximately 1350 °C in the centre of the container when the sampling point at the die inlet was set to 1250 °C. Hence heat losses of the preheated steel billet in the container were inhibited.

11.4.2
Extrusion Performance and Bar Analysis

Laboratory-scale thixoextrusion tests yield very low values of the pressure required for forming, being less than 22 N mm^{-2} during the entire experiment. This indicates smooth and effortless forming behaviour of the semi-solid slurry and confirms the efficiency of this tool concept in preventing heat losses of the work alloy. A characteristic load curve of a laboratory-scale isothermal thixoextrusion test is observed, showing a steep exponential increase in forming forces until the maximum values are reached at the end of extrusion. Intermediate peak values observed in laboratory tests (Figure 11.6) are ascribed to flow irregularities resulting from the relatively hard oxide shell at the bottom of the steel billet that is cracked and deformed during extrusion.

This is contradictory to typical load curves recorded in conventional extrusion processes showing a yield stress at the onset of forming and nearly constant forces during forming (Table 11.2). However, comparison of the extrusion forces applied in the present experiments with relevant work on steel thixoextrusion by other workers [5–8] using cold dies substantiates these findings, revealing a significant reduction in the required process forces.

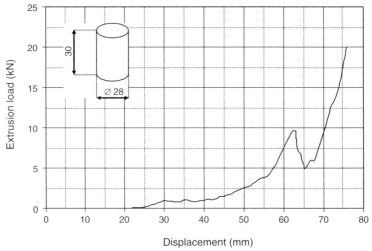

Figure 11.6 Typical extrusion load versus displacement curves obtained in laboratory-scale isothermal thixoextrusion tests.

Table 11.2 Comparison of required forces and pressures for semi-solid extrusion of steels.

Steel grade	Container diameter (mm)	Load range (kN)		Pressure (N mm^{-2})		Source
AISI 304	30	n. s.	n. s.	29	49	[5]
AISI 1080	30	40	140	56^a	198^a	[6]
carbon steel	40	147	196	117^a	156^a	[7]
AISI 1020	22	196	589	516^a	1549^a	[8]
AISI D6 (X210CrW12)	34	1	20	1^a	22^a	Experiments

^{a}Values derived from force values.

The steel billets are completely extruded to bars. The die cross-section is reproduced well except for one arm not being filled (Figure 11.7). This is ascribed to scale residues partly blocking the die inlet and might be prevented by inductive preheating under flushing argon, as carried out in upscaled forming tests. The contour of the die geometry of extruded bars is well defined, indicating smooth material flow during extrusion. Although the bars maintain their shape after cooling and quenching in water, solidification is incomplete after leaving the die, with the result that extruded bars drip down due to their own weight.

However, the excellent formability of steel X210CrW12 observed during isothermal thixoextrusion experiments is confirmed by optical and chemical analysis of extruded bars according to the sampling scheme given in Figure 11.8.

Optical micrographs of different cross-sections along the bar length reveal only small amounts of residual porosity and excellent homogeneity (Figure 11.9).

Figure 11.7 Cross-section of X210CrW12 bar thixoextruded using self-heating alumina dies under isothermal conditions.

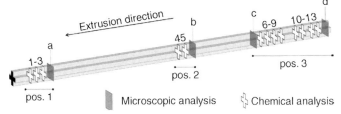

Figure 11.8 Sampling scheme for analysis of extruded bars.
Cross-sections denoted a–d correspond to optical micrographs
depicted in Figure 11.9. Cross-sections numbered 1–13
correspond to results of chemical analysis given in Figure 11.10
Nanoindenter measurements were carried out at positions 2 and
3 (cf. Table 11.3).

The former liquid phase is evenly distributed between the globulitic primary
phase, which is consistent with carbon and chromium contents measured along the
bar length by chemical pulping of respective samples (Figure 11.10). The solubility of
these constituents in the liquid phase is significantly higher than in the solid phase.

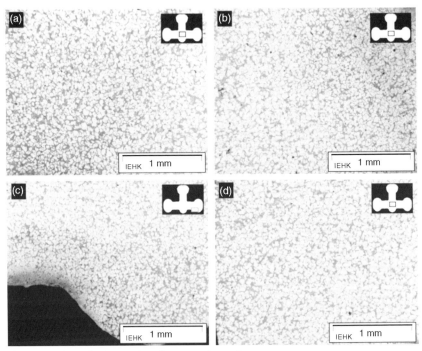

Figure 11.9 Optical micrographs of cross-sections of
as-extruded bars at different locations in the longitudinal
direction from (a) bar tip to (d) bar end. Sampling positions
are given in Figure 11.8

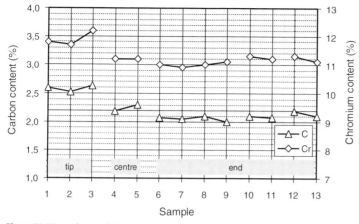

Figure 11.10 Carbon and chromium contents of as-extruded bars. Numbering of sampling points 1–13 refers to the sampling scheme given in Figure 11.8.

Owing to the difficulties in determining the former liquid-phase content of quenched samples, as discussed by Püttgen *et al.* [11], these elements act as a reliable indicator of phase segregation.

The values for both alloying elements are within the fluctuation range of X210CrW12 as given by the material supplier, indicated by the grey areas in Figure 11.10, except for the carbon content at the bar tip.

Hardness and Young's modulus measured by nanoindentation on straight lines at positions near the middle (position 2) and the bar end (position 3) revealed two levels of values. Correlation of both properties is high, as indicated by the averaged results given in Table 11.3.

Moreover, accompanying scanning electron microscopic analysis revealed a correlation of nanoindentation results with microstructural characteristics. Low-level

Table 11.3 Hardness and Young's modulus values [GPa] calculated from nanoindentation experiments[a].

Nano indention data		Cross-section		Average
		Position 2	Position 3	
Hardness	Overall	7.0 (2.7)	7.7 (3.5)	7.4
	Base level	5.7 (0.8)	6.4 (0.6)	6.1
	High level	11.2 (2.2)	14.2 (4.8)	12.7
Young's modulus	Overall	228 (17)	238 (23)	233
	Base level	218 (11)	233 (11)	226
	High level	246 (9)	280 (19)	263

[a]Standard deviations are given in parentheses. Sampling positions are given in Figure 11.9.

Table 11.4 Comparison of hardness and Young's modulus values [GPa] of X210CrW12 during processing.

Sample state	Hardness	Young's modulus
untreated	5.2 (0.5)	220 (34)
thixoextruded	7.4 (3.1)	233 (21)
thixoextruded and annealed	7.8 (0.5)	220 (8)

Standard deviations are given in parentheses.

values for hardness and Young's modulus are determined if the primary globulitic phase is indented, whereas high-level values are measured in the intergranular phase.

According to metallurgical investigations on microstructure formation of X210CrW12 processed in the semi-solid state (see Chapter 3), samples of extruded bars are heat treated at 540 °C for 10 h in air. Comparison of the respective mechanical properties of untreated X210CrW12, thixoextruded bars and thixoextruded bars after annealing (Table 11.4) provides evidence that a significant improvement in hardness may be achieved by thixoforming this alloy. Hardness values are preserved after annealing, which is encouraging with a view to the industrial application of X210CrW12 parts produced by thixoforming. Young's modulus values increase to a lesser extent and decrease to the initial values after heat treatment.

11.4.3
Upscaling of Isothermal Thixoextrusion

Results on thermal and mechanical process parameters obtained in thixoextrusion experiments on the INSTRON laboratory testing machine were transferred to an upscaled and improved tool setup implemented in an SMS Meer open die forging press. The tool and billet dimensions and also the geometric complexity of the die were considerably increased. The applied tool setup allows for higher die temperatures, owing to the improved heating system. Additionally, a water cooling tube was provided directly below the forming table in order to quench the extruded bars rapidly. A mechanism to carry the weight of the extruded bar was not installed. The tool setup is depicted in Section 8.6.4.2, tool dimensions are given in Table 11.5 and the die cross-section is shown in Figure 11.11. The thixoextrusion trials were performed at a punch velocity of 15 mm s^{-1}. The billets used had a diameter of 76 mm and a height of 100 mm.

The temperature evolution of the tool during preheating is shown in Figure 11.12. The die land temperature was selected as the working set point with a target value of 1250 °C, leading to temperatures of approximately 1340 °C in the container.

In addition to the tool heating, a reproducible and homogeneous temperature distribution in the billet for thixoextrusion experiments (isothermal and the following non-isothermal – see Section 11.5) of the steel alloy X210CrW12 used is

Table 11.5 Tool characteristics of self-heating thixoextrusion tool.

Container		
Material		Al_2O_3, 99.7%
Inner diameter	[mm]	85
Outer diameter	[mm]	125
Cross-sectional area	[mm²]	5674.5
Wall thickness	[mm]	20
Height	[mm]	250
Die		
Material		Si_3N_4
Cross-sectional area	[mm²]	301.8
Extrusion channel length	[mm]	300
Feeding angle	[°]	11.3
Extrusion ratio		18.8

required. With an inductive heating device, the billets were heated with time–power strategies into the semi-solid state with maximum power first, followed by homogenization phases to ensure a homogeneous temperature distribution in the billet (Figure 11.13). During heating, the billet was prevented from oxidation by the inert gas argon. Thermocouples measured the temperature development in the billet. With the three-stage strategy, a final temperature of 1270 °C was realized. This corresponds to a liquid fraction of approximately 40%, according to differential thermal analysis as described in Chapter 3. With this strategy, the billets were homogeneously semi-solid but still dimensionally stable and subsequently extruded.

After the heating procedure, the billets were manipulated within 7 s to the extrusion tool and subsequently extruded by the 6.3 MN SMS Meer open die forging press.

Figure 11.11 Si_3N_4 thixoextrusion die for upscaled extrusion tests.

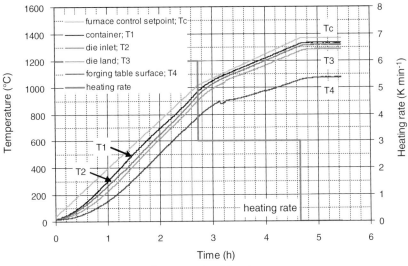

Figure 11.12 Tool temperature evolution in industrial-scale isothermal thixoextrusion experiments prior to forming.

During extrusion, a load curve similar to those in laboratory-scale tests was observed (Figure 11.14). The required extrusion forces were very low; peak forces corresponded to an extrusion pressure of $9\,N\,mm^{-2}$.

The extruded bars showed that the complex shape of the die cross-section was reproduced (Figure 11.15). However, along the bar length local distortions were observed. Despite the cooling tube positioned beneath the die outlet, solidification of the extruded bars was insufficient to prevent drip-like rupture of the thread. Metallographic examination of as-extruded bars yielded no evidence of liquid-phase separation but few pores.

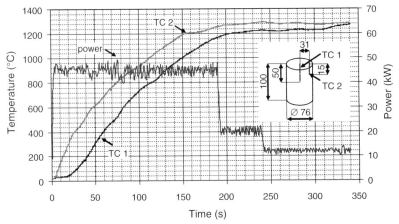

Figure 11.13 Heating strategy to realize a liquid fraction of approximately 40%.

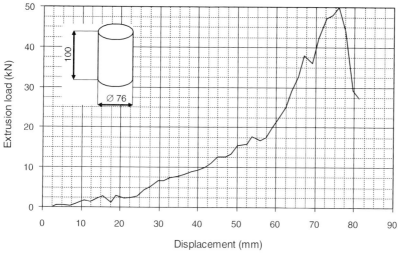

Figure 11.14 Typical load curve versus displacement for upscaled steel thixoextrusion experiments.

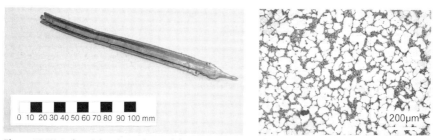

Figure 11.15 Industrial-scale isothermal thixoextrusion extruded X210CrW12 bar and metallographic analysis.

11.4.4
Summary and Discussion

Isothermal steel thixoextrusion experiments were successfully performed by applying the tool concept of self-heating ceramic dies. Analysis of extruded bars proved the beneficial influence of isothermal semi-solid processing on microstructure and properties of work pieces, as depicted in Section 11.4.2. Moreover, on the example of steel X210CrW12, the realization of an improvement in highly application-relevant work alloy properties such as hardness and Young's modulus, as predicted in metallurgical investigations (cf. Chapter 3), was demonstrated. Results prove that not only may forming forces be reduced and the geometric complexity of the formed parts be increased, but also that the homogeneity and thus uniformity of mechanical properties of the work alloy might be improved by isothermal semi-solid processing.

However, tool construction and process control still remain critical aspects for isothermal thixoextrusion of steels. The present experiments revealed a pronounced sensitivity of ceramic tool setups to thermal gradients and unplanned mechanical impacts. Extensive finite element method simulations of temperature profiles and process-induced stresses are essential for defect-free long-term application of ceramic dies in order to benefit from the excellent corrosion and wear resistance of these materials.

This also includes effective control of the solidification step during thixoextrusion. A stable and preferably flat solidification front is required in order to retain the shape of extruded bars and minimize residual porosity by solidification at least under low pressure. Potential solutions include providing guide rollers, as depicted in the tool concept scheme in Figure 11.2, not available in the experiments reported here, on the target to stabilize the as-formed bars until solidification is fully accomplished. Alternatively, a secondary, non-isothermal die exhibiting a defined temperature profile adjusted to the work alloy and extrusion velocity may be provided below the isothermal forming die to control solidification. In the latter case, independently controlled heating cores and a segmented die construction using tailored die materials are essential.

In this respect, the tool setups and experimental results obtained hitherto merely mark the beginning of possibilities for the isothermal semi-solid processing of high-melting alloys. Similarly to other complex, but established, metal forming processes relying on accurate solidification control such as continuous casting and thin strip casting, the transfer of laboratory-scale results to a table process on an industrial scale remains a challenging task.

11.5
Non-isothermal Thixoextrusion

11.5.1
Experimental Strategy, Tools and Process Parameters

The non-isothermal thixoextrusion experiments were carried out on the SMS Meer open die forging press. Due to the press design, the extrusion experiments were carried out in a vertical direction. The lowest press velocity possible was $10\,\mathrm{mm\,s^{-1}}$. For this series of experiments, the version of direct extrusion was chosen. The punch moved in the same direction as the extruded bar. The dimensions of the container (inner diameter 85 mm) were based on the calculations of the maximum hydrostatic pressure. The extrusion tool was designed in a modular way (Figure 11.16a). The forming die holder could be opened to remove the forming die, the discard and the dummy plate afterwards. The tool was designed to extrude billets with a diameter of 76 mm and a height of 100 mm (Figure 11.16b).

The extrusion tool was actively heated to 300 °C whereas the forming die was heated to 600 °C in an external furnace to prevent early solidification of the billet material after coming in contact with the forming die. The forming die was inserted in the container before the extrusion process. This allowed the height of the forming die and also the geometry of the extrusion channel to be changed but prohibited integration of an active cooling system. Additional cooling tube was not installed

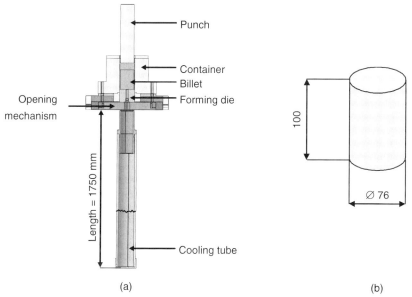

Figure 11.16 Schematic diagram of the extrusion tool for the non-isothermal extrusion process (a) and billet dimensions (b).

from the beginning but was in the later experiments (see Section 11.5.4). This cooling tube was filled with water and had a total length of 1750 mm.

The billets were heated into the semi-solid state according to the already presented heating strategy with three power stages to ensure a homogeneous temperature distribution (see Section 11.4.3). However, in contrast to the isothermal extrusion experiments, two different liquid fractions and related different final temperatures were realized by two different heating strategies that differed in the duration of the first (maximum) power stage. The first heating strategy resulted in a homogeneous temperature distribution in the billet of about 1270 °C ($f_L \approx 40\%$). With the second heating strategy, a liquid fraction of approximately 10% ($T_B \approx 1240$ °C) was chosen to reduce the time required for solidification in the extrusion channel.

11.5.2
Extrusion Performance and Evaluation for Experiments Without Cooling

The experiments were carried out with variation of the main process parameters as shown in Figure 11.17. The active extrusion channel length of the dies with a round exit was 25 and 40 mm. Some experiments were carried out using uncoated or physical vapour deposition (PVD)-coated forming dies. The liquid fraction content was varied between 10% ($T_B \approx 1240$ °C) and 40% ($T_B \approx 1270$ °C) and the first experiments were performed without the cooling tube.

The load stroke curve – for experiments using straight forming die channels and a liquid fraction of 40% – represents the influence of velocity and temperature

Press velocity	[mm s⁻¹]	10, 20, 50, 75
Forming die height	[mm]	35, 50
Liquid fraction	[%]	10, 40
Minimum extrusion channel diameter	[mm]	7.5, 10, 15
Extrusion channel geometry		Straight
		Opening
		Closing
Forming die surface		Uncoated and PVD-coated
Cooling tube		No

Figure 11.17 Variation of process parameters for experiments without cooling tube.

(Figure 11.18). In contrast to conventional extrusion processes, where the required force decreases during the process due to the reducing friction length (Figure 11.18) [12], a continuous rise of the required force was detected in thixoextrusion. As was to be expected, an increase in the press velocity resulted in lower required forces, because there was less time for cooling and solidification.

The friction and the deformation force increased during the process due to the solidification of the material and reached the maximum at the end. The average extrusion force required in the thixoextrusion experiments is very low compared with the conventional extrusion process using the same forming dies and billet dimensions. A rough estimate can be calculated using the elementary theory, that is, the Siebel equation [1]. At a solid-state extrusion temperature of about 1145 °C, the flow stress would be about 70 N mm^{-2} and the force would reach an order of magnitude of

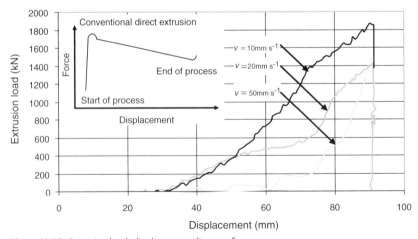

Figure 11.18 Extrusion load–displacement diagram for conventional and semi-solid direct extrusion processes for experiments carried out with a liquid fraction of 40%.

Shell

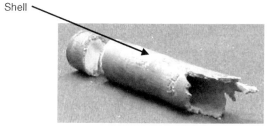

Figure 11.19 Shell formed while extruding in a special case with reduced liquid fraction.

5800 kN, thus being significantly above the thixoextrusion force. However, the maximum extrusion pressure required at the end of the described thixoextrusion experiments is between 254 and 325 N mm^{-2}. This is up to 10 times higher than for the isothermal thixoextrusion experiments, which indicates that a considerable fraction of the force is required to compress the solidifying shell of the billet.

For the chosen process parameters, the material does not solidify in the extrusion channel and exits the forming die still semi-solid, forming clews with constant diameter. Hence no significant influence of the forming die geometry is observed. To achieve solidification in the extrusion channel, the liquid fraction of the billet is reduced from 40 to 10% while keeping the other process parameters constant. For these conditions, another defect – massive shell formation – occurs. The reduced billet temperature results in an immediate shell formation when the billet material comes in contact with the extrusion channel wall. In one special case a massive formed shell is shown in Figure 11.19.

This shell reduced the diameter of the extrusion channel and the following material was extruded through the reduced diameter channel, which allowed the production of a very thin bar with a total length of 450 mm and a diameter of 2 mm on average (Figure 11.20). PVD coatings on the extrusion channel did not influence the

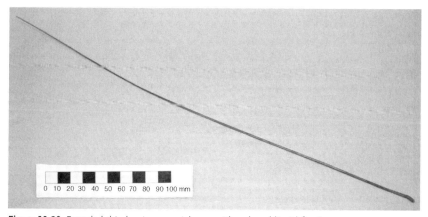

Figure 11.20 Extruded thin bar in a special case with reduced liquid fraction.

shell formation but allowed easier removal of the shell and the discard compared with uncoated forming dies.

11.5.3
Estimation of the Material Dwell Time Required in the Extrusion Channel

As the first experiments did not show sufficient solidification of the billet material in the extrusion channel, a first estimation of the solidification time of the material required in the extrusion channel was performed with a one-dimensional numerical model also allowing the determination of the influence of the forming die material (Figure 11.21).

The simulation consists of the stationary billet material with a homogeneous temperature distribution of 1240 °C ($f_L \approx 10\%$), a total diameter of 15 mm and a simplified forming die with an initial temperature of 300 °C. By assuming adiabatic boundary conditions, the time required for the complete solidification of the material is estimated by transferring the heat from the material into the forming die. The complete solidification is reached at 1220 °C. The forming die material was varied between the hot working steel X38CrMoV5-1, the copper–beryllium alloy CuCoBe, the nickel-base alloy TZM and an Si_3N_4 ceramic to determine their influence on the solidification time. As shown in Figure 11.22, the time for complete solidification can be assumed to be 2.5 s using a heat transfer coefficient of $15\,000\,W\,m^{-2}\,K^{-1}$. Simulations with higher heat transfer coefficients do not influence the total solidification time significantly. Lower heat transfer coefficients are not suitable according to [13]. The solidification time is almost independent of the forming die material. used For the three metallic forming die materials, the solidification time remains almost constant. Only the Si_3N_4 ceramic shows a slower cooling behaviour regarding longer time periods.

With the simulation shown and a required solidification time of 2.5 s using a 15 mm extrusion channel, the required ratio between the length of the extrusion channel and the bar extrusion velocity can be estimated as

$$\frac{l_{channel}}{v_{bar}} > t_{solidification} = 2.5\,s \qquad (11.1)$$

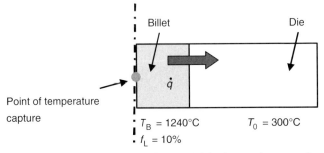

Figure 11.21 Schematic diagram of the model for the one-dimensional cooling simulation.

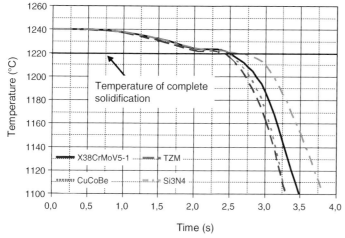

Figure 11.22 Cooling behaviour of billet material in extrusion channel regarding the centre of the bar.

With Equation 11.1, the required press velocity can be estimated by

$$v_{press} = \left(\frac{A_1}{A_0}\right) \frac{l_{channel}}{t_{solidification}} \tag{11.2}$$

As the lowest press velocity achievable with the available forging press is $10 \, \text{mm s}^{-1}$, an extrusion channel length of 800 mm would be required for complete solidification, which is not realizable with the present extrusion tool. Therefore, the following experiments were carried out using the cooling tube of the isothermal extrusion processes to achieve complete solidification after the forming die.

11.5.4
Extrusion Performance and Evaluation for Experiments with Cooling

For the experiments with a cooling tube, reduced press velocities compared with earlier experiments were chosen (Figure 11.23). The press velocity was varied

Press velocity	$[\text{mm s}^{-1}]$	10, 15, 20
Forming die height	[mm]	50, 65, 100
Liquid fraction	[%]	40
Minimum extrusion channel diameter	[mm]	10, 15
Extrusion channel geometry		Straight 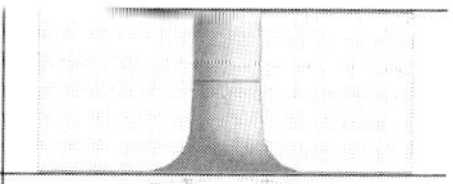
		Closing
Forming die surface		Uncoated
Cooling tube		Yes

Figure 11.23 Variation of process parameters for experiments with cooling tube.

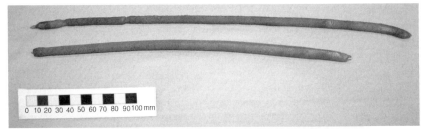

Figure 11.24 Extruded bars.

between the lowest press velocity possible (10 mm s^{-1}) and 20 mm s^{-1}. For better cooling conditions in the extrusion channel, the extrusion channel height was increased to 100 mm. All experiments were carried out considering a fraction of 40% ($T_B \approx 1270\,°C$). The extrusion channel diameter was varied according to the first experiments but no opening forming die geometry was considered. No coatings were investigated in this series.

The installation of the cooling tube resulted in the production of bars. The presented variations of the process parameters (here mainly the reduced press velocity and the elongated extrusion channel length) showed no significant influence on the process: the material exited the forming die in a semi-solid state and solidified in the cooling tube.

Independently of the selected press velocity, extrusion channel height, diameter and geometry, the maximum bar length was limited to 460 mm (Figure 11.24). The weight of the extruded bar increased until the bar cracked due to its own weight. All extruded bars showed such cracks. In Figure 11.25, two different kinds of cracks – dripping cracks and cracks without contraction – are presented. Due to cooling of the

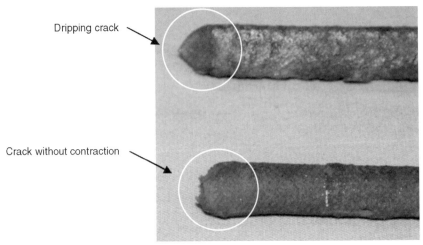

Figure 11.25 Different cracks that occurred.

$\varnothing_{die}$ = 15 mm
h_{die} = 50 mm
v_{press} = 10 mm/s

$\varnothing_{die}$ = 15 mm
h_{die} = 65 mm
v_{press} = 20 mm/s

$\varnothing_{die}$ = 10 mm
h_{die} = 100 mm
v_{press} = 10 mm/s

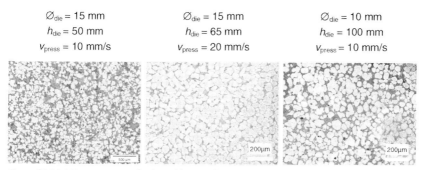

Figure 11.26 Microstructure of selected bars and extrusion conditions.

material on entering the cooling tube, dripping cracks could be observed at a temperature of 1230 °C due to a zero strength temperature. Cracks without contraction occurred in the temperature range 1190–1229 °C [14].

Shell formation occurs when the billet material comes in contact with the extrusion channel wall. The grade of shell formation was independent of the present process parameters but it could be observed in every experiment. The material solidified immediately on entering the extrusion channel and stuck to the channel wall. Hence it reduced the extrusion channel diameter. The diameter of the extruded bars was always reduced by 5 mm compared with the extrusion channel diameter. Hence the shell in the extrusion channel had an average thickness of 2.5 mm. The shell formation could not be prevented in the experiments.

Concerning the microstructure, all extrusion results showed a parameter-independent fine globular microstructure in the radial and longitudinal directions (Figure 11.26). A homogeneous microstructure could be found for every parameter condition. The microstructure analysis was characterized by a homogeneous material distribution.

Pores in the extruded bars were revealed by X-ray images (Figure 11.27). The pores resulted from an unsatisfactory pressure during solidification and occurred mainly in the centre of the extruded bars.

In the extruded bars, the hardness (HV 0.3) differs in the primary grains and the eutectic according the results of Uhlenhaut *et al.* [15]. All over the extruded bars and

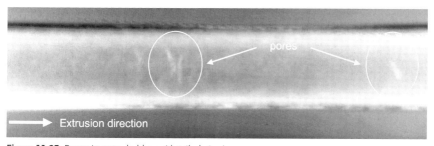

Figure 11.27 Pores in extruded bars (detailed view).

Table 11.6 Pyramid diamond hardness of extruded bars for different sample location and an impact time of 15 s.

Primary grain (HV 0.3)	Eutectic (HV 0.3)	Macro hardness (HV 10)
333 ± 63	722 ± 195	409 ± 33

independent of the extrusion direction, the eutectic had hardness values (HV 0.3) of 722 on average, which were about double those for the primary grains, which had an average hardness (HV 0.3) of 333 (Table 11.6). With the hardness test HV 10, an average macro hardness of 409 for all bars could be observed.

11.5.5
Numerical Simulation of the Thixoextrusion Process

The estimation of the dwell time of the material required in the extrusion channel in Section 11.5.3 shows the requirement for an extremely long extrusion channel or extremely low press velocity. To determine this required press velocity that ensures complete solidification, the process of thixoextrusion was simulated with the simulation program Forge 2005, which uses a Lagrange approach and a rigid plastic material law. For the implicit two-dimensional axis symmetric simulations, triangle elements were used. The material data and boundary conditions were chosen according to the investigations of Shimahara *et al.* [16]. For a complete description of the process chain, the simulations of the process steps 'cooling after heating into the semi-solid state' and 'cooling during manipulation of the billet' revealed the temperature distribution of the billet before the forming process. The measured temperature distribution of approximately 1270 °C directly after heating was used as initial temperatures for the first cooling simulation. Due to radiation and heat transfer during the manipulation of 10 s, the cooling simulations showed a total temperature decrease until inserting the billets in the container of about 20 °C.

The simulation model of the bar extrusion process is shown in Figure 11.28 and is an illustration of the real extrusion tool. It consists of the billet (diameter 76 mm, height 100 mm), the forming die, a die holder and the container with an inner diameter of 85 mm. The initial temperature of the billet was 1250 °C. The forming die temperature was varied from room temperature up to 600 °C to investigate if this could prevent shell formation. The extrusion channel had a diameter of 15 mm and a length of 50 mm.

The simulated press velocities were varied from 0.5 to 10 mm s^{-1} to achieve complete solidification of the material on the one hand and to simulate real experiments on the other. The extrusion was performed in a vertical direction and gravity force was considered. The temperature development of the extrusion material and the forming die is shown in Figure 11.29 for a simulation of the real experiments with a press velocity of 10 mm s^{-1}. After coming in contact with the forming die, the billet cooled immediately. During upsetting the billet up to the container wall, the material continued to cool. When the material was extruded through the extrusion

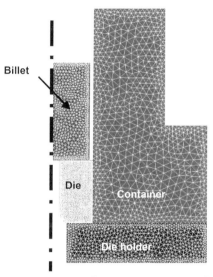

Figure 11.28 Model for extrusion process simulation.

channel, the already solidified material came in contact with the extrusion channel wall and shell formation, as observed in the experiments and mentioned in the literature several times, could be observed. In the experiments, the already solid billet material stuck on the channel wall and the following material flowed by. This resulted in a reduced bar diameter compared with the extrusion channel diameter using a tempered forming die.

When observing the material flow, the development of a dead zone as in the conventional extrusion processes could be detected (Figure 11.30). The material flowed from the generated surface into the forming zone. There, the velocity of the material increased to 300 mm s^{-1}.

With an extrusion channel length of 50 mm and a press velocity of 0.5 mm s^{-1}, solidification in the channel was realized. As shown in Figure 11.31, the bottom of the

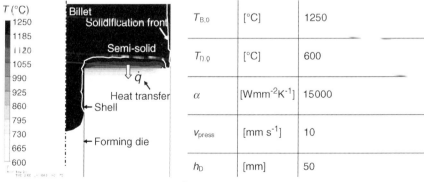

Figure 11.29 Temperature development of the extrusion process.

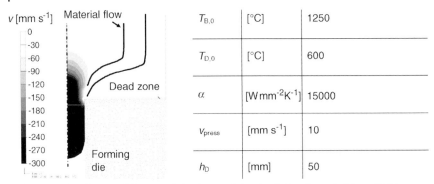

$T_{B,0}$	[°C]	1250
$T_{D,0}$	[°C]	600
α	[W mm^{-2}K^{-1}]	15000
v_{press}	[mm s^{-1}]	10
h_D	[mm]	50

Figure 11.30 Velocity and forming of dead zone of the simulation shown in Figure 11.29.

billet cooled significantly during upsetting and extrusion due to the forming die temperature of 25 °C. The formed shell was massive enough to be extruded through the extrusion channel.

Due to the required extrusion time, the temperature of the forming die increased and reached temperatures which made an active cooling of the forming die essential. With such a cooling system, the press velocity could eventually be increased while realizing complete solidification the die. Figure 11.32 shows an extrusion simulation with integrated active cooling in the forming die. The press velocity was set to 3 mm s^{-1} and the forming die height was 65 mm. The forming die temperature was set to 25 °C and the water temperature was 20 °C. Figure 11.32 shows that the core of the bar is still semi-solid on exiting the extrusion channel.

The simulation of the experiments showed, as estimated, the incomplete solidification of the billet material in the extrusion channel. Furthermore, the phenomenon of shell formation which was observed in the experiments could be simulated. With the chosen reduced press velocities, complete solidification in the extrusion channel could be realized in the simulations. Using an active cooling system, complete

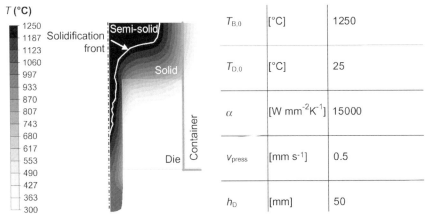

$T_{B,0}$	[°C]	1250
$T_{D,0}$	[°C]	25
α	[W mm^{-2}K^{-1}]	15000
v_{press}	[mm s^{-1}]	0.5
h_D	[mm]	50

Figure 11.31 Realized complete solidification in the extrusion channel.

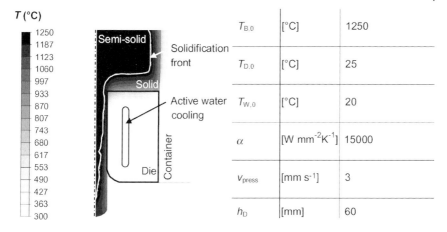

$T_{B,0}$	[°C]	1250
$T_{D,0}$	[°C]	25
$T_{W,0}$	[°C]	20
α	[W mm^{-2}K^{-1}]	15000
v_{press}	[mm s^{-1}]	3
h_D	[mm]	60

Figure 11.32 Extrusion simulation with integrated active cooling in the forming die.

solidification could be realized by increasing the press velocity. Shell formation, the origin of which has not yet been clarified, should be prevented.

11.5.6
Summary and Discussion

The thixoextrusion experiments were carried out with a modular tool where all relevant process parameters can be modified. With this extrusion tool, the forming die, the discard and the dummy plate can be removed by using the opening mechanism integrated in the tool. For the first experiments, the parameters press velocity, extrusion channel length, diameter and geometry were modified using a constant liquid fraction of 40%, which did not result in extruded bars but rather in extruded clews. A reduction in the liquid fraction by changing the other parameters in the same range results in massive shell formation or even blocking of the extrusion channel. Hence a liquid fraction of 10% is not suitable for these experimental procedures. The massive shell formation allows the formation of very thin bars due to a proper press velocity/heat transfer ratio. The investigated PVD coatings simplified cleaning of the forming dies but had no significant influence on the bar extrusion. In comparison with the conventional extrusion process, the average extrusion force required is lower. For thixoextrusion processes, the resulting press force increases towards the end of the process because the material (surface area of the billet) cools and solidifies during the process.

For a first temperature development estimation of material located in the extrusion channel, a simulation under adiabatic and ideal contact conditions was performed in stationary conditions. These simulations revealed that the cooling behaviour of the material in the extrusion channel is independent of the die material.

With the forging press and the designed extrusion tool, the dwell time of the semi-solid material required in the extrusion channel could not be realized. To achieve solidification of the material, experiments using the cooling tube of the isothermal thixoextrusion processes were performed. Using a liquid fraction of 40% and a cooling tube after the tool resulted in successful bar extrusion. The variations of process parameters showed almost the same results. The maximum length of the extruded bars was limited to 460 mm due to the increasing weight of the bar while the resistance of the semi-solid material remained constant. When the gravity force increased due to increasing weight, the bars dropped down. The extruded bars were characterized by a reduced diameter compared with the extrusion channel diameter because of shell formation, which could not be prevented. Independently of the chosen process parameters, which have no significant influence on the solidification of the semi-solid material, the solidification was realized only by the cooling tube. Regarding the microstructure analysis, the thixoextrusion experiments performed show a process parameter- independent behaviour. For the chosen process conditions, the extruded profiles show a similar homogeneous globulitic microstructure. Concerning Vickers hardness, the eutectic shows doubled the values for the primary grains.

Further simulations of the thixoextrusion process permit the complete solidification using low press velocities of 0.5 mm s^{-1}. Active cooling systems integrated in the forming die allow an increase in the press velocity to 3 mm s^{-1}. Shell formation, the origin of which has not yet been clarified, has to be prevented for these conditions. The simulation results show that a tool combination where shell formation is prevented and complete solidification is realized with an active cooling system achieved simultaneously would be a solution for successful applications.

11.6
Conclusion

Ensuring the solidification of the semi-solid material after forming the material is the most important challenge of semi-solid extrusion processes. This solidification can be realized in or after the forming die, which results in two different tool concepts as presented previously: the isothermal and the non-isothermal tool concepts. Both concepts require specific tool setups and process strategies. The general feasibility of both concepts has been shown but open challenges still remain.

With the tool for the isothermal semi-solid extrusion process, forming of the semi-solid material is possible but handling of the bars after the forming die has not yet been solved. The extruded bars drip and limit the bar length in the experiments. Similarly to other complex, but established, metal forming processes, such as continuous casting and thin strip casting, accurate solidification control or bar suspension has to be integrated in the tool.

The non-isothermal extrusion tool setup has not so far achieved the goal of complete solidification of the material in the extrusion channel. Either the press velocity would have to be extremely low or the extrusion channel length extremely long to ensure complete solidification. In addition, the material which comes in contact with the extrusion channel began to stick to the wall, forming a non-extruded shell. Therefore, for the non-isothermal extrusion process, prevention of such sticking shells and ensuring complete solidification are the main challenges.

Solving these problems could eventually be achieved by a combination of both concepts. A first ceramic zone where the material is formed under (nearly) isothermal conditions, followed by a long cooling channel with an integrated active cooling system, could ensure that all requirements are met for successful semi-solid extrusion. An additional mechanism to carry the bar weight is also an option.

Abbreviations and Units

HPDC	high pressure die casting
e.g.	for example
A_0/A_1	extrusion ratio
A_0, A_1	cross section
T_s	solidus temperature
T_B	billet temperature
T	temperature
F	press force
v	velocity
f_L	fraction liquid
T_0	initial temperature
$T_{B,0}$	initial billet temperature
$T_{D,0}$	initial die temperature
$T_{W,0}$	initial water cooling temperature
$l_{channel}$	extrusion channel length
v_{bar}	bar extrusion velocity
v_{press}	press velocity
$t_{solidification}$	time for solidification
Ø	diameter
h	height
h_D	die height
α	angel alpha
TC	Thermocouple
PVD	Physical vapour deposition
mm/s	velocity
s^{-1}	strain rate
mm	length
mm_2	cross section
°	angle
°C	Celsius temperature
K	Kelvin temperature
MPa	pressure
kN/MN	force
N/mm^2	pressure

GPa	pressure
W/m²K	heat transfer coefficient
Hz	frequency
W/kW	power
%	percent

References

1 Lange, K. (1985) *Handbook of Metal Forming*, Society of Manufacturing Engineers, Dearborn, MI.

2 Kiuchi, M., Sugiyama, S. and Arai, K. (1979) Study of metal forming in the mashy-state; 2nd Report. Extrusion of tube, bar and wire of alloys in mashy-state, Proceedings of the 20th International Machine Tool Design Conference.

3 Kiuchi, M. and Sugiyama, S. (1994) Proceedings of the 3rd International Conference on Semi-solid Processing of Alloys and Composites, pp. 245–257.

4 Möller, T. (2004) Verfahrensentwicklung und Ermittlung des thermischen Prozessfensters beim Thixostrangpressen, Dissertation, RWTH Aachen.

5 Miwa, K. and Kawamura, S. (2000) Proceedings of the 6th International Conference on Semi-solid Processing of Alloys and Composites, pp. 279–281.

6 Rouff, C., Bigot, R., Favier, V. and Robelet, M. (2002) Proceedings of the 7th International Conference on Semi-solid Processing of Alloys and Composites, A63, pp. 355–360.

7 Abdelfattah, S., Robelet, M., Rassili, A. and Bobadilla, M. (2000) Proceedings of the 6th International Conference on Semi-solid Processing of Alloys and Composites, pp. 283–288.

8 Sugiyama, S., Li, J.Y. and Yanagimoto, J. (2004) Proceedings of the 8th International Conference on Semi-solid Processing of Alloys and Composites, TMS Publications, Warrendale, PA.

9 Knauf, F., Baadjou, R. and Hirt, G. (2008) Proceedings of the 10th International Conference on Semi-solid Processing of Alloys and Composites.

10 Münstermann, S., Telle, R., Knauf, F. and Hirt, G. (2008) Proceedings of the 10th International Conference on Semi-solid Processing of Alloys and Composites.

11 Püttgen, J.T.W., Hallstedt, B., Bleck, W. and Uggowitzer, P.J. (2007) *Acta Materialia*, **55**, 1033–1042.

12 Knauf, F., Hirt, G., Immich, P. and Bobzin, K. (2007) Influence of tool geometry, tool coating and process parameters in thixoextrusion of steel, Proceedings of ESAFORM.

13 Shimahara, H. (2007) Personal communication.

14 Thome, R. (1998) Einfluss der chemischen Zusammensetzung und der Abkühlbedingungen auf die Rissanfälligkeit von Werkzeugstählen beim Stranggiessen und bei der Warmumformung, Dissertation, RWTH Aachen.

15 Uhlenhaut, D.I., Kradolfer, J., Püttgen, W., Löffler, J.F. and Uggowitzer, P.J. (2006) *Acta Materialia* **54** (10), 2727–2734.

16 Shimahara, H., Baadjou, R. and Hirt, G. (2006) *Solid State Phenomena*, **116–117**, 189–192.

Index

Thixoforming: Semi-solid Metal Processing. Edited by G. Hirt and R. Kopp
Copyright © 2009 WILEY-VCH Verlag GmbH & Co. KGaA, Weinheim
ISBN: 978-3-527-32204-6